Rice Bowls and Dinner Plates

Ceramic artefacts from Chinese gold mining sites in southeast New South Wales, mid 19th to early 20th century

Virginia Esposito

BAR International Series 2674
2014

Published in 2016 by
BAR Publishing, Oxford

BAR International Series 2674

Rice Bowls and Dinner Plates

ISBN 978 1 4073 1316 0

BAR Publishing is the trading name of British Archaeological Reports (Oxford) Ltd. British Archaeological Reports was first incorporated in 1974 to publish the BAR Series, International and British. In 1992 Hadrian Books Ltd became part of the BAR group. This volume was originally published by Archaeopress in conjunction with British Archaeological Reports (Oxford) Ltd / Hadrian Books Ltd, the Series principal publisher, in 2014. This present volume is published by BAR Publishing, 2016.

Printed in England

BAR
PUBLISHING

BAR titles are available from:

BAR Publishing
122 Banbury Rd, Oxford, OX2 7BP, UK
EMAIL info@barpublishing.com
PHONE +44 (0)1865 310431
FAX +44 (0)1865 316916
www.barpublishing.com

Preface

This research was submitted as a PhD thesis to the Australian National University in December 2012 and the award was conferred in 2013. Funding was provided by an Australian Post-graduate Award and an ANU College of Arts and Social Sciences fieldwork grant. Office and laboratory facilities were provided by the ANU School of Archaeology and Anthropology, and scanning electron microscope facilities by the ANU Centre for Advanced Microscopy.

This research would not have been possible without the help of many people. In particular, I'd like to thank, Ian Farrington for his guidance and support; Anthony Flynn for his expertise in ceramics; the late Lindsay Smith for his exploration of Chinese sites; Damian Tybussek and Christine Gant-Thompson for sharing their knowledge of Kiandra; Po-yi Chiang for his translation of Chinese characters; Margaret Parkes for assistance with sorting and cleaning sherds; and Doreen Bowdery, Mary Casey, Paul Macgregor, Trelle Morrow, Dennis O'Hoy and Priscilla Wegars for taking the time to answer all my queries. Thanks to others who provided information, Anggraeni, Lana Arun, Emma Beckett, Alasdair Brooks, Penny Crook, Laura Davies, Melissa Dunk, Emily Dutton, Catherine Fitzgerald, Geoff and Kerrie Ford, Philip Fountain, Gordon Grimwade, Sarah Hayes, Jarvis Hyman, Sarah Kelloway, Tom Knight, Susan Lawrence, Randall Moir, Tim Murray, Kevin Rains, Lisa Solling, Lorna Tilley, Alan Walden and Rowan Ward. Thanks to the staff at the ANU, School of Archaeology, in particular Sue Fraser, David McGregor and Liz Walters; to Frank Brink and Hua Chen from the ANU Centre for Advanced Microscopy, for their guidance in using the scanning electron microscope; and to the staff at the ANU library for reference materials. Thanks to Doug Williams for allowing the use of ceramics from sites near Bungendore; to Diana Osborne for the ceramics from Narrandera; to Penelope Allison for access to the ceramics from Old Kinchega Homestead; and to all the site supervisors and students who participated in the Australian National University field schools at Kiandra. In addition, I'd like to thank Anne-Louise Muir at Heritage Victoria; Sophie Couchman at the Museum of Chinese Australian History; and Louise Holt and John Naumann at Flinders University. Lastly, a special thanks to my family for all their support.

Unless otherwise stated, all photographs and drawings are by the author. Several map illustrations have been adapted from Smith (2006) as many of the examined ceramic sherds are from sites originally surveyed and excavated by Lindsay Smith 1996-2003. Map details from the ANU field schools have also been used. Individual recognition of the origins from which illustrations have been adapted is included in the figure captions, as are acknowledgements of copyright, where appropriate.

Abstract

This research examines ceramic collections from overseas Chinese mining settlements in southeast New South Wales, Australia, including those from Jembaicumbene, Flanagan's Point, Upper Adelong, Adjungbilly, and Kiandra. Traditional and non-traditional methods of ceramic analysis are used to answer major questions, which expand the archaeology of the Chinese in Australia. The analyses enable conclusions to be drawn about the active role of vessels in everyday life, not only within the domestic sphere but also in communal aspects of food and feasting. On a broader scale, the research considers the nature of Chinese supply networks and demonstrates how Western-style ceramics became appropriate substitutes, as those systems changed.

This study is the first intra-site analysis of Chinese camps in Australia and also the first to compare contemporary ceramic assemblages from Chinese and non-Chinese sites in the same region. Furthermore, compositional analysis of Chinese sherds has added another dimension. It has revealed that, while the majority of the brown-glazed storage jars were made in southern China, some may have been made in northern China. It has also demonstrated that visual similarities in vessels do not necessarily relate to chemical similarities. Overall, this research has highlighted short and long-term occupation sites. These camps were not homogenous or static settlements, they changed over time.

Contents

List of Figures

List of Tables

CHAPTER 1

The Chinese in nineteenth century New South Wales

The Chinese, arriving in New South Wales (NSW), were part of a large influx of migrants in the second half of the nineteenth century. Most of them came from the Pearl River Delta of the Guangdong Province and the southern part of Fujian Province (Shaoqing 2002:42). Social and economic pressures, natural disasters and civil wars, which occurred during the nineteenth century, prompted many to migrate (Chan 1997:197). A few Chinese travelled to Australia in the early 1900s (Smith 2006:2) but the first major boatload arrived in 1848 to fill a labour void created by the cessation of convict transportation to NSW in 1840 (Choi 1975:18). In the 1850s promises of wealth lured more migrants and enticed many of the original Chinese farm labourers to leave their Australian employers to work on the goldfields (Choi 1975:19). Those heading to the goldfields from China travelled on European or American ships from Hong Kong, Macao (Macau) or Canton (Guangzhou). They often arrived at ports outside NSW, for example, in 1857, 32 ships arrived in Robe, South Australia, carrying 14,600 Chinese (Smeaton 1865). The groups of Chinese then walked overland to the goldfields in Victoria and NSW.

While the majority of Chinese immigrants in the 1860-1870s were miners, others were storekeepers, fishermen, market-gardeners, butchers, bakers, cooks, carpenters and road labourers (Crawford 1877:3). Each Chinese man arrived in Australia 'carrying his own bamboo-pole brought from China', on which were 'slung baskets of clothes, a roll of matting, a few humble necessities, and perhaps a box of tea or preserved eggs to barter' (Crawford 1877:27). They were expected to be frugal so that they could send a portion of their money back to their families in China (Choi 1975:13).

The decision of who would travel overseas was taken collectively by the family and the money required for the passage was often raised within the clan network. Kwok Bun Chan (1997:195) argued that 'dispersing the patrilineal family' was seen as a way of preserving it and 'a resourceful and resilient way of strengthening it'. The interests of the family and their lineage prevailed over those of the individual so there was an objection to females emigrating because their absence would weaken the lineage (Choi 1975:13). Many rural villages in southern China were occupied by only one or two clans and this influenced the way that migration operated as labourers recruited from one area would be under a headman from an important lineage in that area (Choi 1975:13-14).

The Chinese migrating to Australia brought aspects of their country's administration with them. The domination of Chinese villages by clans was beneficial to the Qing government as it utilised them in their approach to rural control (Hsiao 1960:323). The size of China meant that the government had to adopt a sub-administrative system and this was run through individual villages (Hsiao 1960:43). As villages were usually 'controlled by the heads of the most important families', the headman represented the entire village (Hsiao 1960:262). The government controlled from afar, giving villages the appearance of communities that were self-governing, democratic or even autonomous (Hsiao 1960:261). However, the Qing government took advantage of the fact that most villagers were illiterate. It exercised ideological control over rural Chinese 'which stressed social duties and human relationships' (Hsiao 1960:184).

Aspects of Chinese rural networks were also brought to Australia. Their rural social and market networks were intertwined (Skinner 1964). They were based on central villages which had administrative control and provided resources to smaller towns. Skinner (1964:32) suggested that the 'social field' of Chinese peasants was influenced by the boundaries of their 'standard marketing area' and it was within these areas that social events such as festivals were held. Features of this organisation extended to rural Australia. Smith (2006) identified a hierarchical pattern of community organisation featuring central Chinese settlements which were self-contained, self-sufficient and located on a major goldfield (Smith 2006:256). Each one was associated with a second tier of smaller, self-contained, but not self-sufficient, settlements or work camps. While the central camps had a temple, store and oven, along with numerous dwellings, the smaller camps 'contained only the rudimentary necessities of community life', that is, huts and a communal oven (Smith 2006:256).

When the groups of labourers arrived on the goldfields, they were controlled by a headman under a credit-ticket system which was backed by merchants (Choi 1975:14). This led to a 'lack of autonomy' for individuals (Lawrence & Davies 2011:230). The labourers were made to work long hours for very little return, as described by the newspaper correspondent for Kiandra. The Chinese miners:

> Resemble nothing more intimately than a hive of bees who, under the command of a boss (their queen), toil and strive from morning till night, in many instances, too, for his exclusive benefit. These bosses are generally clever fellows in their way – are possessed of some capital, and as a *sine qua non* have scraped together sufficient English to enable them to act as interpreters. Generally they have received a somewhat better education than their fellows, and have sufficient sabre so to use it as to work upon them bodily and spiritually. Thus, for instance, a boss here has some 150 men under him. He buys claims with their money and his own; puts ten, twenty or thirty men into each claim, and regulates their *modus operandi*, usually securing a fifth or fourth share to himself. He also acts as provedore, charges then

from twenty-three to thirty shillings a-week each for board, which he takes good care to deduct from the weekly earnings (The Sydney Morning Herald, 11 May 1861:7)

Although it was the individual who migrated, it was the family who were active in negotiating with domestic groups, and 'socio-economic, and political forces in both country of origin and country of destination' (Chan 1997:199). As Chan Kwok Bun (1997:200) described, 'the lone migrant…experiences the family in his everyday sojourning life as a real factor, sometimes as a liability, a constraint, other times as a source of strength and enablement'. Henry Chan (2001:9) perceived the 'ever presence of 'the family' as part of the migrant's 'Chineseness''. Chinese poems often focus on 'departure, absence and separation' and these are 'central motifs of Chinese culture' (Chan 2001:9). Although they were away from their homelands, the migrants operated within a Chinese community in their new lands.

The term diaspora has been adopted to explain 'the development through migration of dispersed communities that relate not only to their nation of residence but also to a homeland, or the idea of a homeland, or to each other', so that there are 'multiple senses of belonging' (Curthoys 2001:18). The Chinese who migrated to the goldfields in the nineteenth century retained the bond with their homeland while 'developing their economic and social ties' in their new land (Curthoys 2001:21). These migrants were considered to be sojourners intent on returning home, while at the same time, Europeans were settlers planning to stay. However, recent research in America has shown that similar numbers returned or stayed regardless of nationality (Voss & Allen 2008:9).

Although it was predominantly males who migrated to Australia, there were a few family groups (Chan 2001:8). Research in America has shown that it was not common for Chinese wives to accompany their husbands (Wegars 1993:230). However, some Chinese women travelled by themselves to America, while others were brought by the secret societies and ended up working as prostitutes. These women were later in demand as wives for Chinese labourers (Wegars 1993:236). European women were a common site on the Australian goldfields (Curthoys 2001:26) and many Chinese men married non-Chinese women, a custom that the Chinese migrant household upheld (Chan 2001:9).

As the Chinese workers came to Australia for the gold rush, so did the Hung League, a secret society, formed in China during the Qing dynasty with a slogan 'To overthrow the Qing and Restore the Ming' (Shaoqing 2002:40). It is suggested that the society could have come to Australia through two sources: from Southeast Asia where it was referred to as 'Yee Hing Company'; or from mainland China whose pamphlets used transformed Chinese characters to depict the name of the Heaven Earth Society (Shaoqing 2002:33-34). In China the Hung League had been an association to aid and support peasants and labourers. Shaoqing (2002) stated that the Hung League developed in Australia as the immigrants arrived during the early gold rush period. The society 'arranged jobs for them, mediated their disputes, and assisted with the everyday difficulties of birth, old age, sickness and death and so on' (Shaoqing 2002:40). The society also helped out in times of anti-Chinese violence.

Troubles between Chinese and non-Chinese people on the goldfields have often been emphasised, such as the racism and riots at Lambing Flat, NSW, in 1860-1861. The difference in religion and values often meant that the 'Europeans thought the Chinese were barbarians, and the Chinese, in their turn, thought of the Europeans in the same way' (Curthoys 2001:25). However Crawford (1877:12) had reported that wherever the Chinese 'settled for any length of time in Australia, and become known, they are considerately treated'. Although overseas Chinese are often thought of as a homogenous group, there is a diversity of local cultural and social conditions within Guangdong and Fujian which would have been brought with them (Chan 2001:8). For example, Crawford's (1877:7) report noted ongoing rivalry in Australia between the Chinese from Guangdong and Fujian provinces. Consequently, when the Chinese came to Australia, they were 'embedded in a web of customs, rules and obligations' (Lawrence & Davies 2011:230).

Among the customs brought to Australia was the building of temples. These were generally constructed by people from the same village, a tradition which appears to have continued in NSW (Wilton 2004:85). Temples were a focal point for the Chinese community. In addition to religious and ancestral worship, temples were used as meeting places and for feasts (Grimwade 2003; Wilton 2004). The core of the temple contained a room dedicated to the primary deity (Wilton 2004:87). Other rooms or outbuildings could include a caretaker's residence, meeting hall, kitchen, storeroom or rooms for temporary accommodation (Grimwade 1995, Wegars 2003:71; Wilton 2004).

Another common feature of Chinese settlements in Australia was a communal oven. Although there is no mention of these in contemporary documents, oral evidence relates to their use for cooking meat, particularly at festive occasions (Bell 1995:220; Grimwade 2008). They are circular in shape but some variation in design has been noted (Bell 1995:214). These were used for communal feasts, and may have been used for communal eating at male dominated camps, as sharing food would have reinforced a sense of community and belonging to a group.

Food and feasting played a major role in the lives of the Chinese in Australia. Food is a revealing cultural subject that 'ranks in importance with other major cultural foci' (Yentsch 1990:27). Foodways and their associated material goods, such as ceramic vessels, are 'intimately involved in the creation, maintenance, and manipulation of identity' (Bray 2003:3). Temples and communal ovens were places where these values were reinforced. Chinese food also made an impression on a reporter in 1860, who described a restaurant at the Abercrombie Chinese Camp, near Bathurst, NSW. He noted that 'you may regale yourself on roast duck á la Canton, a pig's head á la Hong Kong, and sponge cake á la Macao'(The Sydney Morning Herald, 10 January 1860:2).

The overall numbers of Chinese in Australia were influenced by legislation. In 1861, NSW limited the number able to enter the colony by imposing a tax and a cargo restriction. This was repealed in 1867 because fewer Chinese were migrating because of limited employment opportunities (Markus 2001:70). Permanent legislation restricting Chinese migration to NSW was introduced in 1881, and the right to naturalisation was withdrawn by the late 1880s. From 1901, there was a ban on Asian immigration with no permanent settlement allowed (Markus 2001:72).

ARCHAEOLOGICAL RESEARCH ON OVERSEAS CHINESE IN AUSTRALIA

Although there is a considerable amount of historical literature on the Chinese in Australia, archaeological studies are scarce in comparison. In 1996, Peter Bell (1996:16) summarised archaeological research to that date describing it as still being in its 'infancy' with many reports inaccessible to the public. In 2006, Lindsay Smith (2006:6) noted that 'only a handful of academic works' had been completed in the prior decade, and that those works were 'site specific or addressed a particular element of the Chinese in Australia', including gold and tin mining and ethnicity. Smith (2006) surveyed and excavated a number of gold mining sites in southeast NSW, investigating the elements that comprise rural settlements. Gant-Thompson (2008) built on that work, surveying further settlements around Kiandra. More recently, Lawrence and Davies' (2011) book has summarised research on overseas Chinese in Australia. They highlighted the research beyond mining, including urban life, market gardens and fish curing. While all these studies have increased our knowledge of Chinese in Australia, none relate to individual rural households and few focus specifically on ceramics. This research fills that gap by re-examining ceramic artefacts collected by Smith (2006), Gant-Thompson (2008) and the Australian National University field schools at Kiandra, NSW.

Ceramic studies

Archaeological studies of overseas Chinese began in America in the 1970s. The first published report on Chinese ceramics from American sites was by Paul Chace in 1976 on material from Ventura Chinatown, California. His paper was essentially a descriptive catalogue of the artefacts, establishing a typology for food containers and tableware. He concluded that these ceramics, along with opium pipes and cans, glass vials and Chinese coins, were the most common elements of material culture which typified the overseas Chinese settlements in America. A series of studies followed in America, including those by Olsen (1978), Pastron, Gross and Garaventa (1981), Hellmann and Yang (1997) and Praetzellis and Praetzellis (1997), expanding and refining the typology of Chinese-made vessels. While some of these list numbers of non-Chinese made ceramics, their focus is generally on the Chinese-made vessels.

The earliest comprehensive study of Chinese in Australasia was by Neville Ritchie (1986). His doctoral thesis on Chinese sites in the Otago goldfields, New Zealand, presented a detailed description of the ceramics found there, both Chinese and non-Chinese, and his results have been used as a reference for many interpreting Chinese sites in Australia. Ritchie (1986:647, 659) concluded that the Chinese in New Zealand were 'in an involuntary acculturative situation' and that they 'strongly maintained their traditional lifeways where possible'. He found that non-Chinese made ceramics were used more widely in urban areas than in rural mining camps and that the use of Chinese-made ceramics generally decreased over time.

The level of detail provided about ceramics in studies of Chinese sites in Australia varies. The first professional archaeological excavation of a Chinese site in Australia was undertaken by Ian Jack in 1982 in Cape York and the first published paper was by Jack, Holmes and Kerr (1984), Ah Toy's Garden: A Chinese Market-Garden on the Palmer River Goldfield, North Queensland. The ceramic artefacts had not been analysed at the time of that report and only rice bowls were noted as being present (Jack et al. 1984). Bell's 1996 survey found that over half of the studies to that time had been conducted in Queensland and a quarter in the Northern Territory. Many early studies of Chinese sites were done by cultural heritage management consultants or as environmental impact studies and lacked the depth of focused academic studies (Bell 1996), therefore details of ceramics are not always present.

Australasian studies which focus solely on ceramics are rare. Exceptions are Anne-Louise Muir's 2008 Master's Thesis, a post-graduate diploma by Lara Arun (2012), and Honours Theses by Jennifer Chandler (2005), Melissa Dunk (2010) and Laura Davies (2012). Some other studies have included Chinese ceramics as a part of their overall research (e.g. Lydon 1996; Bowen 2007). Muir (2008:4) explored Chinese-made vessels from sites in Melbourne's Chinatown, Victoria, and also provided numbers of non-Chinese made vessels although these 'were not examined in detail'. Jane Lydon's (1996) research on the Chinese in Sydney's Rocks area is the only other urban study of artefacts, however her focus was on the relationship between the Chinese and the white population rather than the artefacts themselves. While Chandler's (2005) study of ceramic vessels from the Upper Ovens goldfields in Victoria focused on the artefacts, there was no archaeological context for them as the research was based on collectors' vessels. Another rural study was that of Dunk (2010:2) who examined Chinese artefacts from Atherton Chinatown in North

Queensland, focusing on 'creating a catalogue' of the artefacts. Although Alister Bowen's (2007) PhD Thesis lists ceramic artefacts from Chinaman's Point, in Victoria, his focus was on the Chinese fish curing industry. Arun (2012) and Davies (2012) both examined ceramics from Lawrence Chinese Camp in New Zealand. Arun looked at opium pipe bowls, and Davies, vessels from Sam Chew Lain's house.

Few details are known about the supply of Chinese-made vessels to Australia. It is known that early miners were supplied with provisions by their headmen through Chinese networks. The number of Chinese merchants and from how many sources they obtained their stock is unknown. The only archaeological research on Chinese supply networks is Kevin Rains' PhD thesis, in which he concluded that 'Chinese importation and exportation exhibited a complex pattern of separation and interconnection in relation to the European market' in Cooktown, Queensland (Rains 2005:302).

Scant literature regarding the technical aspects of Chinese-made ceramics from the nineteenth century has been published in English. Few samples of post-1850 Chinese vessels were analysed prior to the 1980s when researchers began scrutinising blue-underglaze decorated vessels in order to determine the difference between genuine antiques and fake copies. Technical analyses that directly relate to overseas Chinese sites include those of Steele (1993), Stenger (1993) and McCrae (2001). These are detailed further in Chapter 2. However, apart from these examples, little technical analysis has been carried out on mid to late-nineteenth century Chinese ceramic vessels. Until the first investigations of overseas Chinese sites in America, there had been very little interest in Chinese ceramic ware from the nineteenth and twentieth centuries. While Ho (1988:3) noted that the 'relative neglect on the part of scholars' does not mean that the wares are 'of no significance', Steele (1993:308) stated that wares from this era have 'been regarded as technically and chronologically unimportant' in the overall history of Chinese ceramics.

The Chinese in southeast New South Wales

Lindsay Smith (2006:209) identified nine Chinese settlement systems within three regions in southeast NSW: around Braidwood, Tumut and Kiandra.

Braidwood Region

The Braidwood goldfields encompass an area measuring about 30 km (east-west) by 50 km (north-south) (Smith 2006:4). Smith (2006) identified two major Chinese settlements around Braidwood, namely Jembaicumbene to the south and Mongarlowe to the north.

Although the first official report of Chinese in the Braidwood area was not until 1859, McGowan (1996:3-4) suggested that it was likely they were present earlier as gold was first discovered at Jembaicumbene Creek in 1851. The goldfields in the Braidwood area 'were the most important in the colony' during the main period of Chinese settlement in NSW, 1858 to 1870 (McGowan 2004b:35). By February 1861, there were 400-500 Chinese on the goldfield at Jembaicumbene (Smith 2006:74). The NSW census of 1871 records 158 Chinese men and two Chinese women living at 'Jembaicumbene Gold Field' (HCCDA 2010a:365). Overall, Chinese were recorded as living in Jembaicumbene from the 1850s to the early 1900s, 'albeit in ever-decreasing numbers' (Smith 2006:287).

McGowan's (2004b:36) study of documents relating to the Braidwood area found cooperation between the Chinese and European miners, not only in mining activities but also in other areas such as market gardening. He pointed out that, although the Chinese and Europeans in Jembaicumbene lived in separate areas, it would have been impossible for them to avoid each other. Several contemporary accounts record social interaction between Chinese and Europeans. In 1861, at the opening of the new temple, the Europeans were invited to celebrate along with the Chinese (McGowan 2004a:329). In the 1870s, on the Mongarlowe goldfield, a European family taught the alphabet to the Chinese (McGowan 2004b:57). There were incidents of violence, but McGowan (2004b:39) suggested that the relationship between Chinese and Europeans should be viewed as 'part of a spectrum of experiences, rather than a continuum of harassment and violence'.

Although there were a 'considerable number of Chinese at work on the Mongarlowe River' (Smith 2006:394), the exact numbers at Flanagan's Point, a second-tier camp in the Mongarlowe system, are unknown. There was a 'sizeable' group of Chinese there in 1862 as shown by archival (McGowan 2004b:42) and archaeological evidence (Smith 2006) and they remained there until at least 1895.

Tumut region

The Tumut goldfields are located about 150 km west-northwest of the Braidwood goldfields. There were a number of Chinese settlements in the Tumut area with camps at Upper Adelong, Middle Adelong and Adjungbilly as well as minor camps elsewhere (Smith 2006). By 1858, there were at least 250 Chinese at the Upper Adelong settlement (Smith 2006:138). A newspaper report described the camp in 1861:

> The Chinese encampment at the Upper Adelong now numbers some 300 souls. They have three stores, two butchers' establishments, two restaurants, two gambling houses, and a variety of minor trading places, all conducted by the Celestials. They have also a hydraulic machine to work the bellows of the forge, a rice mill upon a large scale, all constructed by themselves, and we understand that the place on Saturdays and Sundays presents a very lively picture of Chinese interior life, without the trouble of visiting China (The Maitland

Mercury & Hunter River General Advertiser, 27 April 1861:6).

The first recorded Chinese death in Upper Adelong was in 1860 and the last in 1901 (Smith 2006:514).

Adjungbilly was proclaimed a goldfield in July 1861, with mining continuing intermittently until the early twentieth century (Smith 2006:138). The exact number of Chinese at these goldfields is unknown.

Kiandra Region

These goldfields are located about 150 km west-southwest of the Braidwood and about 90 km south of Tumut, on the Eucumbene River in the Snowy Mountains within Kosciuszko National Park. The Chinese arrived in Kiandra in 1860 and their main camp, inhabited from 1860 to about 1900, was located about one kilometre east of the township. The Kiandra gold rush was 'one of the most hectic short-lived gold rushes in the history of Australia' (Moye 1959:ix). Within a few months of their arrival, numbers of Chinese increased to about 700 or about 20 per cent of the total population (Smith 2006:166). However, in the winter of 1860 most miners were unable to work because of snow falls and the Chinese in the district were engaged to carry goods overland as it was not possible to carry them by packhorse (Preshaw 1860). In 1861 the newspaper correspondent from Kiandra reported 'It is astonishing how quickly they settle down; a little village springs up as if by magic, and where yesterday not a trace of habitation was to be seen, to-morrow will reveal a small township teeming with a busy swarm of Mongolians' (The Sydney Morning Herald, 11 May 1861:7). The 1871 NSW census details 63 Chinese men living in the environs of the village of Kiandra, while none lived in the township itself (HCCDA 2010b:376). In 1877, the number of Chinese in Kiandra was 106, but from then on numbers decreased (Smith 2006:607). The first Chinese person to purchase land in the township of Kiandra was Tom Ah Yan, a storekeeper, in 1882 (Smith 1998:198). A second store, later owned by Ping Kee, was a boarding house for work groups from Nine-Mile when they came into town for supplies (Hueneke 1987:53).

Relations between the Chinese and Europeans were changeable. In 1860, the Gold Commissioner appointed ground for the Chinese camp away from the main township 'to lessen the chances of collision with Europeans, who complain of the Chinamen on account of their dirty habits' (The Sydney Morning Herald, 21 August 1860:7). By 1861, the Kiandra newspaper correspondent reported that the outcry against the Chinese was 'a great deal of bunkum' and that 'the only cause of complaint against them (the Chinese) seems to be that they work too well' (The Sydney Morning Herald, 11 May 1861:7). He stated that the storekeepers in Kiandra were happy to serve the Chinese as they always had money for their purchases. Later that year the Europeans were 'loud in their complaint against the Chinese' and a petition was signed by about 300 miners 'for the entire removal of the Chinese' (The Sydney Morning Herald, 31 August 1861:5). The correspondent then wrote, 'it is hoped that from the Chinese themselves showing a disposition to clear their own community of its scum, any further disturbances will be avoided' (The Sydney Morning Herald, 31 August 1861:5). The Chinese became an important part of the community at Kiandra sharing many experiences with non-Chinese people.

Ceramic artefacts from sites in southeast NSW

The ceramic artefacts, detailed in this book, are from camps at Jembaicumbene, Flanagan's Point, Upper Adelong, Adjungbilly, and Kiandra, along with several smaller camps in the Kiandra Valley (Figure 1.1). In addition, the ceramic assemblages from two Township stores in Kiandra were compared to those of the camps. In total, 76 individual sites from within these locations were examined. This research fills gaps in our knowledge and understanding of Chinese settlements by using ceramics to explore social and economic issues. It identifies variation within and between settlements and is the first intra-site analysis of Chinese camps in Australia. In addition, it adds to the current knowledge of the technical aspects of ceramics found at overseas Chinese sites. The sites in this study had previously been placed within a Chinese settlement pattern and were thought to have functioned as 'homogenous and segregated communities' until the end of the nineteenth century (Smith 2006:2).

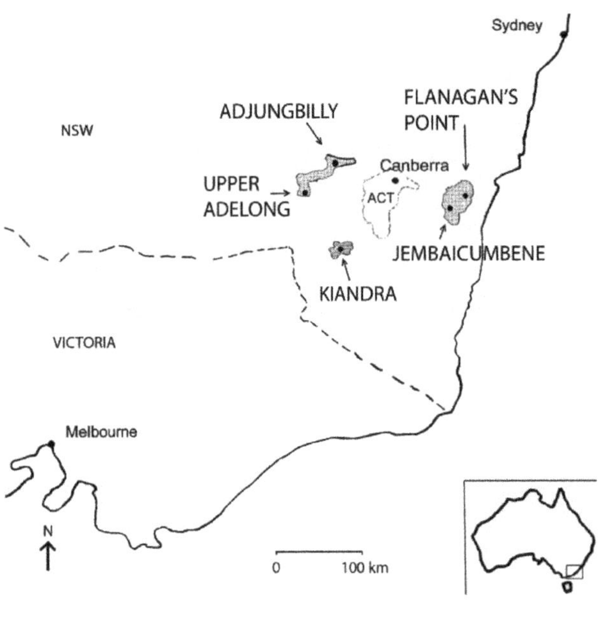

Figure 1.1: Location of sites in this study (shaded areas from left: Tumut, Kiandra, and Braidwood regions after Smith 2006:4)

To date, no pattern of Chinese supply has been established within Australia. This study gains more information regarding this through an investigation of forms and decorative types recovered from different regions. Furthermore, no previous Australian studies have compared Chinese assemblages to contemporary non-Chinese assemblages. Brooks & Connah (2007:143) stated that the topic of ceramic supply in general is currently 'a weak point in Australian historical archaeology'. The inclusion of all ceramics, not just those of Chinese origin, adds knowledge to the overall study of nineteenth century ceramics in Australia. The Chinese sites in southeast NSW have numerous ceramics of non-Chinese origin and this research explores what vessels were used and how these met the needs of the Chinese. The availability of contemporary non-Chinese assemblages from southeast NSW allowed a unique opportunity to compare the ceramics from those sites and address questions on the accessibility of non-Chinese ceramic vessels for the Chinese.

The first section of this book, including this chapter and Chapters 2 to 5, provides background information necessary to interpret the results. Chapter 2 offers a brief overview of archaeological ceramic study, including research on ceramics from Australian sites and technical analyses of Chinese ceramics. It also includes some characteristics of clay and aspects of microscopic analyses which are relevant to this research. Chapter 3 provides a brief overview of the origin, making, marketing and export of Chinese-made ceramic wares, found at overseas Chinese sites from 1850 to the early-twentieth century, and Chapter 4 follows with a description of these vessels. Data from other researchers are used to establish the type of Chinese-made vessels found at other sites. Many studies on Chinese sites have highlighted the Chinese-made ceramics while only making passing reference to others. The Chinese sites in southeast NSW have numerous ceramics of non-Chinese origin and Chapter 5 details the forms and decorative types available from the mid-nineteenth to the early-twentieth centuries in order to provide a context for these vessels.

The next section of the book presents the results of the re-examined ceramic assemblages. Chapter 6 details the assemblages and outlines the methodology used in cataloguing and technical analyses, and the rationale for this approach. Chapters 7 to 10 provide a description of the recovered ceramic artefacts and a brief analysis of sites within each region. Chapter 7 presents the results from the Braidwood region, including Jembaicumbene and Flanagan's Point Chinese Camps. The Tumut region is covered in Chapter 8, including the results from Upper Adelong and Adjungbilly Camps. Chapter 9 details the Kiandra Chinese Camp results and Chapter 10 those from the outlying camps in Kiandra Valley and from the Township stores.

The final section of the book, Chapters 11 to 14, presents further analyses and discussion. This research aimed to establish more knowledge about the number of sources of Chinese ceramic vessels through a compositional analysis of the sherds, the results of which are presented in Chapter 11. In any archaeological study of ceramics, vessels are usually categorised according to their physical similarities and differences and this has proved successful when forms and decoration change over time. The Chinese-made vessels follow a ceramic tradition in which forms and decoration remained stable for hundreds of years. Grouping vessels according to physical similarities only, may not be the most appropriate method of determining their source. The scarcity of similar technical analyses on nineteenth century vessels commonly found on overseas Chinese sites makes this study significant. Chapters 12, 13 and 14 present an overall discussion and a conclusion of the research. Chapter 12 summarises, collates and compares all the ceramic data from each site in this study with specific data from other overseas Chinese sites. Each vessel type is analysed separately within the broad functional categories of storage, tea and tableware, and smoking. While these compilations provide useful knowledge about each form, they do not, of themselves, allow an understanding of the communal, domestic and economic issues which are discussed in Chapter 13.

Since individual Chinese are often more difficult to research through documents than other immigrant groups, archaeologists have tended to approach them as a single homogenous group (Praetzellis et al. 1987:39). Documents relating to the Chinese at these camps or to their ceramics are scant. The few Chinese individuals known from newspaper articles or court proceedings have generally been involved in illegal or tragic events, for example, Cum Tow, a Chinese interpreter, who murdered another Chinese man at Kiandra (The Sydney Morning Herald, 11 April 1871:2) or Mrs Wee Drieg, the Chinese doctor's wife, found drowned near her home in Jembaicumbene (The Sydney Morning Herald, 10 July 1877:5). Little is known about the everyday life of the people in these camps. As McGowan (2004b:58) noted, material evidence is important in understanding the Chinese and their living arrangements, given 'the paucity of most archival records'. Time constraints mean that many studies have only considered ceramic assemblages as a whole. For example, Smith's (2006) Chinese settlement study of southeast NSW interpreted the origin of ceramic artefacts using the total weight from each settlement. This has given the impression that 50-80 per cent of the ceramics from sites in southeast NSW are of Chinese origin (Lawrence & Davies 2011:232). However, analysis of individual sites within each settlement produces a different outcome. Rather than considering the ceramics from settlements as a whole, this research goes beyond the total and looks at individual sites. This allows differences to be seen which reveal changes over time in social and economic circumstances and produces a clearer picture of which ceramic vessels individuals were using in their everyday life.

CHAPTER 2

Ceramics and archaeology

ARCHAEOLOGICAL CERAMIC STUDY

Ceramic vessels are part of everyday life for most people. Deetz (1977:68) summed up their essence when he described them as 'fragile yet indestructible' making them 'among the most informative kinds of material culture'. Vessels contain information about where and how they were made, the site where they were found and provide an insight into the lives of the people who used them. Food and drink containers, cooking and serving vessels, vessels for recreation can all be made of ceramic fabric.

Physical aspects of ceramic sherds recovered from archaeological sites have the capacity to provide information about economies, trade, and site formation. The combination of elements in a ceramic vessel may be traceable to a source as each geological area has its own fingerprint. Once clay has been heated to a certain temperature the process is irreversible and the method of forming a vessel leaves an imprint making it possible to differentiate, for example, between a hand-formed and wheel-formed vessel. This allows inferences to be drawn about the sources of ceramic vessels recovered from archaeological sites and the economies that made them. The condition of the recovered sherds, for example breakage patterns, weathering and burning, can suggest events contributing to the site formation processes.

The forms and decorative types found can provide information about diet, the availability of ceramics in any area, consumer choice, status and identity. The presence of any vessel is related to the household size and composition, and their need and ability to purchase that vessel. The form of tableware used is directly related to the diet of the people using them. Form and decoration change over time according to fashion so that they might be able to be assigned to a particular time period suggesting a date when they were made. While popular culture changes rapidly over time, 'folk culture is traditional and conservative' and therefore less likely to change (Deetz 1977:65). British-made ceramics underwent many changes in both form and decoration during the nineteenth century, but Chinese-made vessels, found on overseas Chinese sites, changed little in hundreds of years. Deetz (1977:73) stated that artefacts must be understood in their 'functional and symbolic role'. He added that each culture has its own set of rules for putting together elements within that culture.

Unfortunately every breakage loses information about the form and decoration of a ceramic vessel but it may add to the overall knowledge of the site and its inhabitants. If there are no diagnostic parts remaining, for example, rim or foot, the form may not be identifiable. Partial decoration may also be difficult to identify, but small sherds may provide information about how they were introduced into the site, for example being dropped, lost or discarded. Conjoining sherds can suggest how a site has been disturbed.

Archaeological evidence is important to the study of ceramics even if they are from the recent past as documentary sources relating to the production of ceramic vessels are often incomplete. For example, in Britain, the source of the majority of nineteenth century ceramic tableware in Australia, only a small number of factory and merchants' records exist (Barker 2001). In addition, museums tend to focus on higher quality and more unique vessels than on everyday mass-produced vessels. Ceramics are found on most archaeological sites making their study central to site interpretation. In Australia, ceramics and glass are usually the largest groups of artefacts at historical sites.

Orton, Tyers and Vince (1993) summarised the history of ceramic studies in their book, Pottery in Archaeology, outlining 'three broad phases': art-historical, typological, and contextual (Orton et al. 1993:4). The art-historical period dated from 1500 and was concerned with whole pots rather than sherds. As excavations unearthed greater amounts of pottery, there was increasing pressure for a classification system, but the true typological phase did not start until the 1880s when Pitt-Rivers developed a typology for other artefacts. At the same time relationships between pottery and stratigraphic sequences were being developed by Flinders Petrie. In effect the typological period was concerned with chronology through stratigraphy and defining cultural areas through spatial distribution (Orton et al. 1993:9).

The contextual period, dating from 1960 onwards, used ethnography, quantification and microscopic analyses to interpret assemblages (Orton et al. 1993:5). Anna Shepard's 1956 seminal work, Ceramics for the Archaeologist, coincided with the introduction of scientific techniques into archaeological ceramic studies. Her book outlined the properties and sources of clay, practical methods of making vessels, analysis of ceramics, and evaluation of ceramic data. Since the 1950s ceramics have been 'a major focus of archaeometric or archaeological science studies' (Tite 2008:216). Scientific studies have been used for dating, assigning provenance and determining function through residue and use wear studies.

This last period of ceramic studies also coincided with the development of historical archaeology. The Post-Medieval Ceramic Research group, founded in Britain in 1963 and later changing its name to the Society for Post-Medieval Archaeology, was initially formed to focus on material dating between 1500 and 1750 AD (Barker &

Majewski 2006). Researchers in Britain have only recently begun investigating nineteenth century pottery sites, uncovering evidence of wares dating from the 1850s (Barker 2001). Pottery wasters, i.e. wares damaged during firing, have provided insights into vessels which were made specifically for export and were uncommon in Britain, including white granite, flown-blue and flown-mulberry vessels (Barker 2001).

Until the 1980s American historical archaeologists' studies focused on eastern American sites that predated 1800 (Majewski & O'Brien 1987). This changed with the initiation of the United States government funded cultural resource management. Majewski and O'Brien (1987) argued that although this opened up new types of sites, such as small farmsteads and inner-city houses, archaeologists lacked knowledge relating to material culture, particularly ceramics. They described ceramics as being 'one of the most misused categories' of artefacts found on historical sites (Majewski & O'Brien 1987:99).

Different approaches in Britain and America led to two methods of cataloguing. Traditionally, historical archaeologists in Britain described their finds by grouping them in categories such as ceramics, glass, metal and bone. At the same time in North America, the emphasis was to group objects in functional categories such as personal or architectural (Cochran & Beaudry 2006).

There are no standard guidelines for artefact catalogues throughout Australia. Crook, Lawrence and Gibbs (2002) commented on the need to produce reliable and consistent catalogues in the absence of guidelines. They put forward the role of catalogues as the 'record of attributes associated with each artefact' and designated a database as a computerised version of the catalogue (Crook et al. 2002:26). They stressed that cataloguing and analysis are distinct: cataloguing details the raw data; analysis then orders that data. The primary purpose of a catalogue is to 'record artefact attributes to facilitate the holistic analysis and interpretation of the site assemblage, or selected parts of it' (Crook et al. 2002:31).

Previous research on ceramics from Australian sites

In addition to research on Chinese-made ceramics outlined in Chapter 1, a number of studies have been carried out in Australia on Australian-made ceramics, assemblages containing mainly British-made ceramics, and on assemblages containing Chinese export porcelain. Lawrence and Davies' (2011) book, An Archaeology of Australia Since 1788, details ceramic analyses associated with a range of topics from convict origins, aboriginal dispossession and gold rushes, to shipwrecks and the archaeology of cities. While many of these studies include ceramic analysis, it is not necessarily the focus of the research.

A study focusing on Australian-made pottery is that of Birmingham (1976) who oversaw the excavation of Irrawang Pottery in the Hunter Valley, NSW, in the 1960s. In 2008 Kelloway investigated the chemical composition of milk pans made by this pottery as part of her Honours Thesis. Another study of Australian-made ceramics was by Casey (1999) who examined the vessels used in dairying at the Brickfields site in Sydney. A number of studies have focused on assemblages which mainly contain British ceramics. Atkins' 1991 Honours Thesis, Not to be Excelled for Elegance or utility: A Study of the Availability of Ceramics in Sydney 1803-1868, investigated the availability of ceramics, most of which were British, using advertisements from two Sydney newspapers, The Sydney Gazette and New South Wales Advertiser (1803-1842) and The Sydney Morning Herald (1831 onwards). Briggs's (2005) PhD Thesis used the ceramic assemblage from working-class residences to determine attitudes towards respectability. Crook, Ellmoos and Murray (2005) evaluated the lives of the working-class residents of Cumberland and Gloucester Streets at The Rocks in Sydney. Crook (2008) used those artefacts in her PhD Thesis, which examined the role of cost, quality and value in nineteenth century ceramics. Beale's (2006) Honours Thesis examined the ceramic assemblage from the Woolgar goldfield in Queensland. Hayes' (2008) PhD Thesis used ceramics to investigate middle-class gentility in nineteenth century Australia. Corcoran (1993) placed Chinese export porcelain within a social context in Sydney as part of her Master's research.

TECHNICAL CERAMIC STUDIES

Technical analysis of archaeological ceramic sherds has been conducted since the 1950s on particular aspects such as the raw materials used, and the methods of making, decorating and firing ceramics. Technical examination of ceramic sherds can be macroscopic or microscopic. While macroscopic examination can determine physical properties of clay, including colour, hardness, porosity and texture without the use of expensive equipment, microscopic analysis may be necessary to determine provenance, firing methods and temperatures, glaze compositions and identification of organic residues. As Orton, Tyers and Vince (1993:146) pointed out, the choice of which techniques to use may be limited to what is available.

Macroscopic characteristics

Clay, the main component of ceramic vessels, is formed by the decomposition of rocks through chemical weathering over long periods of time. Its grain size and characteristics are determined by the way in which the clay was formed, i.e. if it remains in its place of origin or is transported elsewhere. Clay which remains in its place of origin is usually white and has fewer impurities than clay which has been transported elsewhere by the action of water or ice.

A clay body is categorised according to the nature of the research. Geologists or ceramic engineers, for example, may consider the minerals present and their particle size in defining a clay material. In archaeology, potters' terms are often used, classing the clay as earthenware, stoneware or porcelain. Definitions of these types are related to their firing temperature. Traditionally, the lowest-fired clay-type is earthenware which is fired to a temperature of less than 1100°C (Hamer & Hamer 1997:115). Stoneware is usually fired above 1200°C and porcelain above 1300°C (Hamer & Hamer 1997:327, 258).

Although many naturally occurring types of clay can be made into vessels, additional materials may have to be added in order to get a body that is both plastic and workable. Clay minerals contribute to plasticity but non-plastic additives, including quartz, calcite, plant material and shells, can increase workability by countering shrinkage and decreasing cracking during drying and firing (Shepard 1956:24-25; Rye 1981:31). Non-plastic additives are commonly referred to by archaeologists as temper. Rye (1981:31) stated that although the term 'temper' is used in archaeological literature there is 'no equivalent term in the field of ceramic technology' because of different research goals. Archaeologists try to reconstruct human behaviour while ceramicists explore the behaviour of minerals (Rye 1981:31). Archaeologists may distinguish between the ingredients of a ceramic body (clay and temper) to provide more information about the source of materials, clay preparation and production methods in order to understand more about the people who made the wares.

The plasticity and workability of any clay will directly affect what forming methods and firing temperatures are used. Common making methods include throwing on a wheel, hand forming by the use of slabs and coils, and slip casting using liquid clay into a mould. Many factors will ultimately determine what clay is used, how a vessel is made and if it is decorated. These can include the accessibility of raw materials, technical knowledge, availability of labour, tools and equipment, the economic system and market distribution (Tite 2008:222).

The properties of clay are very different before and after firing. Clay becomes plastic when mixed with water which allows it to be worked into a form that can retain its shape. Firing, the process of heating and hardening the clay, will affect the strength, colour and porosity of a ceramic fabric. As clay is heated it undergoes decomposition of organic matter, ceramic change, oxidation of carbon and sulphur, and vitrification. The start of any firing must be slow to allow any steam that has built up, through the heating of water in the clay body, to escape without breaking the vessel. Organic matter begins to break down at about 200°C. During the ceramic change phase, the bound water in the clay crystal structure is driven off (Hamer & Hamer 1997:130). This process starts between 350°C and 450°C and is finished by 700°C. During this phase sintering also begins. Clay particle edges soften and fuse together causing shrinkage and reducing porosity, but only insignificantly before 1000°C (personal communication, Dr Anthony Flynn, 27 February 2011). After 700°C, any carbon and sulphur present will combine with oxygen and be expelled as gases. Carbon will usually be burnt out by 900°C and sulphur by 1150°C (Hamer & Hamer 1997:131). As vitrification of the clay body begins, the clay starts to become impervious and loses its ability to absorb water. If organic material is not fully removed before the commencement of vitrification, gas will be trapped in the body and the vessel wall will be bloated (Hamer & Hamer 1997:27-28). Vitrification occurs as the potassium and sodium oxides within the body interact with the free silica and alumina, filling pore spaces with melted glass.

The colour of any fired clay can have subtle variations resulting from the way it has been fired, therefore the only reason for recording colour would be to provide information about firing atmosphere or to differentiate between ware types, for example yellow, red or white earthenware. An oxidising atmosphere, i.e. one with adequate oxygen, will produce clear colours, while a reducing atmosphere, i.e. one that lacks sufficient oxygen to permit complete fuel combustion, can produce greys. As Shepard (1956:16-17) pointed out there is no 'simple' relationship between the colour of clay before and after it is fired. White pottery is only produced from white clay and coloured pottery from coloured clay, but white clay can also produce fired grey bodies. For example, if clay is low-fired then grey or black colouring may signify carbon material in the clay. Clay is coloured by impurities, principally iron and carbon compounds. Iron compounds are converted to oxides during firing and the resultant colour will depend on the amount of iron, its particle size, distribution and the atmosphere in which it is fired. Raw buff-coloured clay can produce cream, yellow, grey or black bodies when fired, while red and brown clays can produce yellow, red, brown, grey and black ceramic bodies (Shepard 1956).

Colour may have aesthetic appeal but hardness and strength determine 'how well a vessel will stand usage' (Shepard 1956:101). Hardness, another clay property, can be determined through a scratch test but is again of limited use in classification due to the variation within types of clay and the similarities between different types (Shepard 1956).

All ceramic bodies are porous to some extent. Porosity can be defined as 'the total proportion of the air space contained between the solid particles of which the body is composed' (Grimshaw 1971:417). Pores can be either open (not sealed) or closed (completely sealed). The true porosity of any ceramic body is a ratio between the volume of all pores, whether open or closed, and the volume of the body. Apparent porosity, a measure of the unsealed pores of a body, is 'the ratio between the volume of water or liquid capable of being absorbed into it and the total volume of the article' (Grimshaw 1971:419). Porosity may be expressed as a percentage by weight or volume, and is often given in the form of water absorption, which can be determined by a simple test that compares the dry and wet weights of vessel sherds (Singer & Singer 1963:343).

Porosity reduces as firing temperature increases and vitrification occurs. The porosity of a vessel affects its 'density, strength, permeability, degree of resistance to weathering and abrasion, extent of discolouration by fluids...and resistance to thermal shock' (Shepard 1956:126). Porcelain, traditionally the highest-fired ceramic body, will be denser and stronger, and more resistant to weathering than earthenware. Porcelain will have a porosity close to 0 per cent as the body becomes increasingly vitrified when it is fired to 1300°C or above (Hamer & Hamer 1997:53). Stoneware has a variable porosity which is usually less than 5 per cent and often below 2 per cent (Hamer & Hamer 1997:327). Earthenware has a porosity greater than 5 per cent as it is traditionally fired to lower temperatures (Hamer & Hamer 1997:115).

When describing the texture of a pot, archaeologists usually refer to the physical properties observed in the cross-section of a pot. Hargrave and Smith (1936:34) defined texture as the 'visual or tactile character of the core of a pottery vessel, which results from the arrangement, size, shape, and relative quantities of the individual particles composing the core, regardless of color or chemical composition'. They proposed a series of descriptions to assist archaeologists in the field: very coarse, coarse, medium, fine, very fine (Hargrave and Smith 1936).

Ceramic bodies may be glazed to make them waterproof, or purely for decoration. Many methods can be used to glaze a vessel including brushing, pouring, dipping or spraying. A vessel may be glazed in its raw state, i.e. before firing, or after an initial firing. A glaze is essentially a layer of glass which is fused onto a ceramic body and will contain a glass former, a flux and a stabiliser. The most important ingredient in a glaze is the glass former. This is an oxide that retains its molten property when it solidifies through cooling; i.e. it does not return to a crystalline state (Hamer & Hamer 1997:154). The most common glass former is silica; another one is phosphorous. A flux is 'an oxide which promotes ceramic fusion by interaction with other oxides' (Hamer & Hamer 1997:141); i.e. it helps melt the other ingredients. The most common fluxes are sodium, potassium, calcium, magnesium and lead. The stabiliser determines the viscosity (stiffness) of the glaze (Hamer & Hamer 1997:326). The most common stabiliser is alumina which acts to stabilise the combination of the glass former and fluxes ensuring that the glaze does not run off the vessel. The glaze must have the same degree of thermal expansion as its ceramic body, otherwise faults such as peeling and crazing will occur.

The colour of a glaze is a consequence of the oxide colorant used and the heat work produced during a firing. Heat work is a combination of the temperature reached and the length of firing and determines whether a glaze will be shiny, satin or matte. This results in the potential for the same glaze to appear to be a different colour under different firing conditions (personal communication, Dr Anthony Flynn, 28 September 2011).

Any clay can be used as a glaze if sufficient heat is available. Temmoku, a Chinese brown glaze, is basically earthenware clay used as a glaze on a body capable of being fired to a higher temperature (Kaplan 1952:781). Additions of other clays, organic or mineral fluxes (e.g. ash or lime) or feldspars can vary the appearance of a clay-based glaze, as can the firing atmosphere, and as a consequence an infinite variety of glazes are possible. This means that visual distinction alone cannot determine a region or time period to which a glaze can be attributed (Kaplan 1952:8).

Microscopic analyses

Although availability and access to microscopes may have increased in the past twenty years, the cost and location of facilities is still a major consideration. Many methods of microscopic examination that have commonly been used in provenance studies, including electron microprobe analysis, X-ray fluorescence, instrumental neutron activation analysis, proton-induced x-ray and gamma ray emission and inductively coupled plasma - optical emission spectrometry. These techniques all have advantages and disadvantages. Some techniques analyse surface or near surface samples while others analyse bulk samples, either by solution or solid samples.

Electron microprobe analysis (EMPA), the method used to analyse samples detailed in this book, focuses an electron beam onto a sample. This excites the electrons of target elements, resulting in the emission of characteristic x-rays. The concentrations of elements are determined by comparing the intensity of the x-rays emitted with those of standards, which are either compounds of known composition or pure elements. Images produced on a screen result from 'scanning the beam in a television-like raster' (Reed 1996:1). One of the advantages of EMPA is that apart from the original cut to the sample, the process is not destructive. Accuracy to about 1 per cent can be achieved routinely and 'all elements above atomic number 10 can be determined with fairly uniform accuracy' (Reed 1996:3). EMPA is suited to surface or near surface samples (Hancock 2000:17). It can be useful for analysing concentrations of elements and obtaining a chemical profile, which is the 'weighted average of all the mineralogical components of a ceramic specimen' (Bishop et al. 1982:293-294).

Many archaeological studies have made use of the fact that any ceramic vessel will be made from clay and glaze ingredients which come from a particular area and have a particular elemental signature. Using this signature may be a more reliable way of determining the source of ceramic vessels than relying on visual attributes such as shape or decoration. Bishop, Rands and Holley (1982:276) stated that compositional analysis of a ceramic body can distinguish between sources of raw materials either directly by 'establishing probable relationships of pottery to geographically localized raw materials', or indirectly by confirming differences in ceramic samples which are 'sufficient to indicate the existence of geographically isolable resources'. This is based on the 'Provenience Postulate' of Weigand, Harbottle & Sayre (1977:24) which was originally applied to the study of turquoise but has since been extended to the study of ceramics. The postulate dictates that differences exist between sources of raw material and that variation within a source will be less than that between sources.

Although the provenance postulate has been applied to ceramics, any pottery may be the combination of multiple sources of raw materials or many potteries may use raw materials from the same source. In addition, elemental composition may vary within a single clay pit as particle size will not be even throughout, or source clay may have added temper or colorants. In archaeological samples, it is not always known if inclusions in the body are part of the natural clay or whether they have been added as temper. Components within a body may be distinct when that body is unfired or low-fired but will begin to melt together as the firing temperature increases. While some researchers examine clay and temper separately to answer specific questions, others prefer to analyse a vessel's composition as a whole. Burton and Simon (1996:408), for example, stated that they were 'interested in assessing the variability within and among sets of ceramics, as finished vessels, not as lumps of clay'. Despite associated problems, Arnold, Neff and Bishop (1991:88) support:

> A notion that "source" or provenience has important chemical and behavioural (cultural) components and pottery thus encodes both chemical information from the source and behavioural information from the potter. In spite of the problems of relating pottery to its constituent raw materials, pottery made in the same community and drawn from the same set of sources would thus be expected to be similar in chemical composition.

There have been debates about how many elements to test for and whether the major, minor or trace elements provide the best indication of provenance. Some techniques, such as those using powdered samples will automatically choose the elements which are tested for while in other techniques, such as EMPA, the researcher may have to manually select the elements to be tested for by examining graphed spectral peaks.

An argument for using major and minor elements rather than trace elements to determine provenance is that those most abundant in any group of samples, are less likely to be 'susceptible to sampling heterogeneity' (Burton and Simon 1993:46). Major and minor elements have been used successfully to determine provenance by examining the ceramic body as a whole or by separating the clay and temper. For example using EMPA, Summerhayes (2000:38) found that major and minor elements including magnesium, aluminium, silicon, potassium, calcium, titanium and iron were more useful than trace elements in determining similarities and differences between sample groups of Lapita pottery. Abbott, Lack and Moore (2008:59) used EMPA to distinguish between the clay body and the temper, analysing for sodium, magnesium, aluminium, silicon, potassium, calcium, titanium and iron.

Multivariate analysis is usually used to determine the grouping of samples after obtaining chemical profiles. Principal component analysis (PCA) is appropriate to use if 'the structure of the data set is initially unknown' (Baxter & Jackson 2001:253). Discriminant analysis is used to test the probability of group membership once groups have been established. There has been much debate about how many elements to include in the statistical analysis. Some researchers have suggested that all the elements should be used (e.g. Bishop et al. 1982). Grave, Lisle and Maccheroni (2005) too, concluded that high dimensional datasets were more sensitive to discrete compositional groups. These findings conflict with Baxter and Jackson (2001), who suggested it may not be necessary to use all elements measured in the analysis. They stated that those who advocate for the use of the maximum number of elements possible, for example 'Harbottle 1976; Pollard 1986; Glascock 1992', are quoting 'Sneath and Sokal's first axiom of Numerical Taxonomy' which is appropriate to biological classification rather than archaeological classification (Baxter and Jackson 2001:253-254). Based on propositions from Krzanowski (1987) and Jolliffe (1972), Baxter and Jackson (2001:254) suggested selecting a subset of the variables that 'preserve the structure revealed by a principal component analysis (PCA) of the full data'.

Previous technical analyses of Chinese ceramics

Although much research has been carried out on Chinese-made ceramics from the Neolithic period onwards, there is a distinct lack of research on post-1850 samples. Studies have been conducted on compositional analysis of raw materials and/or finished vessels, and provenance determination (e.g. Hall et al. 1973; Pollard & Hatcher, 1986, 1994; Yanyi 1987; Leung et al. 2000; Cheng et al. 2006; Yin et al. 2011).

Ceramic vessels from south and north China have different chemical compositions. This results from the proportions of clay minerals present from the different geological sources. Percentages of the main ingredients were established by Hall, Schweizer and Toller in 1973. They stated that while silica and alumina make up 85-95 per cent of the ceramic bodies, southern wares have 10-24 per cent alumina and 69-83 per cent silica whereas porcelain from northern China has 27-38 per cent alumina and 47-65 per cent silica. Potassium also varies from the south, 2.0-6.6 per cent, to the north, 0.9-2.3 per cent (Hall et al. 1973:65-66).

Early research on blue-underglaze decorated wares (also known as blue-and-white wares) was carried out in the 1950s (Young 1956) and the 1960s (Banks & Merrick 1967) using X-ray fluorescence (XRF) to determine the manganese/cobalt (Mn:Co) ratio of the blue-underglaze. It was found that ratios varied according to the age of the ware and the source of the cobalt. All but four of these samples analysed were made prior to the nineteenth century. In the 1980s researchers began analysing the Mn:Co ratio of underglaze decoration to resolve the differences between original and fake blue-underglaze decorated Chinese wares by analysing a number of vessels from the nineteenth and twentieth centuries (Yap & Tang 1985; Yap 1986, 1988). Building on the work of previous researchers, Yu and Miao (1996:258) used energy dispersive XRF to analyse blue-underglaze decorated wares comparing 27 samples from pre-Qing dynasties, 29 from the Qing dynasty (1662-1908), two from the Republic period (1912-1949) and ten made in 1995, purchased from Hong Kong markets. They reported that differences could be seen in wares dated to different dynasties (Yu and Miao 1996:261).

Research that directly relates to overseas Chinese sites includes that of Steele (1993), Stenger (1993) and McCrae (2001). Steele (1993) used optical emission spectroscopy (OES) to determine Mn:Co ratios present in the blue-underglaze used on Chinese vessels found at American sites occupied by Chinese in the nineteenth and twentieth centuries. He confirmed a change in the underglaze composition over time which could help in the dating of vessels. Stenger (1993) used OES along with XRF to determine the elemental composition of celadon glazes. She found that criteria could be established to differentiate between nineteenth and twentieth century celadon-glazed wares and controversially suggested that celadon-glazed wares found on some American overseas Chinese sites may have been made in Japan (Stenger 1993). Thin section analysis of six Chinese samples (three brown-glazed jars, two olive-green glazed jars and one blue-underglaze bowl) recovered from Cooktown, Queensland, was carried out in 2001 to demonstrate that the wares were not made locally (McCrae 2001). The analysis revealed that the sherds were structurally similar 'derivatives of a volcanic tuff' (Curtis 2001 cited in McCrae 2001:48). They were all stoneware and probably fired between 1200°C and 1280°C. The materials used in the wares were not alluvial, implying 'that a prior mining and manufacturing process would have been necessary before manufacture of the containers' (McCrae 2001:49).

The next chapter provides background information on the origin, production and marketing of the Chinese-made ceramic wares recovered from overseas Chinese sites dating from 1850 to the early-twentieth century.

CHAPTER 3

Chinese ceramics – from kiln to campsite

Chinese-made vessels, found on overseas Chinese sites, have been described as 'the pottery of the working class Chinese' (Chace 1976:528), 'durable rather than particularly beautiful' (Olsen 1978:4). They are the most common element of material culture that typifies overseas Chinese settlements in America, Australia and New Zealand.

Ceramic vessels were an important part of the lifestyle for all Chinese. From the earliest times, ceramic vessels were more highly thought of than vessels made of precious metals, pewter or wood (Medley 1976). The Chinese have two words for ceramic fabrics: *t'ao* refers specifically to earthenware or to pottery in general, while *tz'ŭ* refers to the higher-fired bodies of stoneware and porcelain (Medley 1976:13). During the late-thirteenth and early-fourteenth centuries kilns in China changed from being privately owned 'craftsmen's kilns into a series of industrial complexes' (Medley 1976:171). A highly specialised workforce was developed. Wares for everyday use were made in provincial kilns, or people's kilns, *min-yao*, as opposed to the Imperial kilns, *kuan-yao* (Quellmalz 1976:289).

General surveys of Chinese-made ceramics rarely feature the everyday pottery of the working-class. Honey's (1946) book, The ceramic art of China and other countries of the Far East, which covered ceramics from its known beginning to the nineteenth century, detailed the more highly-valued wares but only briefly mentioned everyday vessels. He noted that:

> The coarse grey porcelain made for the islands of the Pacific, boldly painted in blackish blue...is far more admirable than the laboriously careful copies of classical wares which remained fashionable into the 20th century. Such rough blue-and-white has often been mistaken for early-Ming ware, and has unmistakable vitality (Honey 1946:156).

Quellmalz (1976:289) agreed that these wares are 'exceedingly attractive' but that collectors 'scarcely thought them worth their attention'. He stated that there are a 'multitude of provincial kilns' but 'almost nothing' is known about the wares they produced in the Qing dynasty (Quellmalz 1976:289).

ORIGIN OF WARES

Although provincial kilns are found throughout China, it is likely that the vessels, exported to be used by the overseas Chinese, were produced in the southern Chinese provinces of Guangdong, Fujian, Zhejiang and Jiangxi (Quellmalz 1976). A number of factories operating in the Qing dynasty within the southern provinces have been identified in Western literature and Hetherington (1921) hinted that more pottery centres were written about in Chinese literature. He listed two factories in Zhejiang, ten in Fujian, one in Hunan, three in Jiangxi and eight in Guangdong (Hetherington 1921:9-15). Some of these potteries continued production into the 1930s (Harrison 1995). Most brown-glazed wares that contained native foods and beverages came from Guangdong (Hellmann & Yang 1997; Pastron et al. 1981). Quellmalz (1976:292) stated that the bulk of brown-glazed wares came from Fatshan, a commercial centre near Guangzhou, in Guangdong, which was a distribution point for wares from Shiwan kilns. The Guangdong potteries supplied the needs of the local people as well as making utilitarian items for export (Quellmalz 1976). The majority of blue-underglaze decorated wares found at overseas Chinese sites were made at kilns in Fujian, Hunan, Guangdong and the non-government factories in Jingdezhen, Jiangxi (Ho 1988:3). Archaeological work in China at the end of the twentieth century uncovered areas of manufacture in Fujian which were previously unknown. With the exception of Jingdezhen, areas in Fujian were the largest exporter of blue-underglaze wares from the mid-sixteenth to the beginning of the twentieth centuries (Ho 1988).

Various marks have been found on Chinese-made ceramic artefacts from archaeological sites although many do not indicate a place of manufacture. While some marks have been translated, others appear quite abstract. Marks may have degenerated through the process of being copied (Pastron et al. 1981:436) or be illegible due to weathering, fragmenting or bad application (Olsen 1978:21).

Utilitarian brown-glazed wares are generally unmarked but some marks have been found on liquor bottles. Pastron, Gross and Garaventa (1981:400) stated that four of the five bottles recovered from their site had different marks on the base. They suggested that if these marks signified a particular kiln or factory, then at least four factories supplied the San Francisco Chinese community from 1880 to 1885.

Tea and tableware vessels are usually marked. Some marks indicate a date only, for example several marks on Four Seasons decorated vessels recovered from Tucson, Arizona, display the reign title of T'ung Chih (1862-1873) in red on the base, but individual makers could not be established (Olsen 1978:21). Chace (1976:523) suggested that the blue-underglaze marks found on celadon-glazed bowls were 'imaginary reign marks' and Olsen (1978:18) speculated that they may only have been decorative. Pastron, Gross and Garaventa (1981:436) posited that marks may have related to a factory, region or an individual potter. Occasionally, marks found on wares recovered from overseas Chinese sites identify the kiln site. For example, a blue-underglaze Rocks and

Orchid pattern recovered in Malaysia has been linked to the Cha-p'u factory at Zhejiang, which produced blue-underglaze decorated wares imitating those of Jingdezhen (Willetts 1981:10).

Production methods

A number of reports describing locally-made ceramics from Guangdong, Zhejiang, Jiangxi and Hopei provinces were written in the early-twentieth century, mostly by Western visitors to China.

Two accounts outlined the pottery industry in Shiwan, Guangdong, one by Clinton Laird, from the Canton Christian College, in 1918, and the other by Hoh Shai-Kwong, an Assistant Professor of Chemistry at Lingnan University, in 1933. Pottery has been produced in the town of Shiwan in Guangdong province in southern China and the surrounding districts since Neolithic times (Lau 1979). In 1528 various pottery guilds were established when the 'Temple of the Master Potter' was built on Lotus Mount, Shiwan, resulting in specialisation of production (Lau 1979:10). The residents of this district largely made their living from making pottery as there was no arable land for farming. These kilns were never supervised by the court but belonged to the people (Lau 1979). Hoh (1933) reported that the pottery industries in Shiwan were the only ones still using traditional methods. At that time Shiwan was the largest pottery centre in southern China, having made pots for about 700 years. Hoh (1933) stated that the original clay deposits were exhausted by about 1870, but that sand added as temper still came from the local region.

The clay used for pots, in Shiwan, came from two sources. One was a grey colour and the other 'a creamy white streaked with red' (Hoh 1933:58). The final body was a mixture of these two clays and sand, the relative proportions determined by the type of vessels being made (Laird 1918:569). The mixture was prepared by trampling, each batch taking a man four hours to mix. Vessels from these potteries were either thrown on a wheel, moulded or hand-modelled. Simple dishes, measuring about 40 mm (1.5 in) deep by 150 mm (6 in) in diameter, could be made on a wheel in eight seconds (Laird 1918:569). The glaze used on wares made at Shiwan was composed of river mud, water, rice straw ash and lime (Laird 1918:569; Hoh 1933:60). This 'black glaze', which, when diluted was 'brown to brownish yellow', furnished the 'basis for all glazes' (Hoh 1933:60). The majority of vessels made in this region were raw-glazed so that only one kiln-firing per vessel was needed. The vessels were either dipped into the glaze (Hoh 1933) or if only glazed on the interior, the glaze was poured in, swirled around and then the excess poured out (Laird 1918:569).

Kilns for firing the vessels were built on the side of hills. They could be thought of as long inclined chimneys and were about 45-60 m (150-200 ft) in length (Hoh 1933). Their internal dimensions varied from about 1 m (3 ft) in width and height at the front, or lower end, to about 2 m (6-7 ft) in width and height at the upper end. The fuel used to fire the kilns was pine wood. Thinner vessels were placed near the lower end and heavier vessels in the upper end. The temperature increased more quickly at the lower end and this would have caused problems with cracking for heavier items which required a slower increase in temperature. The firings took about 12 to 15 hours (Laird 1918:571; Hoh 1933:64).

The only industries in Shiwan in the early-twentieth century were the potteries and those to support the potters. The people, mostly men, who worked at the potteries were organised into guilds according to the category of ware they produced. There was a six year apprenticeship which had to be served before wages would be paid (Laird 1918:571). A large kiln at Shiwan could fire about 10,000 items at once and if 80 per cent of these were marketable the firing was considered a success (Laird 1918.). About a third of the production was used by local people, one third for people in Hong Kong and the rest were exported for use by overseas Chinese (Hoh 1933).

Rudolf Hommel (1937), a Christian missionary, lived in China in 1921-1926 and 1928-1930 and visited potteries in Zhejiang and Jiangxi. He observed that glaze was applied to jars before they were fired 'with brushes on the outside and a hempen tassel on the inside' (Hommel 1937:351). The tassel was coated with glaze and swung around inside the pot. One of the kilns in Zhejiang measured over 120 m (400 ft), was fired for 24 hours and opened four to seven days later, when it had cooled (Hommel 1937:354).

A contemporary account of the pottery industry in Hopei, northern China, was written by Professor J.B. Tayler in 1930. He tabled the annual value of pottery and porcelain in China for 1925, showing that the south-eastern provinces of Jiangxi, Guangdong and Fujian accounted for about three-quarters of China's total production value (Tayler 1930). The local industries in Hopei used materials that were found within a short distance of their kilns, a practice that occurred in Guangdong before the local materials were exhausted. The Hopei industry consisted of small individual businesses which produced bowls and jars. Whilst the individual businesses were small their aggregate was large. A skilful potter could throw 500 bowls in a day and kilns in Hopei would be capable of firing 60,000 medium bowls at one time (Tayler 1930). These bowls were placed in saggars, which are coarse receptacles to protect the wares, before being loaded into the kiln (Tayler 1930).

Frank Lentz (1920) wrote about the procedure for producing porcelain at the time of his visit to Jingdezhen, in Jiangxi province, in the early-twentieth century. The Chinese referred to the bones (kaolin) and flesh (a mixture of feldspar and quartz) of porcelain. A process of cleaning, sifting and refining the clays occurred before they were kneaded together and placed next to a potter's wheel (Lentz 1920). The potter continuously threw almost identical vessels, placing them on a tray for the next artisan. If the vessels required handles, they were added next and then the vessels were dried for the addition of underglaze decoration, and finally glaze. Wood and straw were used for firing, making it necessary for the vessels to be placed in saggars to protect them from ash. The kilns were about 15 m (50 ft) long and 3.5 m (12 ft) high and were large enough to fire wares from many factories (Lentz 1920). If over-glaze decoration was required then another firing in a smaller kiln would take place. Chinese porcelain vessels were classified according to their shape: round ware (plates, bowls, cups and saucers); irregular rounds (teapots, vases, ink and paint boxes); and irregulars (images and statues). Factories usually specialised in the production of one type of ware. Most of the porcelain produced at that time in Jingdezhen was for domestic use (Lentz 1920).

Marketing and exporting wares

Skinner's (1964) study of markets within China found a pattern which existed over several centuries. His model was based on one originally proposed by Christaller in 1933, which addressed the distribution and size of towns and their relationship to supply and demand. Skinner stated that there were so many markets in China that all households had access to one where they could sell their produce and buy goods for themselves. Each city or town was a 'central place' which had either a standard, intermediate or central market, depending on the size of the town (Skinner 1964:5-9). The standard market provided a place where locally made ceramics were sold. Most of the central market towns also served as administrative areas. Skinner suggested that the market towns were distributed regularly within the landscape, roughly in a hexagonal fashion. He reported that the majority of market towns had six neighbouring market towns. There was a regular flow of goods between the standard, intermediate and central market towns. Sellers could either sell their products directly to consumers or dealers based in their town, or to agents visiting from higher level market towns. Consequently, markets took the form of 'interlocking networks' (Skinner 1964:31). Skinner also suggested that the 'social field' of Chinese peasants was influenced by 'the boundaries of (their) standard marketing area' (Skinner 1964:32). For example, social events such as festivals were held within the market area (Skinner 1964).

While jars were marketed at local potteries, bowls and finer wares were sold through commission houses run by brokers who sold the wares to merchants from the big cities (Tayler 1930). Travelling merchants belonged to guilds which were set up all over China to cope with differences in dialects or hostility towards outsiders from neighbouring states (Douglas 1895). These guilds served to promote trade and members paid a small percentage of their sales revenue to cover the guilds' expenses. The large amount of trade ensured that the balance of the guilds was constantly in surplus (Douglas 1895:142-143).

Hong Kong became a regional trade centre and major shipping port because of its proximity to China. It was an important centre for distribution of goods throughout Asia and there were benefits for both Chinese and Western merchants conducting business from there. In the second half of the nineteenth century nearly two million people left China via Hong Kong (Hui 1997:118). While the Chinese secret societies controlled most of the trade in labourers, their transportation was carried out by European shipping companies. However, the Chinese controlled regional trade as they already had an extensive business network throughout Asia before European colonial expansion in the region. European businessmen could not have succeeded in East Asia without their collaboration with the overseas Chinese and their business networks. Chinese merchants 'played the role of middlemen' between the Europeans and the native Chinese and established 'networks of Chinese agents and petty traders' rather than forming an 'independent economic entity' (Hui 1997:123). In 1842 the Treaty Ports were opened up to the West. In the 1850s many foreign firms moved from China to Hong Kong as a result of the Taiping Rebellion and Sino-British wars. Western merchants needed Chinese middlemen who had access to inland China. Over the second half of the nineteenth century Chinese middlemen in Hong Kong became leaders of the Chinese community there and among the wealthiest people. That allowed some to become independent of the Western firms and to set up their own businesses. Some businesses were the result of joint investment ventures between Chinese merchants and foreign firms rather than the employment of Chinese by those firms. There was a mutual dependence. Chinese merchants needed 'the protection and connections with the West' while foreign firms needed 'the capital and domestic connections of the Chinese' (Hui 1997:126).

Traditionally China held its merchant class in disesteem and this view did not change until the 1860s when it was realised that Western power lay in its 'superior skills in trade and industrial technology' (Chan 1977:25). In the 1860s and 1870s, merchant groups formed through regional ties still controlled supply sources and transportation networks (Motono 2000). At that time merchants' values were shaped by loyalties to family, clans or other shared membership groups (Chan 1977). The right to conduct business was held to be more important than the 'profits or property of individual merchants' (Motono 2000:166). By the 1880s Chinese society in the treaty ports was being influenced by Western culture and many could speak and write in English. Commercial networks of English-speaking Chinese began working with foreign firms to benefit from the profits obtained (Motono 2000). Chinese merchants place in society became more important as men from gentry or official backgrounds entered the merchant class (Chan 1977). It was not until 1903 however that an imperial edict was released stating that merchants should be more highly regarded because they 'enrich the people's wealth and strengthen the state's foundations' (Chan 1977:26).

Ceramic vessels, similar to those used by local Chinese, have been exported to Southeast Asia, India and Africa since the ninth century and by the tenth century trade extended further into Southeast Asia to Philippines, Borneo, Thailand, Malaya, and Sumatra (Kuwayama 1989:105). Chinese-made wares were exported to Europe soon after Marco Polo saw them being made there in 1280 (Gulland 1928:1). From the sixteenth century wares were made specifically for the Western market. Designs of vessels in materials other than ceramic were sent from Europe to be copied by Chinese potters. This expanded greatly during the Qing period. The second half of the eighteenth century saw the peak period of reproduction when actual English, French and German porcelain and English earthenware were sent to China as designs to be copied (Honey 1946:158).

Early trade goods from China, destined for a Western market, entered Australia via Britain because of the monopoly of the British East India Company until 1833 (Jack 2001:44). However, these vessels are not the type found at overseas Chinese sites, but are known as Chinese export porcelain. They have been found at a range of sites in Australia during archaeological excavations (Corcoran 1993:53). In Sydney, ware has been recovered from the First Government House site, Lilyvale, and Cadman's Cottage. In Parramatta, Chinese export porcelain was found in Barrack Lane and George Street sites which were initially occupied by convicts. It was also found at the rural properties of Elizabeth Farm and Regentville. All these sites date prior to 1850 (Corcoran 1993). Failure to modernise during the mid-nineteenth century when they faced competition from Japan and Europe contributed to the decline of Chinese export porcelain (Harrison-Hall 1997:194).

While much is known about Western trade with China through archival records, not much is known about Chinese exporting goods for an Asian market, possibly due to language barriers. Shipwrecks have proved a source of knowledge about the trade of Chinese-made ceramics in the absence of written records. A contrasting example is that of the ships Diana and 'Desaru' which sank off the Malaysian coast in the early-nineteenth century (Brown & Sjostrand n.d.:21). 'Desaru' was the name given to the ship at the time of its discovery as its real name was not known. The Diana was a British ship en route to Madras when it sank in 1817, 10 per cent of its cargo being Chinese-made ceramics (Brown & Sjostrand n.d.:63). These wares were destined for a market with a Western-style diet and archival records detailing the history of the vessel were kept by the British East India Company. The 'Desaru' was a Chinese wooden ship filled with wares suitable for an Asian market and no known records exist that detail its history or destination. It sank about 1830. It was speculated that its cargo included wares from Jiangsu, Jiangxi, Fujian and Guangdong provinces (Brown & Sjostrand n.d.:65).

Some Chinese potters from Guangdong and Fujian produced their style of wares overseas. For example, in Indonesia, Chinese miners were supplied with stoneware jars by immigrant Chinese potters from around 1830-1840 (Harrison 1995:100). The clay resources for stoneware vessels, and the kilns needed to fire them, were readily available in Borneo (Kalimantan). However, stoneware jars were still being exported to Indonesia from the Shiwan kilns in Guangdong until the middle of the twentieth century (Harrison 1995).

A Chinese kiln was built in Australia, but made bricks rather than jars. In 1859, A'Fok, Fok Sing and Company applied for a licence to build a kiln at Ironbark Chinese Camp in Bendigo (DPCDV 2012). The Bendigo Advertiser reported on the progress of the kiln as it was being built:

> The kiln for burning the bricks is worthy of inspection, both from its novelty and apparent fitness for the purpose of design. It is very neatly built in a conical form, something like a beehive or coke oven, and terraced round, with different openings for the parties to go in to load the kiln. The other portion of the works is most ingeniously arranged, and no doubt when all is completed it will present a most complete plant (Bendigo Advertiser 25 October 1859:2).

The kiln was abandoned in the 1880s. This site was excavated by Heritage Victoria in 2005 and 2012. A number of Chinese-made food/beverage storage jars were recovered, along with other Chinese artefacts and a demijohn that was made in Bendigo (DPCDV 2012).

Overseas Chinese merchants

Pastron, Gross and Garaventa (1981:467) suggested that the Chinese in America did not have access to a full range of goods that were available to the rest of Americans so they depended on Chinese merchants who were involved in 'the lucrative import-export business with China' and who were supplying a 'virtually captive market'. The presence of Chinese merchants provisioning Chinese migrants in America evolved within the first decade of their arrival (Evans 1980:89). Contemporary sources reported that Chinese stores during the period 1851-54 were stocked with various Chinese food products and other supplies (Borthwick 1857 cited in Spier 1958) suggesting that Chinese miners need only buy mining implements and boots from American stores (Spier 1958). While the majority of any shipment from China contained food or beverages, a total of '131 distinct items' were listed as part of one ship's stores (Spier 1958:80). As well as the food, 'chinaware, wooden ware, bamboo ware, lacquer ware, iron and copper pans, chopping knives, chopsticks, ladles, tongs and mills' were listed as being imported (Spier 1958:80). Chinese stores also served as 'post offices, hiring halls, social centres, and opium-smoking establishments' (Wegars 2003:70).

Similarly, a trade network existed in Australia (McCarthy 1988:146). Early Chinese-run stores were set up on the goldfields to supply the Chinese miners. These operated from temporary structures, calico tents or slab huts, and only later operated from permanent structures. Storekeepers in rural New South Wales often networked through Sydney stores to China (Wilton 2004). They sold food, utensils and opium. The stores were also venues for gambling and meeting places for the Chinese miners (Wilton 2004). An analysis of imports into the colony of Victoria for 1859 shows the greatest diversity of goods were shipped from Hong Kong (Macgregor 2012). They included essentials for a Chinese mining population such as preserved foodstuffs, opium, wine, medicines and clothing.

In a report about Chinese immigration to the Australian Colonies, Crawford (1877:18) noted that:

> The bulk of the Chinese business in Australia...is in the hands of two or three great houses, whose representatives, under changeful names, are found at every focus of immigration. Through them immigrants forward letters to their homes, and diggers bound for Hong Kong are delegated to carry...packets of gold dust, addressed by different diggers to their relatives in China...The profits of the Chinese trade, apparently so large, become reduced to an ordinary Chinese business per-centage by being spread over a number of agents, consignees, retailers, and brokers. The merchant buys from his hong brother rather than from a stranger in the cheapest market; and often sells at a loss when mining prospects are gloomy.

Rains' (2005) research into the socio-economic relations of the overseas Chinese in Cooktown, Queensland, in the late-nineteenth century confirmed separate patterns of Chinese and European imports. Both groups displayed preferences for different items. Chinese firms imported larger quantities of rice, fresh and preserved vegetables, and tea, while European firms imported larger quantities of alcohol and other goods. The quantity of Chinese-made imports rose dramatically in the 1870s when more Chinese migrated to Cooktown. By the 1890s the volume of imports reduced as the local economy matured. The Chinese merchants in Cooktown also advertised non-Chinese goods for sale to the general public. Rains posited that there was a linkage between Chinese and European supply systems. There were weaknesses in both 'which necessitated a level of mutual dependency', for example Chinese storekeepers required access to non-Chinese goods while Europeans relied on the Chinese to supply fresh produce and other Asian goods (Rains 2005:303). Rains emphasised that patterns of exchange were not static but changed over the period 1870s to 1890s.

Chinese merchants played an important role by sponsoring and advancing money to their countrymen thereby enabling passage to Australia. They were among the elite of the overseas Chinese (Choi 1975:14), elevated to a higher status through their acquisition of business wealth and their use of guanxi (Lydon 1999:80). Guanxi literally means 'a relationship' between 'objects, forces or persons'; a network of personal and business connections based on 'mutual interest and benefit' (Yang 1994:1). It has been in existence in China for centuries and 'is of unprecedented importance' (Yang 1994:75).

While details of the lives of many Chinese remain unknown, those of some merchants and storekeepers have been documented. For example, Quong Tart, a well-known Chinese merchant from Sydney who also had connections to the Braidwood region, and Sat Wong, a NSW storekeeper who operated a business selling Chinese and European goods at Bolong from the late 1870s to 1916. Sat Wong's store extended credit to their customers and sometimes bartered goods for other products or services (Webber et al. 2003:31). Wilton (2004) detailed collections from Chinese stores which are now housed in museums throughout NSW. Lydon's (1999) study of On Jang, a Chinese merchant living in The Rocks in Sydney between 1916 and 1924, revealed how he used his wealth to acquire status symbols. His ceramic assemblage contained a large proportion of Chinese-made wares, but along with the usual vessels, there was a more unusual sand pot, a vessel rarely found outside China. Lydon suggested that guanxi, material culture and other Chinese traditions were used to manipulate the system and to form strategies to deal with the Europeans. She stated that merchants from George Street, Sydney, 'played a key role in regional trading networks' and that they were part of a network that extended from China and Hong Kong to inland country towns (Lydon 1999:84-86).

Analysis of the business records, 1871-1883, of Kwong Tai Wo Company, a general store in California, by Sando and Felton (1993) has provided an insight into the merchandise available at that time. The store sold a large range of goods including food and beverages, and tea and tableware. Ceramic vessels were listed individually or in sets of ten. The most common vessels listed were bowls, followed by tea and liquor cups, with 'relatively few spoons, teapots or liquor warmers' (Sando & Felton 1993:153). Large and medium bowls had values which were 'considerably higher' than rice bowls (Sando & Felton 1993:163). Bowls generally fell into two price categories, one being about twice as expensive as the other. The cheaper bowls included blue-underglaze decorated designs such as Bamboo and Double Happiness, while the more expensive bowls included Winter Green (celadon-glazed) and Four Seasons, which had an over-glaze decoration. In the early 1870s, cheaper vessels made up the bulk of the stock. By the late 1870s to early 1880s, the more expensive bowls 'were stocked in comparable or greater numbers' (Sando & Felton 1993:165). Bamboo was the most common tableware pattern on 1880s railroad campsites and on other post-1870 rural sites, while in post-1870 urban sites, Winter Green was the most common tableware. Sando and Felton (1993) speculated that the Bamboo pattern replaced Double Happiness in about 1870 as prior to that date Double Happiness had been the most common tableware for low-paid Chinese workers. They stated that the inventory of the Kwong Tai Wo Company may support this theory as there were relatively few Double Happiness bowls in comparison to Bamboo bowls after

1870. A problem arising from the inventory was that cheap 'Green bowls' listed have not been identified in the archaeological record and it is not known how these may differ from the celadon-glazed Winter Green (Sando and Felton 1993).

While the majority of overseas Chinese obtained their wares from merchants, Willetts (1981:11) reported that some Chinese miners going to work in America and Malaysia took with them 'chinaware for personal use', including rice bowls in the common blue-underglaze patterns of Bamboo and Double Happiness. Chinese immigrants to Australia may also have brought their own bowls, along with their other belongings, as described by Crawford (1877:27).

In summary, little is known about the supply of Chinese ceramics to Australia in the 1860s. Chinese immigrants were organised along clan lines and supplied with wares by their bosses. In the 1870s and 1880s Chinese stores were set up. More is known about these enterprises as some records still survive along with the names of many merchants.

CHINESE DIET AND VESSEL CHOICE

The type of ceramic vessels found at any site is related to the diet of its inhabitants. Many of the Chinese-made ceramic containers held food and beverages. The major Chinese food exports in the late-nineteenth century were tea, sugar, fruit, vermicelli, beans and bean cake, preserved vegetables and medicines (China Imperial Maritime Customs 1885 cited in Spence 1977). Chang (1977:11) stated in his book, Food in Chinese Culture, that 'few other cultures are as food orientated as the Chinese'. The preparation and eating of food is usually a social event (Anderson & Anderson 1977); the kinds and amounts of food determined by the occasion and status of the parties involved (Chang 1977).

Anderson and Anderson (1977:319) highlighted the nutritional benefit of the southern Chinese diet as it sustains 'more people per acre than any other diet on earth'. A balanced meal requires appropriate amounts of both *fan* (grains and other starches) and *ts'ai* (meat and vegetables) (Chang 1977:7). In the Qing Dynasty, 1644-1912, a large variety of food was available (Spence 1977). In the late-nineteenth century the daily diet in southern China consisted of 'two bowls of rice with scraps of vegetables or pieces of fish added' (Douglas 1895:357). The general words used for food were that of the staple rice and vegetables (Douglas 1895). Wealthier people included 'pork, fowl, or aquatic animals' in their diet but when animal proteins were not available legumes including soy beans, broad beans, and mung beans, provided protein (Anderson & Anderson 1977:319, 328).

The traditional diet of southern China was often maintained by overseas Chinese. For example, the Chinese in Otago, New Zealand, had a diet very similar to that of their homeland, with rice and vegetables being the major components (Piper 1988). Although the Chinese imported a lot of food they also began producing their own. By 1872 'the Chinese were producing two-thirds of all vegetables eaten in California' (Nordhoff 1872 cited in Spier 1958:81). Research has shown Chinese at a site in Tucson, Arizona grew local vegetables which they incorporated into their traditional cooking practices (Diehl et al. 1998). Similarly in Australia, many Chinese market gardens have been located with over 200 known to 2006 (Smith 2006:32).

Food was part of social transactions, used as a form of communication. The average daily meal for Chinese merchants of the early-nineteenth century was three or four dishes, but, when entertaining, a larger number of dishes were served. Important guests received 16 dishes, second class guests, ten dishes, and third class eight dishes, each with differing food according to the importance of the event (Spence 1977:277). Different food was also served for religious and seasonal festivals. In contrast, research into the dining arrangements at isolated overseas Chinese workers' camps in California has found that the workers were 'communally fed' (Greenwood 1980:114).

The standard beverage in southern China is green tea but wine/liquor is also consumed with meals for all classes. Alcoholic beverages in China are generally referred to as wines regardless of their content, although Anderson and Anderson (1977:342) noted that in southern China all alcoholic drinks are technically either beer, i.e. 'undistilled drinks from grain' or vodka, i.e. 'distilled, unaged alcoholic drinks made from starch bases'. They suggested that a few may qualify as whisky but in general the drinks are not aged long enough. It was customary to heat beverages to avoid infections and this also extended to alcohol. Hommel (1937:147) illustrated a brown-glazed ceramic liquor bottle of similar shape to those found at overseas Chinese sites and related his observation that at 'each meal a child or servant (was) sent to the shop to have the container filled'. The alcohol was then heated in that bottle over a small stove (Hommel 1937). Other containers specifically for heating and serving alcohol, blue-underglaze decorated porcelain pots with a handle and spout, have been recovered from overseas Chinese sites.

Ceramic vessels most commonly used by the southern Chinese for food and beverages include rice bowls, main-dish platters, sauce dishes, teacups and spoons (Anderson & Anderson 1977). 'Ordinary' families would use their cooking pots, usually sand-pots, for serving the main dishes and would have individual rice bowls and a few dishes for sauces, although they would have 'fancy platters for special occasions' (Anderson & Anderson 1977:365). The rice bowls doubled as liquor cups during a feast if people could not afford liquor cups (Anderson & Anderson 1977). The method of cooking and serving a dish was important to wealthier people. Food was warmed in ceramic vessels not metal, cheaper dishes were to be served in small bowls and fried food on platters (Spence 1977).

CHAPTER 4

Chinese ceramic vessels

FOOD AND BEVERAGE STORAGE CONTAINERS

The most common types of vessels found on overseas Chinese sites are brown-glazed storage jars. They are collectively known as 'brown glazed stoneware' or 'brown ware' (Chace 1976; Hellmann & Yang 1997; Pastron et al. 1981) or *Jiàn Yòu* and predominantly used for packing, shipping and storing food and beverages (Pastron et al. 1981:392). Vessel shapes and glazes have remained constant for at least two hundred years, making them difficult to date (Olsen 1978; Hellmann & Yang 1997). The body of these vessels contains mineral inclusions, some protruding through the glaze.

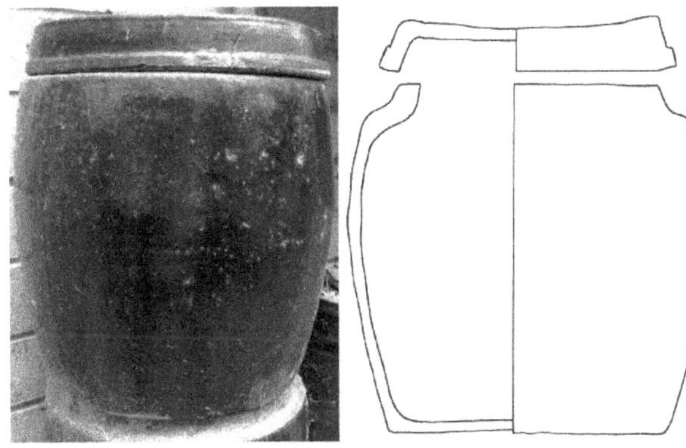

Figure 4.1: Barrel jars and lids (left: private collection)

Barrel jars

These are known in Cantonese as *ngá gong* or 'pottery barrel' (Hellmann & Yang 1997:187); or 'barrel-shaped jar' (Wegars 2012:3). They have been described as having a nearly straight side with an indented rim (Figure 4.1) and referred to as 'huge, or barrel size' straight-sided jars (Pastron et al. 1981:413). Rim diameters range from 260-300 mm, and foot diameters, 330-360 mm (Pastron et al. 1981:657). Height is about 430 mm (Ritchie 1986). The vessels are thrown on potters' wheels. Their exterior has a dark brown glaze, and interior, a thinner version of that glaze which appears brown or red-brown. Sometimes jars are unglazed on the interior.

Barrel jar lids are flat, flanged lids that fit over the gallery around the rim of the jar. They have a slight curve on the side wall and are made from stoneware which varies from a fine paste to one with sand-like inclusions; the body colour is light red to grey. The diameter of the lid is usually equivalent to the foot of the vessel and the height is about 45-55 mm. The lids are thinly glazed, appearing light brown, reddish-brown or blackish brown.

Barrel jars were originally used for packaging goods for transport, for example, food such as *pintong* or sheet sugar (Hellmann & Yang 1997:187) or wares like rice bowls which were packed with sawdust or straw (Ritchie 1986:253). The empty barrel jars were then reused to store rice (Ritchie 1986:253) or to catch rain water (Hellmann & Yang 1997). Ritchie (1986) reported that a barrel jar, found outside the Chinese store at Arrowtown, had been reused as a planter. In addition, barrel jars have been associated with the remains of Chinese who died on the goldfields. A lidded barrel jar, found near a grave site overlooking Jessops Creek, Queensland, contained skeletal remains (Comber 1991:331, 336). In southeast NSW, sherds from a minimum of two jars and three lids were recovered from Mongarlowe cemetery (Esposito 2012:629). The jars were also used to ship bones back to China (Hellmann & Yang 1997:187).

Globular jars

Globular jars are known in Cantonese as *chǐng*, or 'jar' (Hellmann & Yang 1997:182). Pastron, Gross and Garaventa (1981:407) referred to these as 'steep-shouldered jars' and related that the earliest extant reference to them came from the eighth century. The diameter of the base is about the same size as that of the rim, or only slightly larger (Ritchie 1986:255) and the height about the same as the maximum width of the body (Figure 4.2). Jars were made in a number of sizes. Rim/foot diameters range from 80 to 140 mm, and heights from 230 to 450 mm (Chace 1976:521; Pastron et al. 1981:656-657; Ritchie 1986:254).

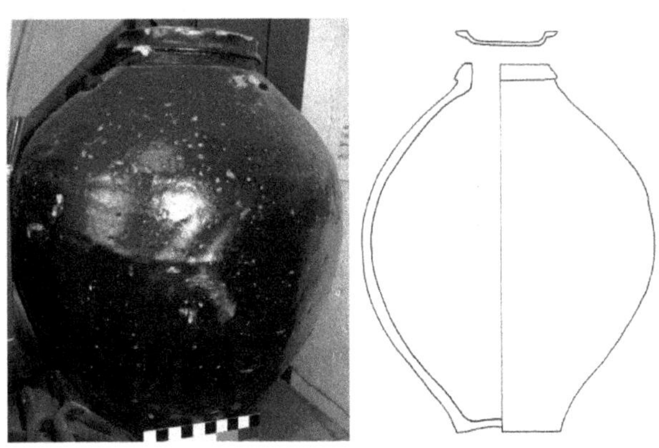

Figure 4.2: Globular jars (left: private collection; right: after Ritchie 1986:255)

Their shape is characterised by 'a constricted triangular flared rim' and 'four splayed lugs' placed about two to three centimetres below the rim (Pastron et al. 1981:408). Variations to the rim have been found, for example 'two triangularly indented tiers' (Pastron et al. 1981:409) and a 'square-section rim with thread-like grooves' (Ritchie 1986:254). The narrow rims and bases allow the vessel to be tipped easily. Globular jars were thrown on potters' wheels and the bases shaped when the jars were leather hard, i.e. a stage when a vessel can be handled without deformation, but before it is completely dry. The concavity in globular jar bases probably results from the firing, as bases are often much thinner than the lower sides of the jar (personal communication, Dr Anthony Flynn, 7 January 2011). The jars have a varying-colour dark-brown glaze which is lighter where thinner; some have a shiny appearance and others matte. The exterior of each jar is glazed to within 5-20 mm of the base allowing room for the glaze to run if it is over-fired. Drips of glaze are often seen extending to the base of the jar. The interior glaze is dark brown. Pin-holing, present in most globular jar glazes, occurs as gas bubbles from the body pass through the glaze and come to the surface, but the glaze cools before the hole created heals over.

Rims generally exhibit 'extreme patterns of wear' and this has previously been explained as being the result of prising off lids (Pastron et al. 1981:409). Research by Esposito (2012) suggests that the chipped rims resulted from the way the jars were stacked in the kiln during firing. Observed globular jar bases have glaze scars which are about the same size as jar rims (Figure 4.3). Unchipped areas of damaged rims appear to have had the glaze wiped off, which would have been done if the jars were stacked vertically in the kiln for firing, however any remnants of glaze would have stuck to the jar above. The over-fired glaze and the unchipped rim on the sherd in the right of Figure 4.3 suggest that this jar was at the top of a stack when fired, i.e. in the hottest area within that section of the kiln. The rim sherd shows the vessel was considerably over-fired, making the holes created by the gas bubbles much larger, giving a lumpy appearance to the glaze (personal communication, Dr Anthony Flynn, 7 January 2011).

The lugs on the jars are used in securing the lid. Lugs are formed by taking a strip of clay about 50-60 mm long, pinching the middle and pressing the ends onto the jar, leaving a hole under the middle section of the clay strip. This allows the lid to be tied onto the jar via the lug. Quellmalz (1976:293) advocated that lugs enabled the jars to be sealed with pigs' bladders. Olsen (1978:33) stated that 'a wooden lid padded with oil-soaked paper' was attached to the jar 'by means of hemp cord or split bamboo'. Ritchie (1986:255) detailed an unglazed earthenware lid that possibly belonged to a globular jar.

These jars contained 'wine, peanut oil, soy sauce or vinegar' (Olsen 1978:32). They were later reused to store many foods including 'soy sauce, pickled carrots, scallions, salted cabbage, melons, cucumbers, ginger and salty duck eggs' (Hellmann & Yang 1997:182).

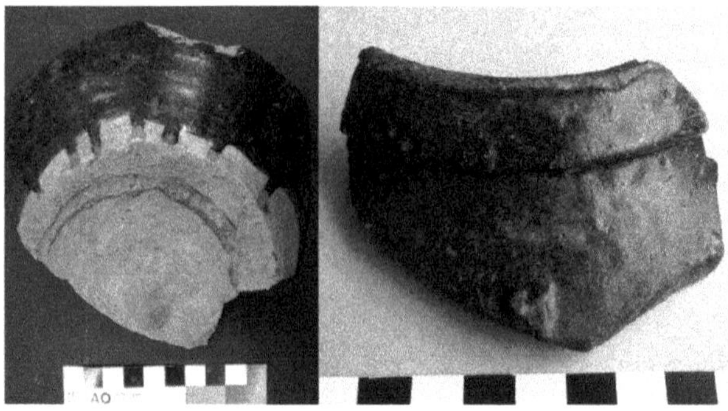

Figure 4.3: Globular jar base (left) and rim

Wide-mouthed jars

These jars are known in Cantonese as *fut hów ngá pěng* or 'wide mouthed pottery bottle', or as wide-mouthed jars (Hellmann & Yang 1997:182), shouldered food jars, bean pots or food jars (Wegars 2012:1).

They have a high-shouldered squat body with an outwardly rolled rim (Chace 1976). Olsen (1978) stated that the shape of these vessels was the same regardless of size and that it could be inferred that they contained varying amounts of the same product. The jars are found in a variety of sizes from rims of 80 mm to 130 mm (Olsen 1978; Pastron et al. 1981). They were thrown on a wheel, their base being splayed at the foot and slightly concave underneath (Figure 4.4). It has been reported that some Chinese potters placed rice straw ash on the wheel before throwing a pot (Hoh 1933; Laird 1918). Imprints of material, such as straw, have been seen in the central area of the underside of the base in the wide-mouthed and spouted jars. The bases, slightly wider than the bodies, were shaped when the jars were leather hard. Laird (1918) suggested that the addition of the rice straw ash enabled a pot to be easily removed from the wheel after being thrown. It may also have been done to save time when shaping the jars (personal communication, Dr Anthony Flynn, 7 January 2011) as only the outer section of wide-mouthed /spouted jar bases observed have been shaped. The thinness of the base may have contributed to its concave shape.

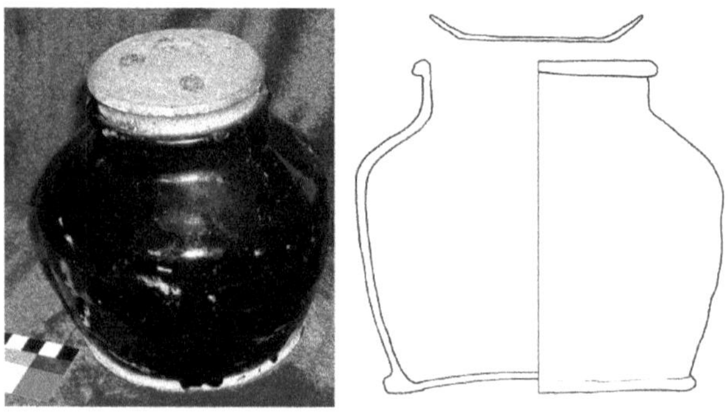

Figure 4.4: Wide-mouthed jars and lids (left: Museum of Chinese Australian History collection, Melbourne, lid may not belong to jar)

The jars have a glaze colour that ranges from dark to light brown and appearance from shiny to matte. Generally the dark-brown colour is lighter where the glaze is thinner. Pin-holing occurs on the exterior glaze and the interior glaze is patchy. The jars were raw glazed (Hoh 1933), i.e. they were glazed when the clay was leather hard. If they were glazed by the method described by Pastron, Gross and Garaventa (1981:405) as 'fill-and-slosh-while-pouring', the glaze would probably have been poured into the jar first whilst it was being turned around to distribute the glaze. The jar would then immediately be dipped into the glaze to coat the exterior, the shaped foot making it easy to hold the jar. Speed was necessary for both these actions or the pot would become too wet and fall apart. This could account for the patchiness of glaze on the interior as there would be no time to go over any missed areas. Alternatively, they may have been glazed by the method described by Hommel (1937:35) who saw hempen tassels being swung inside the pots to coat the interiors with glaze.

Wide-mouthed jar lids are unglazed saucer-like discs which correspond to the jar sizes. Chace (1976:519) suggested some are moulded, however all jar lids observed by the author have been thrown using a potter's wheel, probably off a mound of clay. That is, a larger amount of clay than required is put on the wheel and lids are continuously thrown and cut from that lump of clay. This method is more economical than centring small amounts of clay needed for each lid. The saucer-like lids have cutting wire marks on the base consistent with being removed from a mound of clay.

Chace (1976) detected traces of soft clay on the lids and suggested that this enabled the lid to stick to the jar. Olsen (1978) stated that little was known about the way in which the lid was secured to the jar, but that it could be assumed that the method involved the rolled rim. Wegars (1988:46) stated that lids were 'held onto the jars with a hard plaster-like substance'. During a visit to China, Hommel (1937:349) witnessed jars being 'closed with leaves, over which are placed cup-shaped covers of earthenware smeared with clay, to make the joint airtight'.

The wide-mouthed jars contained 'preserved tofu; sweet bean paste; black, brown and white beans; pickled turnips; cabbage; and shrimp paste' (Hellmann & Yang 1997:182), sweet gherkins, soy bean cheese (Olsen 1978), 'salted garlic, salted radish, salted onion, and pickled lemon' (Pastron et al. 1981:413). Wide-mouthed jars have been reused as cooking pots, as evidenced by their blackened bases (Ritchie 1986:242). Olsen (1978) observed that these jars are roughly made and could be thought of as being similar to the Western culture's Mason jar, both being cheaply mass-produced to contain products and later reused after being emptied and cleaned.

Spouted jars

These jars are called *ngă hú* in Cantonese or 'pottery pot' (Hellmann & Yang 1997:182). Other names include soy sauce pot, soy sauce bottle, spouted bottle and spouted jar (Wegars 2012:1). Olsen (1978:36) noted that referring to these vessels as a 'soy pot' is restrictive as these vessels may have contents other than soy sauce. For example, the spouted jar (Figure 4.5) contains Gee oil from Yue Mar Tee, Hong Kong.

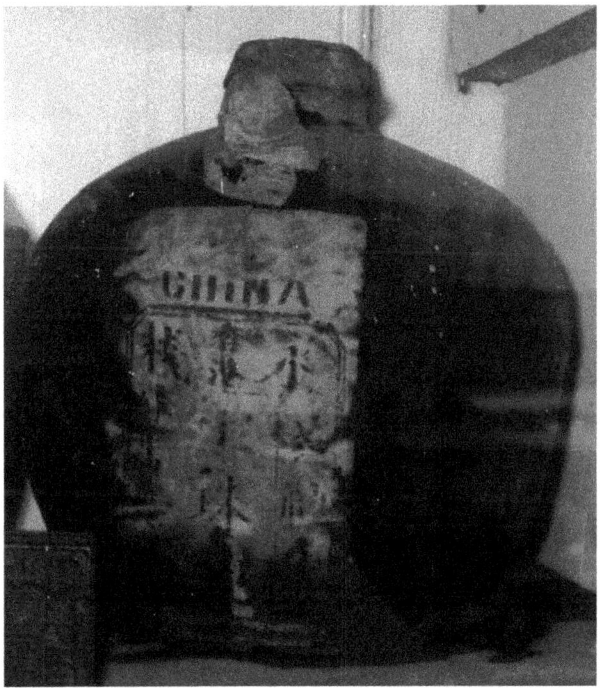

Figure 4.5: Spouted jar with label (private collection)

These vessels are similar in shape to the wide-mouthed jars with a high shoulder but have a short spout between the neck and shoulder and narrower rims with a curved lip (Figure 4.6). The jars are made in a number of sizes. The rim/foot diameters for the small jars are about 43/67 mm. The larger spouted jar rim diameters measure 44-48 mm and the feet, 126-137 mm (Pastron et al. 1981:656). Ritchie (1986:234) pointed out that 'side and base fragments are often very difficult to distinguish from pieces' of wide-mouthed jars.

The jars are wheel-thrown but have hand-modelled spouts. Spouts are formed by joining the ends of small slabs of clay with slip. The base of the jar is shaped when the clay is leather hard and a hole is made in the jar for the spout. The spout is then attached to the jar over the hole using a small coil of clay on the exterior around the base of the spout. Spouts vary in length and are sometimes misaligned as shown on the right of Figure 4.6.

These vessels have a dark-brown glaze, which is often thicker around the rim as there is an overlap from separate exterior and interior glazing. The interior glaze is a diluted version of the exterior glaze, thinner and lighter in colour and sometimes patchy.

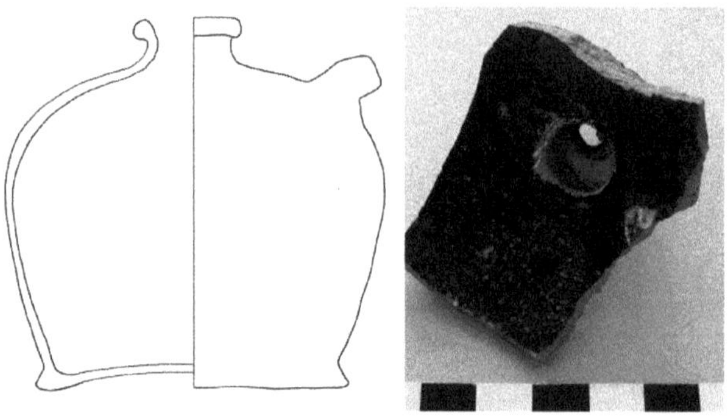

Figure 4.6: Spouted jar and a jar interior showing spout offset from hole

The characteristic spout and lipped rim are sealed during transportation (Olsen 1978). The majority of small spouted jars in Pastron, Gross and Garaventa's (1981) assemblage were corked. Ritchie (1986:234) recovered a 'mushroom-shaped earthenware plug in the neck' of a spouted jar.

Spouted jars were used to transport and store a variety of products including soy sauce, black vinegar, molasses (Olsen 1978), liquor and peanut oil (Hellmann & Yang (1997).

Straight-sided jars

These are known as *jiung* or 'covered cup' (Hellmann & Yang 1997:182). They have roughly parallel sides and are thin walled (Chace 1976) (Figure 4.7). There is a continuous range of sizes, with base diameter from 30 mm to 150 mm and heights from 30 mm to 140 mm (Ritchie 1986:247). The loosely fitted lids are glazed while the seat for the lid is unglazed. The smaller jars were used for medicine while the medium and large jars contained tofu (Hellmann & Yang 1997).

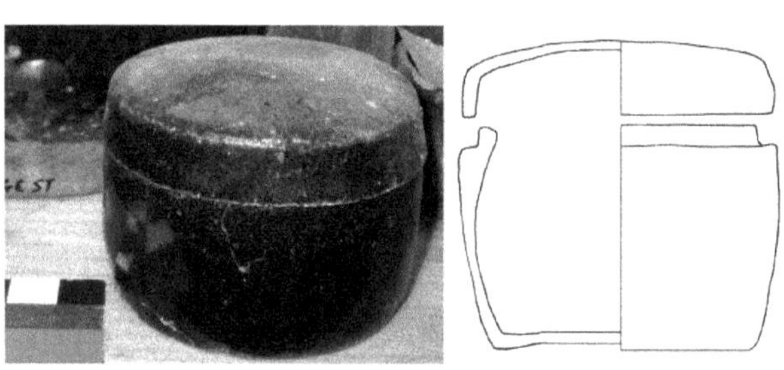

Figure 4.7: Straight-sided jars (left: private collection)

Liquor bottles

These are known by many names including 'wine bottles' (Chace 1976; Olsen 1978) or simply 'beverage' or 'alcohol' bottles (Pastron et al. 1981; Ritchie 1986). The Cantonese name is *tsáo tsun* or 'liquor bottle' (Hellmann and Yang 1997:182).

The bottles have a squat bulbous body which tapers to a flaring neck. The foot ring is unglazed. The traditional shaped bottles have a rim diameter of 56-58 mm and a foot diameter of 83-88 mm (Pastron et al. 1981:656). The bottles are made in a mould, often in three pieces (Figure 4.8) comprising the lower and upper halves of the body and the neck (Chace 1976).

The exterior glaze on bottles varies in colour from light-brown to almost black and may appear lustrous or matte, often exhibiting pin-holing. Bottles which are glazed to the base may show plucking i.e. have chipped edges, around the foot ring, suggesting that they have been forcibly removed from the kiln shelf after the glaze has run and stuck. Sometimes there is sand stuck in the glaze on underside of the base, illustrating the method of placing vessels on sand when firing. The interior glaze varies in thickness and coverage.

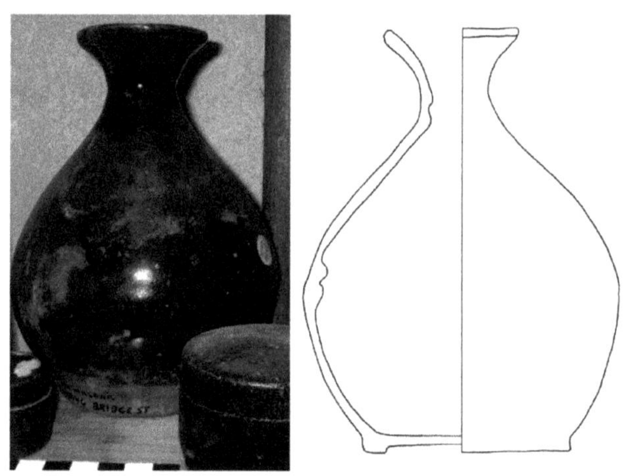

Figure 4.8: Liquor bottles (left: private collection; right after Ritchie 1986:233)

Olsen (1978) suggested that bottles could be assigned dates, according to their form and glaze. He stated that they all have a dark-brown glaze but the earliest ones have an olive-brown glaze which does not extend to the base, leaving an unglazed strip around the bottom. The next bottles in the series are similar but are glazed to the base. The most recent bottles are marked 'Federal law forbids sale or re-use of this bottle' after a 1933 law passed by the United States government (Olsen 1978:27).

Chandler's (2005) study of Chinese artefacts from the Upper Ovens goldfield in Victoria, found some bottles are glazed on the base while others are unglazed. Her analysis showed that the bottles with unglazed bases have slightly thicker walls (Chandler 2005:87-88). It is not known if this difference is temporal or is due to kiln origin. There are two distinct thicknesses of rims/feet on the bottles in this study. Sherds with the thicker foot ring do not appear to have the thick glaze on the base, which is evident on sherds with a thinner foot ring.

There has been much discussion regarding the contents of these bottles and whether it was wine or liquor. Labels found on Chinese bottles indicate that they were used for both wine and liquor. Bottles contained distilled spirits such as *Ng Ka Py* and *Mui Kwe Lu* (Pastron et al. 1981). Liquor was also referred to as medicinal wine. For example, *Wai Sang* medicinal wine had instructions on the label 'to be taken with meals daily' (NAPCVM 2010). Alcohol was also used for cooking (Hellmann and Yang 1997).

Ginger jars

Another common food container found on Chinese sites is the ginger jar. There are two types: green-glazed jars and blue-underglaze decorated jars (Figure 4.9).

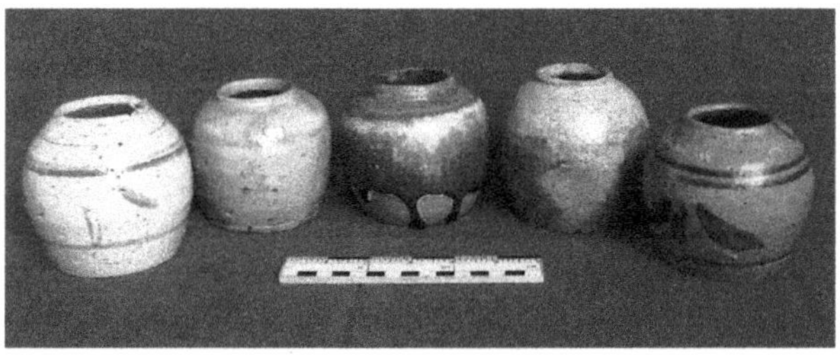

Figure 4.9: Ginger jars. Four blue-underglaze decorated jars and a green-glazed jar in the centre (photograph: Trelle Morrow)

Chace (1976) described the clay of the green-glazed ginger jar as having a yellowish-cream colour with the same mineral inclusions as the brown-glazed wares. Green-glazed ginger jars from the nineteenth century are round in shape and exhibit 'extreme simplicity and no endorsements' (Keddie 2006 cited in Morrow 2009:52). Hexagonal green-glazed jars, made in the twentieth century, may be temporally classified by their side-wall and collar embossing (Morrow 2009). Some jars were mould-made in two halves (Chace 1976) while others were wheel-thrown (Ritchie 1986). Jars were 'produced in numerous sizes' (Morrow 2009:50). Those from New Zealand sites have rims of 73-82 mm, foot diameters 58-71 mm and heights of 77-84 mm (Ritchie 1986:260). Green-glazed jars have an exterior 'iridescent, mottled green' glaze and an interior brownish-red glaze (Chace 1976:522). The glaze covers the top two-thirds of the jar and is 'commonly flawed with pits and mineral inclusions' (Ritchie 1986:259).

Blue-underglaze decorated jars are rounded. Sizes range from rims of 55-70 mm, foot diameters of 115-120 mm and heights of 125-150 mm (Olsen 1978:34; Ritchie 1986:267). They are thrown on a wheel. Chace (1976:522) suggested that the jars were slipped with white clay and 'painted with soft, broad, vertical strokes of blue'. Olsen (1978:34-35) described them as having coarse clay with 'abstract blue swirls and streaks' covered by a clear glaze. Ritchie (1986:267) outlined three types of blue-underglaze decorated ginger jars found in New Zealand, two of which dated after 1910: type A has 'patches of smudgy blue underglaze surmounted by a creamy-beige overglaze'; type B has 'abstract swirls and wavy linear bands of blue underglaze surmounted by a translucent overglaze' and was less lustrous than type A; and type C, which is not known from archaeological contexts and probably dates to after 1920, has a blue underglaze with a translucent clear glaze.

Ginger jar lids are loose fitting. They sit over the unglazed rim and neck of the jar and are probably secured by waxed paper and split bamboo (Olsen 1978). Sometimes impressed seal marks are found on the lid. Ritchie (1986:262-274) stated that it is 'often difficult to match lids' with specific jars; he detailed a range of lids. The green-glazed jars have domed flanged lids, similar to straight-sided jars, and are glazed in green on the exterior and brown on the interior. The blue-underglaze jar lids are flanged with a flat top, the surface of which is covered by the same glaze as the jar (Ritchie 1986).

Olsen (1978) explained that the term ginger jar is misleading as the jars contained a variety of products including green plums, preserved chopped garlic, preserved green onions, preserved sliced turnips, sweet gherkins, and preserved fish. Several of the jars he examined have an impressed mark which translates as 'joyful life' and were thought to contain preserved fish (Olsen 1978:35).

Ovoid jars

These large jars are wheel-thrown and have a constricted rim, as shown on the left of Figure 4.10. They are about 475 mm in height with a rim diameter of 120 mm, a foot of 225 mm and a maximum body diameter of 375 mm (McCrae 2001:19). The glaze varies from light to dark mustard and patterns on the upper part of the jars consist of: 'a vertical concentric comb pattern surmounting three concentric rings placed below the shoulder'; or 'a vertical concentric comb pattern deeply incised with vertical striations extending from the base of the neck through the comb pattern to the concentric circles' (McCrae 2001:19). They were the most common containers found on a site at Cooktown and have also been found in the Northern Territory (Rains 2005). Although Rains (2005:285) suggested that these jars appear to be 'restricted to sites in northern and northwest Australia', they are also found in Bendigo (personal communication, Dennis O'Hoy, 21 February 2012). They may be found elsewhere in Australia but to date no data confirming this have been published.

Spouted liquor storage jars

These jars have not been referred to in archaeological literature. Figure 4.10 (right) shows a jar measuring 280 mm in height. The rim diameter is 135 mm, the foot 250 mm, the maximum width 305 mm and the spout length is 70 mm. Chandler's (2005:32) examination of a collection of ceramics from the Upper Ovens goldfield recorded a vessel as a globular jar but noted that it had a small spout.

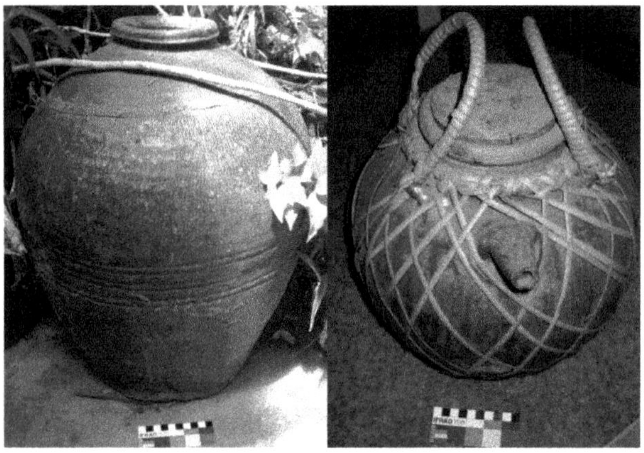

Figure 4.10: Ovoid jar (left) and spouted liquor storage jar (both from private collection)

Archaeological presence of food and beverage containers

Table 4.1 details sites at which brown-glazed vessels or ginger jars have been identified. Only identified vessel-types are included. If reports have shown 'storage jars' without specifying a vessel type, the data have not been included. Many archaeological sites may only have unidentifiable brown-glazed sherds leading to the generic description of 'storage jars'. Data are from Australian, New Zealand and American sites.

Two assemblages are from Lacey Place in Melbourne's Chinatown (Muir 2008). Other Victorian assemblages are from the Upper Ovens goldfield (Chandler 2005) and Chinaman's Point, a Chinese fish curing establishment near Port Albert (Bowen 2007). The Narrandera assemblage is from the Chinatown related to the market garden area on the outskirts of the European town (Esposito 2012: Appendix 7). Emmaville and Tingha, in NSW, were both associated with tin-mining. The Emmaville assemblage comes from a Chinese workers' camp situated about 500 m from the European town and the Tingha assemblages from the Chinatown and temple complex (Yit 2005). The Sydney assemblage is associated with a Chinese merchant living at the Rocks (Lydon 1996). The Queensland assemblages are from Atherton and Cooktown Chinatowns (Grimwade et al. 2007; Dunk 2010).

Other data are from sites in New Zealand and USA. The New Zealand sites include Chinatowns at Arrowtown and Cromwell, and rural camps in the Otago goldfields (Ritchie 1986) and Sam Chew Lain's house at the Lawrence Chinese Camp, also in the Otago goldfields (Davies 2012). The American sites include Tucson (Olsen 1978), Sacramento (Hellmann & Yang 1997), San Francisco (Pastron et al. 1981) and Ventura (Chace 1976).

In 2007, Brabyn analysed stoneware recovered from two sites in Market Street Chinatown, San Jose, USA, dating to about 1860-1880s. She compared the number of brown-glazed sherds as a percentage of total artefact numbers from a domestic and a commercial property, finding that the overall percentage was about the same at each site, 13 per cent and 14 per cent respectively. Differences were seen when the assemblages were broken down into large/small vessels and lids. The domestic property had 20 per cent large vessels, 75 per cent small vessels and 5 per cent lids. The commercial property had 45 per cent large vessels, 27 per cent small vessels and 27 per cent lids (Brabyn 2007).

In Australia and New Zealand, barrel jars are most commonly recovered from Chinatowns or areas associated with commercial activities. For example, all but one of the eight barrel jars and one lid from Ritchie's (1986:245) New Zealand sites were from Chinatowns. They represent the second most common jar type (MNV=15) recovered from a rubbish dump area in Cooktown, Queensland, thought to have been associated with commercial activity (Rains 2005:415).

Globular jars have mostly been recovered from Chinatowns, such as Ritchie's (1986:245) study with a MNV of nine. A MNV of four were recovered from Narrandera Chinatown in NSW (Esposito 2102: 653).

Wide-mouthed jars are commonly found on overseas Chinese sites in New Zealand and America. They are the most common jar type at Ritchie's (1986:245) sites, dating 1860s-1930s, with 84 jars and 97 lids in total. Most of his sites had at least one jar or lid. They are also the most common jar-type from Sacramento Chinatown, dating 1855-1861, with a MNV of 42 (Hellmann & Yang 1997:179). However, they do not appear common in Australian literature. This may be because they have been described simply as storage jars, or because their base sherds are similar to those of spouted jars and their rims have been absent from assemblages. Exceptions are small numbers from Sydney, NSW (Lydon 1996: Appendix 1), Upper Ovens goldfields, Victoria (Chandler 2005:33) and Chinaman's Point, Victoria (Bowen 2007:170).

Small numbers of spouted jars have been found at sites in NSW, at Narrandera (Esposito 2012:653) and at Tingha (Yit 2005:39). Spouted jars appear to be more common elsewhere. For example, in New Zealand spouted jars are the third most common type of jar with a MNV of 26 (Ritchie 1986:245). Most were found in Chinatowns or stores but single jars were recovered from some hut sites.

Table 4.1: Presence of food and beverage containers

Location	Date	Barrel jar	Globular jar	Wide-mouthed jar	Straight-sided jar	Spouted jar	Bottle	Ginger jar	Reference
AUSTRALIA									
Melbourne, VIC	1850s-1910					x	x	x	11
Melbourne, VIC	1886-1928	xx	x		x		x	x	11
Upper Ovens, VIC	1858-1880s		x	xx		x	x	xx	7
Chinaman's Point, VIC	1860s-1900			xx		x	x	x	9
Narrandera, NSW	1870s - early 20th		x			x	x	x	14
Emmaville, NSW	1870s-1900s	x							8
Tingha, NSW	1874-1910					x		x	8
Sydney, NSW	1916-1924	x		x				xx	3
Atherton, QLD	1880-1920					x	x		12
Cooktown, QLD	late 19th-early 20th	xx	x			x	x	x	10
NEW ZEALAND									
Arrowtown	1860s-1920s	x	x	xx	x	x	x	x	6
Cromwell	1860s-1930s	xx	x	xx	xx	x	x	xx	6
Kawarau	1870-1890s	x		xx		x	x	xx	6
Queensberry	1880-1915			xx		x	x	x	6
Luggate	1890-1915			xx		x	x	xx	6
Lawrence	late 1860s-early 20th	xx	x	xx	x	x	x	x	13
USA									
Tucson, AZ	1850s-1960s		x	x		x	x	x	2
Sacramento, CA	1855-1861	x	x	x	x	x	x	x	4
San Francisco, CA	1880-1885	xx	x	x		x	x		5
Ventura, CA	c1895-1907		x	x	x	x	x	xx	1

Key: x = vessel only; xx = vessel and lid; References: 1-Chace 1976; 2-Olsen 1978; 3-Lydon 1996; 4-Hellmann & Yang 1997; 5-Pastron et al. 1981; 6-Ritchie 1986; 7-Chandler 2005; 8-Yit 2005; 9-Bowen 2007; 10-Grimwade et al. 2007; 11-Muir 2008; 12-Dunk 2010; 13-Davies 2012; 14-Esposito 2012

They are the most common jar type from the Atherton Chinatown, dating 1880-1920 (Dunk 2010:49), and from the San Francisco dump site, dating 1880-1885, with a MNV of 38 (Pastron et al. 1981:399). A minimum of 19 were found at Chinaman's Point, Victoria (Bowen 2007:170). Straight-sided jars appear to be uncommon in Australia. In New Zealand straight-sided jars were mainly recovered from Chinatowns rather than hut sites, with a MNV of ten jars and seven lids from Ritchie's sites (1986:245).

Liquor bottles are common, for example, they are the second most common storage vessel, after wide-mouthed jars, from Ritchie's (1986:245) sites with a MNV of 80. They are the most common brown-glazed vessel from Sam Chew Lain's house at the Lawrence Chinese Camp (Davies 2012:60), also the most common at Walnut Grove, California (Costello & Maniery 1988) and at Narrandera Chinatown (Esposito 2012: 653).

Ginger jars are found at many sites. Green-glazed jars are more common from Ritchie's (1986:263) sites (MNV=51) compared to blue-underglaze decorated jars (MNV=6). The majority of those jars were recovered from Chinatown sites.

While other vessel types, such as ovoid jars, have been reported at some sites they are not as common across all sites.

Tea and tableware

This outline of Chinese tableware considers the main decoration types separately: celadon-glazed Winter Green wares; Four Seasons wares; and blue-underglaze decorated wares. The most common tea and tableware found on Chinese sites in Australia is celadon-glazed porcelain and the second most common is the Four Seasons pattern. A variety of patterns are found in blue underglaze decoration.

Celadon-glazed or Winter Green wares

These wares are referred to in southern China as *dong qing*, or 'winter-green' (Olsen 1978:18). Qing celadon-glazed wares imitate those of the Song dynasty 'with more mechanical perfection but less character' (Fournier 1986:38). Hodgson (1905:31) stated that the name celadon was first used in France in the seventeenth century to describe popular Chinese 'sea green' porcelain. Its popularity stemmed from a successful play adapted from Honoré D'Urfe's novel, 'in which Céladon, the shepherd courtier, appeared on stage clad entirely in sea green' (Hodgson 1905:31). Celadon-glazed wares are the colour of jade, the most valued Chinese gem, which symbolises 'excellence and purity' (Williams 1960:133). The colour of celadon-glazed wares has been described differently by researchers. Chace (1976:523) expressed the colour as that of 'winter melons'. Olsen's (1978:18) celadon-glazed wares generally have a pale green exterior and a milky white interior but one exterior is a 'thick moss green'. Pastron, Gross and Garaventa's (1981:432) celadon glaze ranges from 'dark grass green to pale sea green'.

Celadon glazes are produced by using small amounts of iron oxide as the colorant and being fired in a reduction atmosphere, i.e. one that is lacking sufficient oxygen. The glaze varies in colour, but is generally grey-green to blue-green. The thickness of the glaze affects its colour; the thicker the glaze the darker the colour. A thin application of the glaze can appear almost white rather than green.

Observations of sherds in this study reveal the glaze is applied thickly to the exterior of hollowware, but thinly to the interior and base. On flatware, the glaze is applied thickly to the interior as well as the exterior, but thinly to the base. Generally, the exterior glaze on the hollowware vessels is about twice as thick as the interior glaze. The foot ring on all items is unglazed. Rim edges appear to have been wiped clean of glaze and are sometimes an orange-brown colour. This is caused when the porcelain body is re-oxidised during firing (Wood 1978:12). Other factors affecting the appearance of the glaze include the firing temperature, length of firing, and the position of the vessel in the kiln. Visually, the colour of the glaze is altered by the angle of the light reflecting off bubbles in the glaze, which are easily seen under low magnification. These bubbles are formed by gases during firing. Celadon is a 'stiff glaze', i.e., one that does not run, and the trapped bubbles give it 'a frosted opacity' (Hamer & Hamer 1997:27) and 'distinctive quality' (Fournier 1986:29).

The celadon-glazed wares found on overseas Chinese sites have a white porcelain body, but as Chace (1976:523) noted, these wares are 'not particularly thin or fine quality'. Body sherds in this study are only translucent to about 3.5-4.0 mm (when shining a light through) and cracks are present in the thicker parts of vessels, particularly around the foot (Figure 4.11, left). Although Europeans regarded translucency as a distinctive quality of porcelain, the Chinese considered the quality of the sound emitted when a vessel is struck to be of more importance (Lindberg 1947). The celadon glaze on these vessels has many faults including puckering (incomplete coverage of glaze), iron spotting and inclusions of foreign matter (Figure 4.11, right). Iron, occurring naturally in a clay body, can cause black/brown spots on the surface when it is drawn to the surface during a reduction firing, forming 'darker and sometimes slightly crystalline specks' (Fournier 1986:124).

The range of celadon-glazed wares includes plates, serving dishes and bowls, sauce dishes, rice bowls, tea cups, liquor cups and spoons (Figure 4.12), in a number of sizes (Table 4.2).

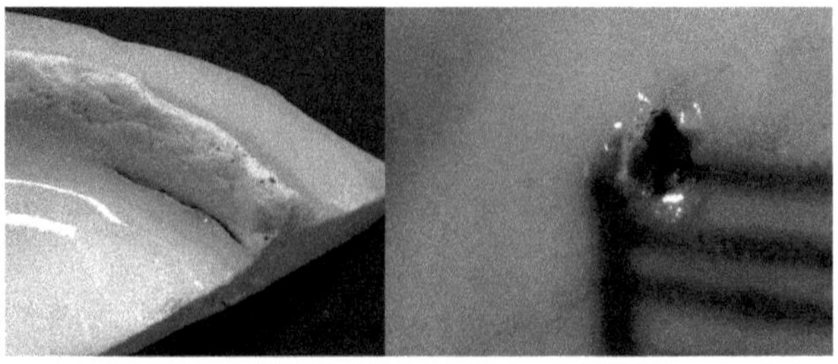

Figure 4.11: Crack in foot of bowl (left: Cat no. KCC2007-1007) and quartz inclusion in base of celadon-glazed bowl (Cat no. KCC2001-1319)

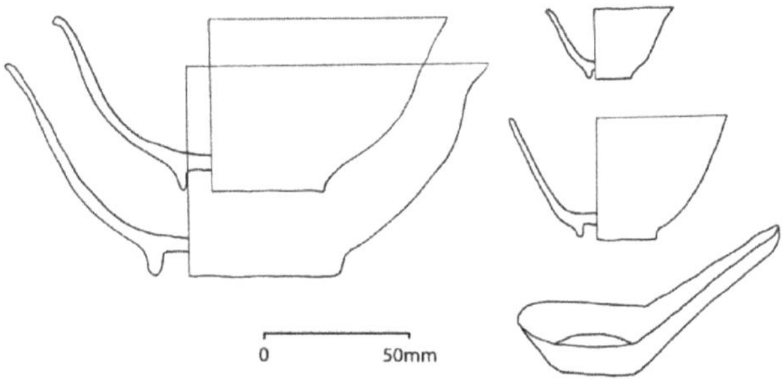

Figure 4.12: Celadon-glazed vessel forms (after Ritchie 1986:208-9)

Table 4.2: Size-range of celadon-glazed vessels (after Pastron et al. 1981:659-660; Ritchie 1986:208-209)

Form	Rim (mm)	Foot (mm)	Height (mm)
Plate	147-250	82-c140	21-29
Serving dish	142-190	105-111	24-33
Serving bowl	160-166	62-64	72
Sauce dish	80-85	37-45	26-29
Rice bowl	120-150	42-61	55-70
Small tea cup	67	24	31
Large tea cup	75-85	24-32	37-44
Liquor cup	42-51	17-21	22-26

These vessels have been found at sites in Australia, New Zealand and USA (Table 4.3). Rice bowls are the most common celadon-glazed form found at overseas Chinese sites, with liquor cups, spoons and tea cups also being relatively common. Celadon-glazed vessels such as plates, serving dishes and bowls, and sauce dishes appear to be found more often in Chinatowns although some have been recovered from rural camps. Some researchers have noted that it is possible to reconstruct sets of Chinese tea and tableware from sherds found at their sites. Pastron, Gross and Garaventa (1981) found a complete celadon-glazed set including liquor and teacups, spoons, rice bowls, and shallow dishes as well as serving dishes in small and large sizes. The small dishes were used for sauces and condiments and the larger plates as serving platters.

Table 4.3: Presence of celadon-glazed tea and tableware

Location	Date	Plate/ serving dish	Serving Bowl	Sauce dish	Rice bowl	Tea cup	Liquor cup	Spoon	Reference
AUSTRALIA									
Melbourne, VIC	1850s-1910				x				10
Melbourne, VIC	1886-1928						x		10
Chinaman's Point, VIC	1860s-1900				x	x	x	x	8
Narrandera, NSW	1870s - early 20th	x			x	x	x	x	13
Emmaville, SW	1870s-1900s				x		x		7
Tingha, NSW	1874-1910				x		x		7
Sydney, NSW	1916-1924				x				3
Atherton, QLD	1880-1920	x			x	x	x	x	11
Cooktown, QLD	late 19th-early 20th			x	x			x	9
NEW ZEALAND									
Arrowtown	1860s-1920s		x	x	x	x	x	x	6
Cromwell	1860s-1930s	x	x	x	x	x	x	x	6
Kawarau	1870-1890s				x			x	6
Queensberry	1880-1915				x		x		6
Luggate	1890-1915				x			x	6
Lawrence	late 1860s-early 20th	x	x	x	x	x	x	x	12
USA									
Tucson, AZ	1850s-1960s			x	x	x	x		2
Sacramento, CA	1855-1861	x			x		x	x	4
San Francisco, CA	1880-1885	x		x	x	x	x	x	5
Ventura, CA	c1895-1907				x	x	x	x	1

Key: x = presence; References: 1-Chace 1976; 2-Olsen 1978; 3-Lydon 1996; 4-Hellmann & Yang 1997; 5-Pastron et al. 1981; 6-Ritchie 1986; 7-Yit 2005; 8-Bowen 2007; 9-Grimwade et al. 2007; 10-Muir 2008; 11-Dunk 2010; 12- Davies 2012; 13-Esposito 2012

Four Seasons decorated wares

This pattern has been referred to by many names including Four Seasons (Chace 1976; Ritchie 1986; Wegars 2012), Four Flowers and 'Enamelled Flower Ware' (Pastron et al. 1981). The pattern appears on the exterior of bowls and cups and on the interior of dishes and spoons (Figure 4.13). It consists of four plants, each one representing a season. Each flower has 'its appropriate meaning and purpose' in China (Gulland 1928:32). The peony is the sign of spring, an emblem of love, symbol of feminine beauty and an omen of good fortune (Williams 1960:317). Lotus, representing summer, is the symbol of purity and perfection and is also an emblem of offspring because of its numerous seeds (Williams 1960:256). Autumn is represented by the chrysanthemum which is the symbol of joviality (Williams 1960:68). The winter flowering plum or prunus is the symbol of longevity (Williams 1960:327). On the centre interior of bowls and dishes there is also a peach or groups of peaches and leaves (Pastron et al. 1981). The peach symbolises immortality and is an emblem of marriage (Williams 1960:312).

The execution of this pattern on vessels from archaeological sites varies from 'bold, crudely drawn' to 'painstakingly executed, finely drawn' examples (Pastron et al. 1981:437). Olsen (1978) suggested that the earlier hand painted designs are more carefully executed while the later ones appear to have been painted hastily because the infill colours are not contained within the painted borders. The infill is often done with a broad brushstroke rather than taking care to stay within the lines (Olsen 1978). Pastron, Gross and Garaventa (1981:437) suggested that there are 'at least three and possibly four sub-pattern varieties':

> 1: Pink or peach-color tones with light aqua-green enamels outlined in brown underglaze
> 2: Black with orange enamel
> 3: Orange and heavy green overglaze in addition to the brown underglaze outlining
> 4: Solid orange flowers outlined in orange with a light, translucent green and black enamel stem outlined with brown underglaze (Pastron et al. 1981:438).

Olsen (1978) described the colour of the decoration being pink, yellow, green, orange and black.

The pattern is painted in enamel, a soft melting glass using oxides as colorants, over the glaze. Oxides commonly used by the Chinese include copper, cobalt, iron, manganese and gold and they could have contained 40-70 per cent lead oxide (Harrison-Hall 1997:198). An additional firing is required for enamels; usually about 300°C lower than the original glaze firing (Hamer & Hamer 1997:229) and this probably accounts for the greater cost of these vessels in comparison to the underglaze decorated items.

The glaze of the Four Seasons ware has been described as a 'white, granular background' (Olsen 1978:19) or as having a 'slight greenish-blue cast' which is rarely blemished (Chace 1976:525). Fissures have also been observed in the Four Seasons porcelain body particularly around the foot of serving bowls (Pastron et al. 1981).

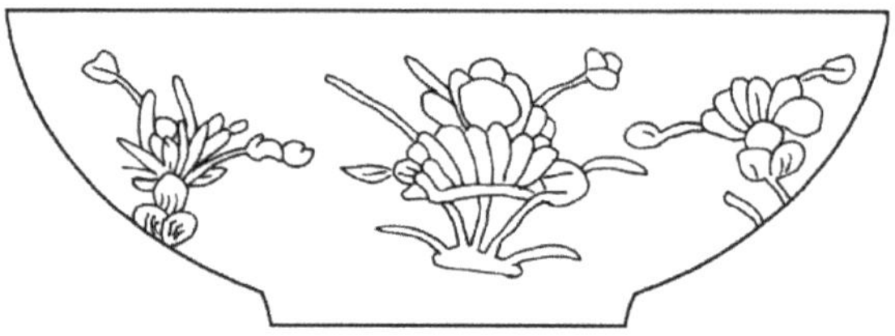

Figure 4.13: Four Seasons pattern

The range of Four Seasons wares includes serving dishes, sauce dishes, serving bowls, rice bowls, tea cups, liquor cups and spoons, in a variety of sizes (Table 4.4). Four Seasons decorated forms have been recovered from sites in Australia, New Zealand and USA (Table 4.5). A variety of other polychrome enamel patterns have also been recorded from Chinatown sites.

Table 4.4: Size-range of Four Seasons decorated vessels (after Pastron et al. 1981:665-666; Ritchie 1986:217)

Form	Rim (mm)	Foot (mm)	Height (mm)
Serving dish	126-226	76-150	19-38
Serving bowl	160-220	67-91	54-75
Sauce dish	78-110	44-65	17-25
Rice bowl	110-130	40-56	39-59
Octagonal tea cup	60-61	24-26	38-40
Circular tea cup	85	40	40
Liquor cup	40-55	17-21	22-27

The most common vessels found decorated in Four Seasons pattern are serving bowls, spoons, sauce dishes and liquor cups. While rice bowls in this pattern have been recovered from sites in the USA, none have been reported from Australia or New Zealand. The Four Seasons pattern is the most common polychrome enamel pattern and found in the widest variety of forms at Sacramento Chinatown (Hellmann & Yang 1997). Chace (1976) noted that the Four Seasons pattern was rare in an 1860's deposit from Donner Pass, but common in later deposits at Tucson. There were no Four Seasons vessels at the Chinaman's Point fish curing establishment near Port Albert, Victoria, which Bowen (2007:238) equates to a second-tier mining settlement in Smith's (2006) Chinese settlement system.

Blue-underglaze wares

These wares are decorated with blue cobalt oxide before being glazed. Cobalt was introduced into China around the fourteenth century (Medley 1976). Cobalt ore contains various impurities including nickel, iron and manganese which results in shades of grey-blue, blue-purple and blue-green (Fournier 1986). The blue-underglaze decorated wares are made in porcelain and stoneware and the body varies from white to dark grey often exhibiting numerous cracks and fissures (Pastron et al. 1981). The clear glaze has a green or blue cast often with many faults. Many patterns were made, the two most common at overseas Chinese archaeological sites being Bamboo and Double Happiness.

Table 4.5: Presence of Four Seasons decorated tea and tableware (and other over-glaze patterns)

Location	Date	Plate/ serving dish	Serving bowl	Sauce dish	Rice bowl	Tea cup	Liquor cup	Spoon	Other patterns	Reference
AUSTRALIA										
Melbourne, VIC	1850s-1910		x							10
Melbourne, VIC	1886-1928									10
Chinaman's Point, VIC	1860s-1900									8
Narrandera, NSW	1870s - early 20th		x	x						13
Emmaville, NSW	1870s-1900s		x?							7
Tingha, NSW	1874-1910						x		x	7
Sydney, NSW	1916-1924			x						3
Atherton. QLD	1880-1920		x					x		11
Cooktown, QLD	late 19th-early 20th	x	x	x			x	x		9
NEW ZEALAND										
Arrowtown	1860s-1920s		x	x		x		x	x	6
Cromwell	1860s-1930s		x				x	x	x	6
Kawarau	1870-1890s									6
Queensberry	1880-1915		x							6
Luggate	1890-1915									6
Lawrence	late 1860s-early 20th		x					x		12
USA										
Tucson, AZ	1850s-1960s		x		x		x	x	x	2
Sacramento, CA	1855-1861	x	x	x	x		x	x	x	4
San Francisco, CA	1880-1885	x	x	x	x	x	x	x	x	5
Ventura, CA	c1895-1907	x	x	x			x	x	x	1

Key: x = presence; x? = uncertain form; References: 1-Chace 1976; 2-Olsen 1978; 3-Lydon 1996; 4-Hellmann & Yang 1997; 5-Pastron et al. 1981; 6-Ritchie 1986; 7-Yit 2005; 8-Bowen 2007; 9-Grimwade et al. 2007; 10-Muir 2008; 11-Dunk 2010; 12-Davies 2012; 13-Esposito 2012

Bamboo pattern

This design was previously referred to by different names including Three Circles and Dragonfly (Chace 1976:523; Pastron et al. 1981:426) and Blue Flower (Olsen 1978:15). By the 1980s the name Bamboo was 'widely adopted to describe this motif' (Wegars 1985 cited in Ritchie 1986:218). Olsen (1978) noted that this design has changed little since the Ming Dynasty. It is hand-painted on the exterior of bowls and includes three circles, a dragonfly, a marsh and a flowering plant (Figure 4.14). The interior usually has blue bands at the rim and lower down near the curve by the foot (Pastron et al. 1981). A slight variation of the design was found in Sacramento on a rice bowl (Hellmann & Yang 1997). This bowl had an interior stacking ring, which is an unglazed area where the foot of another bowl would sit during firing.

The glaze on these wares has been described as a clear-glaze with a light green-blue cast and the underglaze as a blue or dark green-grey colour (Chace 1976). The glaze is often pitted or has small protrusions marring the surface (Chace 1976). In some cases patches are left unglazed (Olsen 1978). Pastron, Gross and Garaventa (1981) stated that the glaze irregularities and rough pitting indicate that these items were cheaply-made.

Rice bowls are the only vessel found in this decoration (personal communication, Dr Priscilla Wegars, March 2013). Two sizes of bowls, with rim diameters of 120-130 mm and 137-150 mm, were used as individual rice bowls by Chinese labourers (Pastron et al. 1981:427-428). An assemblage from San Francisco's Chinatown contains 60 rice bowls, all made of stoneware. Six of the bowls recovered have a green cast to them while the others have the more common blue cast (Pastron et al. 1981). Only three Bamboo rice bowls were recovered by Ritchie from sites in New Zealand. Two are from Chinatowns and have a blue cast, while the other from a rock shelter site has a green cast (Ritchie 1986).

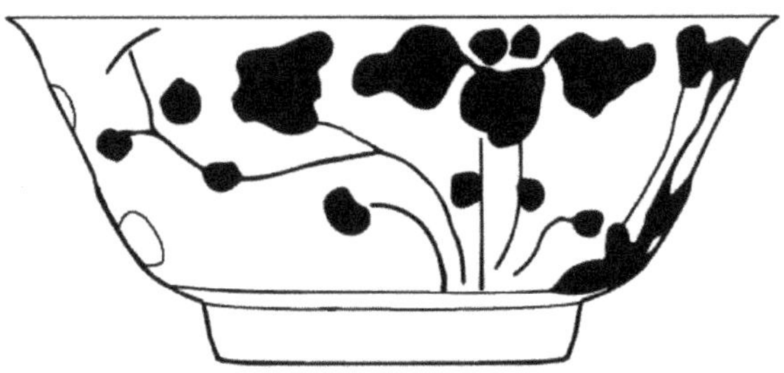

Figure 4.14: Bamboo pattern

Double Happiness pattern

This pattern was also known as Swirl, a term introduced by Praetzellis in 1979 as a 'precise pattern' name because the character for double happiness appeared in several patterns (Pastron et al. 1981:430). However, Double Happiness is now the accepted name for the pattern (Wegars 2012), which consists of the double happiness character, *hsi*, repeated three times and interspaced with swirls or a dragon (Figure 4.15). There is also a single heavy line or two thinner lines on the foot of the vessel and at the rim (Pastron et al. 1981).

The majority of Double Happiness rice bowls found at Pastron, Gross and Garaventa's (1981:424, 430-431) San Francisco site are stoneware and have rim diameters ranging from 110-140 mm, but they have recovered some porcelain bowls. No vessels with this decoration were recovered from Ritchie's (1986) sites in New Zealand.

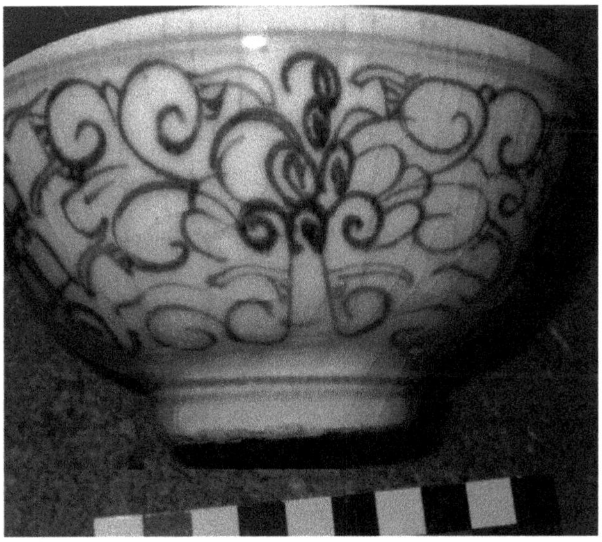

Figure 4.15: Double Happiness pattern (private collection)

Other patterns

Other blue-underglaze patterns found at overseas Chinese sites include Rocks and Orchid, Sweet Pea, Peach and Fungus, Bat and Knot. These are all hand-painted designs featuring motifs of their title. They are found on porcelain or stoneware and have similar glaze/underglaze colours to Bamboo and Double Happiness. These patterns are found on serving vessels such as dishes and plates. Sweet Pea is often found on liquor pots used for warming alcohol.

While the patterns named above are usually hand-painted, another type of decoration using blue-underglaze is block-printing, which allows repetition of simple patterns. The block stamps used for printing are soaked in pigment before being pressed onto the leather hard surface of the vessel (Harrison 1995). There are two sorts of stamps, one hard and one soft. Hard stamps leave an impression in the surface of the vessel so they are mainly used on flat surfaces. The soft stamps allow patterns to be applied on curved surfaces. These soft stamps are made from paper or leather mounted onto a block (Harrison 1995). Another suggestion has been that the stamps are made in a mould from konnyaku, a form of starch from an Arum lily, which hardens after being heated. This process was first used in Japan but was extensively used in China in the nineteenth century (Finch 2010).

Blue-underglaze decorated vessels have been recovered from sites in Australia, New Zealand and USA (Table 4.6). In America there has been discussion as to whether Double Happiness was an older pattern which was replaced by Bamboo in the 1870s. Chace (1976) noted that this pattern is present at Donner Pass sites (1865-1869) but absent from the Ventura site (1870s-1880s). Chace suggested that there may have been changes in the distribution network but Pastron, Gross and Garaventa (1981) argued that the abundance of the Bamboo pattern pointed to it having replaced an old style. Ritchie (1986) found no Double Happiness at his sites in New Zealand and stated that Bamboo was relatively rare. In Australia, both patterns have been recovered from sites dating from 1870 onwards. They have been found at Cooktown, Queensland (Grimwade et al. 2007:38-39) and at Narrandera, NSW, on stoneware rice bowls and Double Happiness on porcelain bowls (Esposito 2012: 653). Bamboo and Double Happiness patterns are found on rice bowls while other blue-underglaze decoration types are found on other forms. Other blue-underglaze patterns are more commonly found in Chinatowns rather than rural camps.

Other patterns, found on rice bowls, at The Rocks in Sydney, include Peach and Fungus, Chrysanthemum and Conch and Om (Lydon 1996: Appendix 1) and in New Zealand, patterns designated 'Crab motif' and 'Bird and floral motif' were found in the Chinatown settlements (Ritchie 1986:222).

Tea and tableware summary

Serving bowls and dishes are more commonly found in Chinatowns than rural sites in America and New Zealand (Evans 1980:92; Ritchie 1986:210, 216). Serving bowls are the most common Four Seasons vessels found in New Zealand (Ritchie 1986:216). This pattern is also commonly found on serving bowls from American sites (Evans 1980:92; Pastron et al. 1981).

Rice bowls are the most common vessel type. They are the most common celadon-glazed vessels at Ritchie's (1986:210) New Zealand sites, dating from the 1860s to the early 1900s (MNV=61). He found no Four Seasons or Double Happiness rice bowls and only three Bamboo decorated bowls. Rice bowls are the second most common celadon-glazed vessels (MNV=42) from a San Francisco, California, dump site dating to the 1880s (Pastron et al. 1981:659). They are the most common Four Seasons decorated vessels (MNV=44). Sixty rice bowls were recovered in Bamboo, and seven bowls in Double Happiness (Pastron et al. 1981:424). The assemblage from Ventura, California, dating to c1900, contains 11 Winter Green and two Four Seasons rice bowls (Evans 1980:92). Bamboo decorated rice bowls dominate the blue-underglaze patterns (MNV=17). Other patterns were also found but not detailed (Evans 1980).

Table 4.6: Presence of blue-underglaze decorated tea and tableware

Location	Date	Bamboo	Double Happiness	Rocks & Orchid	Sweet Pea	Other patterns	Reference
AUSTRALIA							
Melbourne, VIC	1850s-1910						10
Melbourne, VIC	1886-1928		?				10
Chinaman's Point, VIC	1860s-1900	?	?				8
Narrandera, NSW	1870s - early 20th	s/w	porc; s/w				13
Emmaville, NSW	1870s-1900s					?	7
Tingha, NSW	1874-1910						7
Sydney, NSW	1916-1924			x		x	3
Atherton, QLD	1880-1920		x			?	11
Cooktown, QLD	late 19th -early 20th	x	x		x		9
NEW ZEALAND							
Arrowtown	1860s-1920s	s/w					6
Cromwell	1860s-1930s	s/w				x	6
Kawarau	1870-1890s					x	6
Queensberry	1880-1915						6
Luggate	1890-1915						6
Lawrence	late 1860s-early 20th	?	porc; s/w				12
USA							
Tucson, AZ	1850s-1960s					x	2
Sacramento, CA	1855-1861					x	4
San Francisco, CA	1880-1885	s/w	porc; s/w			x	5
Ventura, CA	c1895-1907	s/w				x	1

Key: x = presence (unknown body fabric); s/w = stoneware; porc = porcelain; ? = possible pattern; References: 1-Chace 1976; 2-Olsen 1978; 3-Lydon 1996; 4-Hellmann & Yang 1997; 5-Pastron et al. 1981; 6-Ritchie 1986; 7-Yit 2005; 8-Bowen 2007; 9-Grimwade et al. 2007; 10-Muir 2008; 11-Dunk 2010; 12 Davies 2012: 13-Esposito 2012

While there are relatively few teacups from Australian and New Zealand sites, there are a large number found at some American sites. The majority of Winter Green vessels from a San Francisco dump were teacups (MNV=67) and a further nine Four Seasons octagonal teacups were found (Pastron et al. 1981:659).

Winter green is the most common decoration found on liquor cups in New Zealand (Ritchie 1986) while sites in America have more Four Season liquor cups (Evans 1980; Pastron et al. 1981). Mueller (1987:304) suggested that liquor cups have a ritual association after finding the highest frequency of these cups at the Temple site in Riverside Chinatown, California. He stated that a photograph from a temple in San Bernardino, California, showing liquor cups placed on an altar supported this theory. On the other hand, liquor pots (used for heating alcohol) have been referred to as being 'a common piece of household ceramic' (Olsen 1978:28) and by inference liquor cups.

As with other forms, Winter Green spoons are more common than Four Seasons spoons in New Zealand (Ritchie 1986:210, 216) while the reverse is true in San Francisco (Pastron et al. 1981).

Opium pipe bowls

Opium pipe bowls are made from earthenware or stoneware clay and are bell-shaped, similar to a poppy seed pod (Figure 4.16). Common bowl shapes include a circular rim with a faceted, flared or straight side; hexagonal rim; or octagonal rim. The bowls are attached to a bamboo pipe stem by a brass or copper fitting (Etter 1980). Opium pipe bowls are usually made by pressing clay into a mould but some are wheel-thrown. The most common bowls are made of orange earthenware, commonly called terracotta. They have an orange slip, the colour of which depends on the clay particle size (personal communication, Dr Anthony Flynn, December 2011). The finer particles fire to a dark-orange and appear shiny. The other orange slip is lighter and not as shiny but still has a burnished appearance. Makers' or owners' marks can often be found impressed or scratched into the surface of the pipe bowls.

The top of the pipe bowl is known as the smoking surface and has a small hole in the centre, into which the heated opium is placed (Etter 1980). The neck and connecting flange, or stem, are found at the base of the bowl. Most pipe-bowls measure 50-78 mm in rim diameter and 40-50 mm in height (Etter 1980:99).

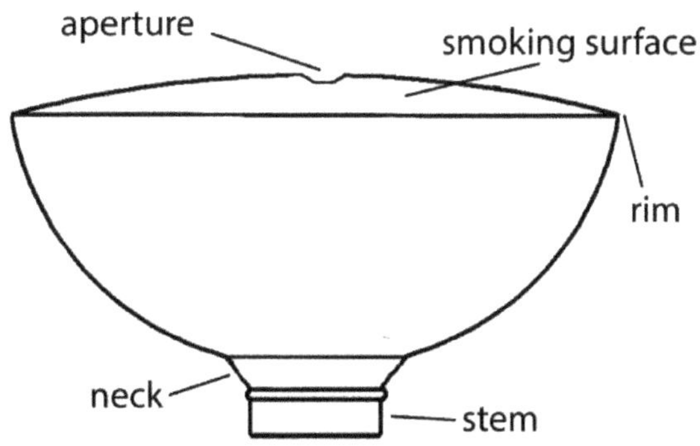

Figure 4.16: Opium pipe bowl detailing diagnostic parts

A greater variety of opium pipe bowls have been recorded from Chinatowns than from rural sites. Etter (1980) stated that Chinese construction workers at Donner Summit smoked mostly plain slipped terracotta pipe bowls, while the residents of Virginia City Chinatown had a more extensive choice of bowl styles. Williams (2004:7) noted that the opium pipe bowl assemblage from Market Street Chinatown, San Jose, is very diverse. The minimum number of pipe bowls is 95 and no two are exactly the same. Williams roughly divided the pipe bowls into expensive (decorated) and inexpensive (undecorated) categories and just over half were inexpensive pipe bowls. He concluded that these pipe bowls were not found exclusively at opium den sites, brothels, clubs or the back of stores, indicating that opium was consumed in a variety of contexts (Williams 2004:10-11).

Ritchie's (1986:371) analysis of opium pipe bowls from sites dating 1870-1925 established 12 basic types which were 'further subdivided into 26 subgroups based on stylistic differences' such as clay colour and decoration. The bowls in his assemblage are orange or grey and basic shapes include nine circular, two octagonal and one hexagonal. Ritchie's (1986:375) most common types are designated: C1 – circular smoking surface with flaring rim, smooth sides with a single ridge; and C3 – circular smoking surface with flaring rim and ten faceted sides (Figure 4.17).

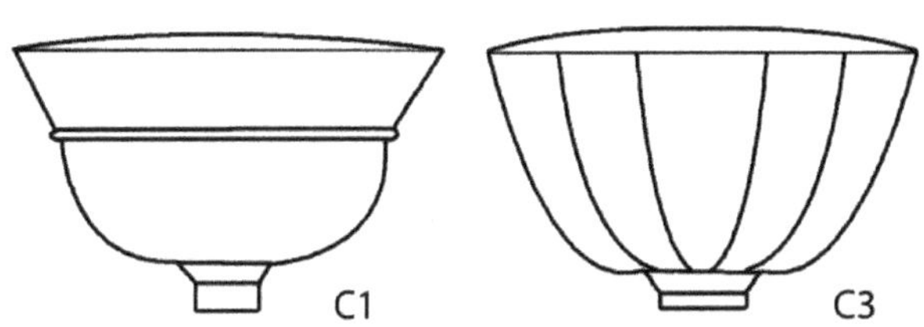

Figure 4.17: Opium pipe bowls designated C1 (left) and C3 (after Ritchie 1986:376)

MISCELLANEOUS VESSELS

Lamps

Chinese lamps from the early-twentieth century are illustrated in Hommel (1937:3, 311, 313-318). Lamps were made of two parts: a stand and a small dish which was filled with oil and held a wick. The oil used before the introduction of kerosene was obtained from soy beans or rape seed. The materials used for the lamps were wood, earthenware, porcelain or metal. Their price depended on the material used, with wood being the cheapest and metal the most expensive. Another simple kind of lamp was a miner's lamp. It consisted of a small porcelain cup resting in a loop of wire, the other end of which was bent to form a hook (Hommel 1937:3).

Miniature Floral Decorated Pot

A small pot with a rim diameter of 22 mm, foot of 20 mm and measuring 25 mm in height was recovered from Ritchie's (1986:224) Upper Chinatown site. Its pattern of pink, green and orange enamel flowers on a white background is reminiscent of Four Seasons.

Toothpowder lids

A possible Chinese-made toothpowder pot lid was recovered from Lacey Place in Melbourne. It is made of porcelain and is the same size and shape as English pot lids but has a label featuring Chinese and English script (Muir 2008).

CHAPTER 5

Ceramics of non-Chinese origin

A visit to any archaeological site in Australia, once occupied by Chinese, presents a variety of non-Chinese made ceramic forms and decorative types. However these are often overlooked during site analysis, or statistics are used to accentuate the Chinese nature of the site, particularly when the artefacts are not the focus of the research. For example, Smith (2006:88) stated that 'Most of the ceramic artefacts' from the Jembaicumbene Temple Complex 'were of Chinese origin', but this is only correct if the artefact weights are considered. The sherd numbers are almost equal and the minimum number of Chinese-made vessels is half that of the non-Chinese made vessels (Esposito 2012:116). Similarly, Bowen (2007:194) used the number of sherds to state 'Ninety seven and a half percent of the ceramics from Chinamen's Point are of Chinese form, decoration and origin; the remaining 2.5 percent is European tableware'. Although still mostly Chinese, he does not include the percentage if the MNV is considered, 84.3 per cent Chinese and 15.7 per cent non-Chinese. Ritchie (1986:319), on the other hand, pointed out that in New Zealand 'many of the Chinese miners regularly used Euro-ceramic' tableware and included an analysis of those vessels in his thesis. This chapter outlines the origin, production, availability, and form and decorative types of non-Chinese made ceramic vessels used in Australia in the mid-nineteenth to early-twentieth century in order to provide a context for those found at the Chinese sites in this study.

ORIGIN OF WARES

Britain

Most of the earthenware tea and tableware recovered from nineteenth century sites in Australia was manufactured in Britain, mainly in North Staffordshire (Brooks 2005). Britain's domination of the refined earthenware market extended from the late 1700s to about 1890 in America (Majewski & O'Brien 1987:115). In Australia, it was the major source of tea and tableware into the twentieth century.

By 1850 Staffordshire potteries were not only dictating what was bought but also influencing what was produced in Europe and North America (Barker 2001). The main pottery towns were Tunstall, Burslem, Hanley, Stoke, Fenton and Longton, with minor centres such as Longport, Cobridge and Etruria. The pottery industry thrived because many types of clay, along with coal needed for firing, were available in the area thus lowering production costs. New products were continually being developed and refined, influencing popular tastes over time. In 1762 about 150 potteries operated in the area, employing 7,000 people. By 1800, the number of people employed in Staffordshire potteries had risen to 15,000 – 20,000 (Barker 2001). A parliamentary report (Birks 2012) into the use of child labour in the pottery industry in 1862 found that there were 180 potteries in Staffordshire, 'employing 30,000 operatives, of whom 6,500 were classified as young persons and 4,500 as children under 13 years of age'.

Potteries in other areas of Britain either changed what they were producing to compete with the Staffordshire potteries or they closed (Barker 2001). The advent of mass production resulted in specialisation in all areas. Smaller potteries, which did not have their own resources, used freelance mould makers and engravers, resulting in uniformity across many forms and patterns. They used tools and kiln furniture produced in Staffordshire and as a result Staffordshire-type wares became the industry standard throughout the nineteenth century. Uniformity of forms and decoration means that it is not possible to identify unmarked vessels. Many potteries were involved in obtaining wares from others potteries to fill their orders, reinforcing the fact that these products are not distinguishable. The influence of the Staffordshire potteries was spread not only by the size of that industry but by the migration of potters to other parts of Britain and throughout the world (Barker 2001).

Throughout the nineteenth century there were many variants of earthenware bodies. In 1805 Spode developed stone china, a high-fired earthenware 'which approximated porcelain in terms of hardness' (Majewski & O'Brien 1987:120). Ironstone, another development, was so called because of the ironstone slag content (Hughes n.d.). Mason's 'Patent Ironstone China' was first marketed in 1813 and became instantly successful because of the popularity of Mason's other products (Godden 1999). An abstract of the patent read 'A process for the improvement of the manufacture of English porcelain, this consists of using the scoria or slag of Ironstone pounded and ground in water with certain proportions, with flint, Cornwall stone and clay, and blue oxide of cobalt' (Godden 1999:72). Over 300 other manufacturers then produced ironstone-related bodies during the nineteenth century (Godden 1999).

British producers and retailers used the terms earthenware, porcelain and china synonymously regardless of body-type and this has caused problems in the classification of nineteenth century white-earthenware ceramic bodies (Majewski & O'Brien 1987). Some of the trade names used for earthenware include '"Stone China", "Ironstone China", "New Stone", "Semi-porcelain", "Opaque Porcelain", "Granite ware" (and) "Kaolin ware"' (Wakefield 1962:19). All these names have been found on bodies produced in Staffordshire but there is little technical difference between them

(Wakefield 1962). A total of 61 different names were listed by Wetherbee (1985 cited in Majewski & O'Brien 1987:122). Members of the American ceramic industry complained about this ambiguity in descriptive terms at the beginning of the twentieth century (U. S. Department of Commerce 1915 cited in Majewski & O'Brien 1987) and the White Wares Division of the American Ceramic Society approved a classification of dinnerware based on 'percent absorption', 'mechanical shock resistance' (related to hardness) and 'translucency' (Watts 1939:314 cited in Majewski & O'Brien 1987:112).

The most confusing terms relating to nineteenth century earthenware bodies are ironstone and white granite. White ironstone was the undecorated version of ironstone which became popular because of its low cost and durability as well as the ease of producing matching sets of items (Godden 1999:160-162). It was particularly popular in America where it became known as white granite. Brooks (2005:30-35) suggested that white granite should be separated from the other white varieties of earthenware for cataloguing purposes. He described white granite as a semivitreous post-1840 earthenware which is often moulded and has a blue-grey tinge to the body rather than the glaze. He noted that American collectors such as Dieringer and Dieringer (2001) refer to white granite as white ironstone. Godden (1999:162) used the terms interchangeably describing 'white granite or undecorated ironstone wares' as 'thickly potted' and 'often relief-moulded', and stated that they 'were probably introduced in the late 1830s' but there is no record of the exact date. He stated that white ironstone wares are 'all but unknown' in Britain or to British collectors (Godden 1999:163). He posited that two types of white granite/ironstone were made for the American market, the whitest body for city buyers and the slightly-blued body for rural inhabitants (Godden 1999). Majewski & O'Brien (1987:115) stated that white granite refers to 'an American-produced, nonvitreous-to-semivitreous body made to imitate the British classic ironstones'.

Wares were also made in other colour clay bodies; red, yellow, brown and buff earthenware were commonly used for utilitarian and food preparation vessels and teapots. While these are naturally occurring clay colours, oxides were also added to clay to obtain other colours, including turquoise, blue, lilac and green (Miller, posted on HISTARCH on February 16, 2004).

The other main ceramic product from Britain in the nineteenth century was bone china, a type of porcelain body (Birks 2007). It was developed by Josiah Spode in the late eighteenth century to compete with Chinese porcelain which was very popular at that time. In the early nineteenth century, this new English porcelain was under the patronage of the Prince of Wales, who influenced tastes in Britain. Its popularity also increased because of the rising cost of Chinese porcelain due to high import duties, which were 108 per cent in 1799. Although bone china is initially fired to about 1250°C the subsequent glaze-firing is in the traditional earthenware range, making it easy for existing earthenware factories to make bone china (Birks 2007).

Australia

The main Australian-made ceramic vessels found at nineteenth century sites are food and beverage containers, and utilitarian items. Bricks, the first ceramic product made in Australia, were being manufactured at Brickfield Hill in Sydney shortly after the arrival of the first fleet in 1788 and the first domestic earthenware was made by potters who arrived from Staffordshire in 1790 (Ford 1995). However, the first successful stoneware firing was not until 1819 when Jonathon Leake, who also made lead-glazed earthenware, successfully produced salt-glazed stoneware (Ioannou 1988). Although some local potteries, for example the Irrawang Pottery, became well known through advertisements (Birmingham 1976) rapid expansion of the industry did not occur until the 1850s (Ioannou 1988). Government policy did not encourage early domestic potters because it was of the opinion that the quality of goods imported from Britain was superior (Ford 1995:9). Nevertheless early potters 'saw themselves as an essential source of both utilitarian and ornamental pottery' and concentrated on manufacturing these products rather than tea and tableware (Casey 1999:23). In 1844, the Irrawang Pottery advertised ewers, wine coolers, water monkeys, wash basins, filters, mugs and milk pans (Birmingham & Fahy 1987:9). Bendigo Pottery, established in 1864, produced 'chimney tops, flower pots, butter pots, preserving jars, wine and spirit jars...and every class of stoneware' (O'Hoy 1989:14). Bristol-glazed wares were first produced in the 1870s at the Magill Pottery, South Australia, and in the 1880s at Lithgow pottery, New South Wales (Ioannou 1988). At that time, stoneware jars and bottles, with salt or Bristol glazes, were common as food and beverage containers. Rockingham-glazed wares, such as teapots, jugs and spittoons, were also produced. In the late-nineteenth century commercial pottery wares were 'similar Australia-wide and closely comparable to those of the North American and British industry' (Ioannou 1988:67). Those vessels included food and beverage containers, teapots, jugs, mixing bowls, baking dishes, chamber pots, basins and ewers. The market for tea and tableware was still dominated by British imports. Around 1900 Bakewell Brothers, a Sydney firm, were the first to produce transfer-printed tableware. They used staff trained in England and patterns and marks similar to English wares in order to compete for market share (Ford 1995:94).

Japan

Japan ended its 'period of self-imposed isolation' in 1854 when a trading treaty was signed with America (Jackson 1992:245). In 1858 a similar treaty was signed with Britain. The exportation of Western-style ceramic vessels began during the Meiji period, 1868-1912, a time when ceramic technology in Japan was greatly advanced (Kondo 1923:212). The Japanese had realised that to compete with the rest of the world modernisation was imperative. They sent students to various Western countries and brought foreign teachers into Japan (Stitt 1974). The availability of cheap labour in conjunction with Western technology produced low-priced goods of variable quality (Stitt 1974).

In 1922, the National Council of the Pottery Industry (NCPI) from Stoke-on-Trent, England, published the results of a survey into the Japanese pottery industry. From 1868, two categories of items were exported from Japan: those of traditional Japanese style; and those of Western-style made specifically for the export market. The first category was not seen to compete directly with goods from other countries but the second category competed with similar types of goods being manufactured elsewhere. Cups, saucers, jugs and 'good quality dinner ware' (plates and sets) were among vessels exported to Australia in the period 1918-1921 (NCPI 1922:18).

Modern factory equipment, coal burning kilns and plaster moulds had been installed as early as 1870 to facilitate mass-production in large pottery-making areas like Arita, near Nagasaki (Stitt 1974) but in general in the late-nineteenth century, Japanese firms could not compete successfully as most of their equipment and methods were outdated (NCPI 1922). At that time in Japan, most ceramics were still being produced by families or small groups (Stitt 1974). In the 1880s the large amount of orders for wares required hand-painted decoration to be done at great speed resulting in poor quality wares. From the beginning of the twentieth century a number of companies developed modern factories which produced large quantities of goods, enabling them to compete on a world market (NCPI 1922). Quality improved as machine-made decoration was used in conjunction with jiggers, which were machines for the shaping vessels (Stitt 1974).

The Japanese classified their ceramic bodies as earthenware, hard-earthenware and porcelain (NCPI 1922). While earthenware and porcelain corresponded to the English equivalents, hard-earthenware corresponded to what was known at that time in England as semi-porcelain or ironstone china. Very few Japanese firms made hard-earthenware items, the main export ware being porcelain (NCPI 1922).

Among the most efficient of the early factories was Nihon Toki Works (NCPI 1922). It was owned by the Morimura brothers who had formed a trading company in 1876 in Japan to export goods to America and by 1878 were trading in New York as 'Morimura Brothers', importing 'china, porcelain and dolls' (Noritake Co. 2010). In 1904 the brothers established the company Nippon Toki Gomei Kaisha (Nihon Toki Works) in Noritake. This company later took the name of the village becoming the Noritake Company and by 1914 they had made 'Japan's first Western-style dinner set' which was exported to America. In 1932 they were the first to create bone china in Japan (Noritake Co. 2010). Early maker's marks featured 'Nippon Hand Painted' and a circle of M's (for Morimura) forming a flower.

One of the firms making earthenware products was Nihon Koshitsu Toki Co. Ltd., established in 1908, whose website described the company as 'a pioneer in the manufacture of semi-porcelain dinnerware', the production of which was first attempted in Japan in 1905 (Nikko China 2010).

Europe

European countries began producing porcelain in the eighteenth century, particularly Germany and Austria, but these products were expensive at that time. Later, in the age of mass-production, porcelain became cheaper and was afforded by most people. One of the main regions making mass-produced porcelain export wares was the Czech Republic, originally called Bohemia. From the late-nineteenth century many European firms exported porcelain (Snodgrass 2005).

PRODUCTION METHODS

In 1908, Reverend Malcolm Graham wrote a contemporary account of the methods of earthenware production in North Staffordshire. He described a series of processes from the preparation of the clay to the final product, detailing the equipment and people needed to carry out each task. The clay used was a mixture of ball clay from Devon or Dorset, china clay from Cornwall and flint or silica. Each part of the vessel making process was done by a specialist. He wrote 'the man whose work it is to make saucers never makes cups; the man who makes cups never puts the handles on' and so forth (Graham 1908:21).

Making methods in Staffordshire included pressing, slip casting and throwing. Pressing tea and tableware vessels involved using a jigger or jolly, a machine which consists of a revolving disc on which a mould is placed to shape the clay. Flatware items were made by pressing the clay onto the mould and shaping with the jigger from the outside. Graham (1908:24) noted that a 'good Flat-presser can make a hundred dozen plates in the day'. For the making of hollowware vessels, clay was pressed inside the mould and then shaped from the inside with the jolly. Handles were made by pressing clay into the bottom half of a two-piece mould and putting the top on resulting in the clay being shaped by the mould. This job was done by boys, who could make about 1,200 handles each per day. Slip cast vessels were made by pouring liquid clay into a plaster mould. A thin layer of clay stuck to the mould as the water was absorbed into the plaster. The remaining liquid clay was then poured out, a process which usually only took a few minutes. This process was used mainly for ornamental ware. Some vessels, for example teapots or vases, were thrown on a wheel. An assistant supplied balls of clay to the thrower; using exactly the amount required for each article so there was no wastage. Graham (1908:35) 'timed a thrower making twelve teapots (without, of course, spouts and handles) in two minutes' and noted 'a quick worker can turn out 300 or 400 half-pint jugs in an hour'.

The most common method of making vessels in Japan was slip-casting (Stitt 1974). Jiggers, initially introduced around 1900, but not in general use until the next decade, enabled an increase in mass-production (Stitt 1974). Most of the material used for making ceramics was obtained from local sources as it was not considered to be economical to import materials (NCPI 1922).

Ceramic vessels often require two firings, the first firing known as the biscuit firing and the second as the glost or glaze firing. The biscuit, or initial firing, hardens the clay although it is still porous enough to absorb the applied glaze. The most common glaze used for this period in Staffordshire was lead based, often producing plumbism, or lead poisoning, among those who handled it. Lead glazes were popular because they produced a shiny glaze which flowed and covered many faults resulting in a 'superior finish' (Graham 1908:43). The dangers of using lead were lessened slightly by replacing the raw lead oxide with fritted lead, that is, a mixture of lead oxide, silica and other ingredients which when fired together, produce a glass that is ground and used in the glaze mixture. Once vessels had been dipped into the glaze and fettled (the process of removing drips and generally smoothing the glaze) they were placed on thimbles, spurs or stilts, small kiln furniture used to separate the vessels so their glaze did not stick to each other. Vessels were placed inside saggars, which are receptacles made of coarse clay, to protect them from the fuel used to fire them. The glaze firing was much quicker than the biscuit, taking only about 25 to 30 hours in comparison to 60 hours for the biscuit (Graham 1908).

Japanese porcelain vessels in modernised factories were twice fired, that is, biscuit and glaze, using coal as a fuel (NCPI 1922:68-69). Plates, dishes and flatware were put on setters in a saggar for firing while cups and bowls were placed on rings at the bottom of a saggar. In smaller businesses items were often raw-glazed and fired once in old style kilns using wood as a fuel (NCPI 1922:90).

Decoration can be applied under the glaze, i.e. after the biscuit stage, or over the glaze, i.e. after the glaze firing (requiring an additional firing). Common under-glaze methods used in Staffordshire were transfer-printing and hand-painting. Over-glaze decoration, using enamel colours, included printing, painting, ground-laying and colour dusting (Graham 1908). Japanese decorative types included 'decalcomania transfer work, printing and filling in, hand painting, ground-laying, gold paste work, gilding and etching' (NCPI 1922:81). Decals replaced hand-painting as mass production increased. (These methods will be explained further in the section relating to decoration.)

EXPORTING WARES

Larger Staffordshire potteries had agents or outlets in the countries that they exported to, but the smaller manufacturers had to rely on merchants (Barker 2001). In the nineteenth century trade items were a large part of their production, particularly exports to America. American research into the ceramic trade has shown that merchants, supplying rural areas, selected packages of ceramic vessels based on their customers' needs. Miller (1984:45-46) noted that vessels from assemblages, for the same time period, showed a 'consistent similarity to each other' and doubted 'whether the merchants themselves could have distinguished differences among each other's stock'.

There were a large number of small manufacturers in Japan who fulfilled orders for dealers who sold to Commission houses who in turn shipped the goods overseas (NCPI 1922). In addition, British firms purchased goods from Japan to export to Australia, South Africa and elsewhere. During the early-twentieth century, the price of vessels in the country of destination was fixed so the quality of goods declined if production costs rose in Japan (NCPI 1922:16). Some larger firms, for example Morimura Brothers, later Noritake, had an advantage over other Japanese firms in that they could export directly to their branches in other countries without the need to deal with buyers and shippers (NCPI 1922). They had a firm in Sydney which handled their imports and those of some other Japanese factories (NCPI 1922:23).

AVAILABILITY OF ITEMS

Archaeologists have focused more on interpretations of sites using ceramics than on where the ceramics were made or how they got to Australia because in the nineteenth century most pottery was imported and most of it came 'from a single national source along known trade routes', i.e. from Britain (Brooks & Connah 2007:144). Different colonies received different classes of ceramic goods and transfer-printed patterns (Brooks 2005:60-61). Evidence has suggested that Australia became a dumping ground for ceramic items intended for the American market (Brooks & Connah 2007:143; Hayes 2008:353-354). Briggs (2005) confirmed that South Australian merchants requested specific patterns from British manufacturers. As mass production increased the number and type of objects on the market increased and it was only the quality and quantity of the goods that differed (Breward 2004). The family home played an important role in all levels of Victorian society with objects used to express identity. Books, articles and catalogues offered advice to homemakers and 'respectable' working class families strived to emulate the middle and upper classes (Breward 2004:108-109).

Advertisements in two Sydney newspapers, The Sydney Gazette and New South Wales Advertiser (1803-1842) and The Sydney Morning Herald (1831-1868) revealed that there were two main methods of purchasing ceramic vessels in Australia, from auctions or retail outlets (Atkins 1991). Auctions of goods for commission took place regularly in the first half of the nineteenth century, along with those occurring as people left the colonies, or through insolvency (Atkins 1991). From the middle of the century large consignments of ceramics were advertised for sale. These were aimed at dealers and seemed 'to represent wholesale activity' implying that goods went through Sydney to other areas of Australia (Atkins 1991:13).

Many names were used for advertising semi-vitreous earthenware including 'ironstone china, stone china, new stone china, white stone, ironstone chip ware, royal ironstone, white ironstone and pearl stone china' (Atkins 1991:51). Commonly advertised wares included transfer-printed, flown, gilt and sprigged vessels. Transfer-printed vessels were available in a wide variety of forms. The main transfer-printed colours advertised for the period 1803-1868 were blue, brown and green, with only a few references to 'purple, mauve and mulberry' in the 1860s, while black was not advertised at all (Atkins 1991:34). The most common colour was blue. In the early nineteenth century the cheapest transfer-printed pattern was Willow and potters classed it as a separate category (Miller 1991:8). Tableware was the predominant form available with dinner sets in this pattern being available from the 1820s (Atkins 1991:46). Sponge-printed wares advertised during the period to 1868 appeared in blue and purple with forms including bowls, basins, wash sets, cups and saucers and jugs. Gilt decorated wares were known as 'white and gold, white and gold edge, white and gilt, W. & G., white and gold plain' (Atkins 1991:56). These were advertised from about 1837 and forms could be embossed or enamelled in addition to the gold decoration and could be on earthenware or porcelain. These wares were available in a variety of forms but appeared most popular on breakfast and tea sets. A few sprigged items, including breakfast cups and saucers were advertised from 1841, but they became more common from 1844 (Atkins 1991:61). The majority of ceramics advertised in Sydney newspapers came from Britain. There were a few references to European ceramics although details of decoration types were not provided by the newspapers. Two-thirds of the advertised ceramics were marketed by their function or fabric, while only one-third made mention of the type of decoration (Atkins 1991:69).

Illustrated trade catalogues also provided a source for advertising ceramic vessels. Australian general store catalogues were first produced in the 1860s, although no copies survive; the oldest known department store catalogue is from Melbourne's Mutual Store, dated to 1892 (Crook 2008:66-67). The Harrod's catalogue of 1895 shows the price of earthenware to be much cheaper than bone china/porcelain (Adburgham 1972). From a selection of cup and saucer sets advertised, the cheapest at that time was the earthenware 'band and line' red or blue at 2s 9d per dozen, followed by floral-decorated earthenware at 3s per dozen. The plain white or 'white and gold' china cups and saucer both sold for 3s 6d, while the more ornate china cups and saucers sold for 5s 6d per dozen (Adburgham 1972:474).

The ability to purchase ceramic vessels has usually been considered to be directly related to their cost. Crook's (2008:283) examination of trade catalogues, dating 1872 to 1907, concluded that 'social interaction and cultural conditioning' may have a greater impact on consumption. She ascertained that 'aesthetic appeal and quality workmanship' are significant aspects of retail promotion and that there is a strong relationship between the cost of an item and its quality (Crook 2008:276). Crook (2008:280) suggested that nineteenth century ceramics typically found on Australian sites 'may be less expensive than we assume' and were probably easily afforded by middle-class and many working-class families.

Industrialisation allowed objects that were once only available to the upper class, such as sets of matching vessels, to become more accessible to the middle and working classes. Ceramic sets expressed the values of the middle class and were aspired to by the working class (Young 2003:191). Definitions of a set for American contexts have varied, for example, 'three or more different forms in the same pattern' (Yamin 2001:160) or 'various forms which have the same body type and decoration' (Worthy 1982:348). In the 1860s in America, groups of vessels with the same shape or motif were considered to be matching sets (Praetzellis and Praetzellis 1992; Fitts 1999; Wall 1999). It was only towards the end of the nineteenth century that matching sets became common in America. Evidence suggests that even into the early twentieth century consumers in remote areas of America were not concerned with purchasing sets (Majewski and O'Brien 1987:179). The definition of a set in Australia is also varied. While some researchers use an American definition (e.g. Briggs 2005), others have suggested 'individual vessels bearing the same pattern, form and maker's mark' (Crook 2008:237) or 'two or more vessels of an identical pattern' (Hayes 2008:142). The general availability of tea and tableware sets throughout Australia during the nineteenth and early-twentieth centuries has not been established. Brooks (2005) suggested that because Australia's market was relatively small in the nineteenth century, it became a place where unfashionable stock would be dumped from England or America, the implication being that consumers may not have been able to purchase sets.

Differing numbers of ceramic sets have been recovered from working class sites in Australia dating to the second half of the nineteenth century. Crook's (2000:23) study of five households at the Rocks, Sydney, found one household had eight sets, two had no sets and the others had two and three sets. The most common patterns were Two Temples and Willow. Most of these sets were tableware; only the household with eight sets had a set of matching table and tea ware. Lydon's (1999) study of two assemblages dating to the 1860s-1870s found no sets present even though evidence suggests that these residents had easy access to the ceramic market at that time. The crockery used at Mrs Lewis' boarding house, occupied from 1861 to 1873, was made up of a number of large breakfast cups and saucers, which while individually matching, did not form a set but could be considered to be complementary (Lydon 1999:48). Research into three households in Port Adelaide, occupied from 1840 to 1900, found the number of sets for each household was six, seven and ten; common patterns on tableware were Albion, Asiatic Pheasants, Rhine and Willow (Briggs 2005). All households had gilt and Chelsea Sprig porcelain tea sets. One household had one combined tea and tableware set and another, three combined tea and tableware sets; all other sets were either tea or tableware (Briggs 2005).

FOOD AND BEVERAGE CONTAINERS

In the time period covered by this study, food and beverages could be purchased in ceramic containers, although the need for ceramic bottles decreased as glass bottle production increased. Many areas in Britain and Australia produced these ceramic vessels which were usually thrown on potters' wheels.

Forms

White-bodied earthenware food jars, which contained foods such as fish paste or jam, were plain or decorated. Jars, with a clear glaze only, often had labels which do not survive in the archaeological record. Some decorated jars, known as Pratt-ware, had multi-coloured transfer-prints featuring portraits, landscapes, pastimes or contemporary nineteenth century events such as battles. Small stoneware food jars, which contained jam, cheese or cream, were decorated with Bristol glaze. These often had lids which sat inside a gallery at the mouth of the jar. Larger wide-mouth food storage jars were decorated with either Bristol or salt glaze.

Bottles for beverages such as ginger beer or stout were made of stoneware and were salt or Bristol-glazed. Various shaped finishes were made over the nineteenth century including the blob top, cork seal, screw top and crown seal. The screw top was invented in 1872 (Noël Hume 1976:79) and the crown seal in 1891 (Vader & Murray 1979:75). Larger beverage containers, known as demijohns, were also salt or Bristol-glazed. Many demijohns found in Australia were made in Scotland or in Bendigo, Victoria (Arnold 2006).

Decoration

The most common glazes on stoneware food and beverage storage jars and bottles are salt and Bristol glazes as both only require a single firing. Salt glaze is produced by throwing salt (sodium chloride) into a kiln when it reaches a sufficient temperature (usually 1100-1200°C). The salt volatises and the sodium reacts with the silica and alumina in the clay body to produce a glaze, while the chlorine is expelled from the kiln as a gas. The thickness of glaze on any vessel depends on the amount of salt added and the position of that vessel in the kiln. Salt glaze only coats the exterior of any vessel, which means that if an interior glaze is required it has to be done separately using another glaze.

In 1835, William Powell of Bristol produced a glaze that could be applied to the interior and exterior of unfired vessels, requiring one firing (Noël Hume 2001). Within a few years it became more popular than salt-glazing as many potters began using what became known as Bristol glaze. The glaze is a whitish colour. Its main flux, an oxide which helps melt other oxides, thereby promoting ceramic fusion, is zinc oxide. The glaze is often applied in conjunction with a golden-brown glaze using iron and manganese oxides as the colorants. The darker glaze features on the top section of a bottle or jar. The recipe for Bristol glaze used at the Bendigo Pottery differs from the original British Bristol glaze. It is a grey colour and does not contain zinc oxide but has calcium oxide as the main flux (O'Hoy 1989:15).

FOOD PREPARATION AND CONSUMPTION VESSELS

The shape of vessels produced is related directly to the diet and customs of those using them. Many etiquette books were written in the nineteenth century, originally read only by the wealthy classes but by the mid-century they were more widely accessed by the expanding middle class (Lucas 1994). The popular mode of dining in the 1850s and 1860s was à la Russe which involved serving many courses, leading to the need for many plates, especially in smaller sizes (Lucas 1994:81-82). Drinking tea became an important social event in Britain, particularly within the timeframe of this study. Queen Victoria popularised afternoon tea which was taken up by all classes, leading to the use of matching cups and saucers.

Forms

Ceramic vessels were used for food preparation, serving and consumption. Food preparation vessels include mixing bowls, baking dishes, pudding dishes, colanders and milk pans. These vessels were made of fine or coarse earthenware, often white-bodied but sometimes yellow-bodied or terracotta. The majority of tableware found on nineteenth century Australian sites is refined earthenware. A dinner service can include dishes for bones, butter, cheese, pickles, vegetables and general serving dishes, compotes, tureens, gravy and sauce boats, platters, drainers and jugs. Platters, made in sizes from 250 mm to 600 mm (10 to 24 in.), are commonly oval in shape but some are oblong or hexagonal (Miller 1991, Neale 2005). A variety of plate sizes were produced (Table 5.1) however they are often found in slightly larger sizes. Potters provided their customers with a slightly larger size for the cost of a smaller plate in order to get around price fixing agreements; for example, 9¾ inch plates were sold as 'Suppers' (Miller 1991:11).

Tea service vessels comprise teapots, creamers, milk pitchers, sugar bowls and cake plates in addition to cups, saucers and tea plates. Tea ware is usually made of refined earthenware, porcelain or bone china. British teapots, cups and saucers were copied from the shapes of Chinese ones imported in the seventeenth century by the East India Company (Buckrell Pos 2004). British cups of that time did not have handles as they were a copy of the Chinese teacups. Saucers had been used from the seventh century in China as plates to hold hot handle-less teacups (Buckrell Pos 2004:59). The shape of teapots and cups changes over time.

Table 5.1: Plate names and sizes (after Miller 1980:23, 1991:11; Neale 2005:160-161)

Plate name	Size (mm)	Size (in.)
Dinner / table	250	10
Soup plate	250	10
Supper	230	9
Cheese / breakfast / twiffler	200	8
Cheese / breakfast / muffin	175	7
Side / nursery / muffin	150	6

Decoration

The most common decorative types are detailed below, starting with body decoration, then underglaze, glaze and over-glaze decoration.

Dyed body

Coloured earthenware bodies were being produced by most 'good earthenware potters' by the 1820s (Griffin 2001:167). The body was coloured through the addition of oxides to the clay. Many colours were produced throughout the nineteenth and early-twentieth centuries, including blue, green and grey. A white slip was added to the underside of plates or the inside of cups, jugs and teapots by some potters resulting in those areas appearing white under the clear glaze while the rest of the vessel was coloured. Miller (2004) stated that these vessels do not appear in the Staffordshire potters' price fixing lists he examined and suggested that not enough were being made to make it worthwhile setting prices. These bodies feature a clear shiny lead glaze or a smear glaze. The smear glaze results from coating the inside of saggars with a glazing material which vaporises during firing and provides a thin glaze coat on the vessel. This type of decoration was used on relief-moulded wares as glazing the vessel directly could obscure pattern detail (Griffin 2001).

Sprigs

Sprigs are small shapes made in a mould and applied as decoration to a vessel while it is leather hard, i.e., it still contains some water in the clay, but the vessel can be handled without deformation. Common patterns found on nineteenth century Australian sites are Chelsea Sprig and Chelsea Grape. Both have added blue sprigs, one featuring flowers and the other grapes, although Chelsea Sprig is often used as the name for either pattern. It was common on British sites from about 1820 to the later nineteenth century (Brooks 2005:43).

Relief-moulding

This term is used to describe a surface with a raised decoration, which is part of the mould in which the vessel was made, rather than being an added sprig. This decoration was particularly popular on white ironstone or granite, where relief-moulding appears around the marly.

Transfer-prints

Transfer-prints were first used in the 1750s (Little 1969). They allowed decoration to be applied quickly and uniformly. Majewski & O'Brien (1987) proposed that their popularity may have been the reason that nineteenth century marketing was based on decoration rather than clay body-type.

The process involves paper designs being copied onto a copper plate via tissue paper (Copeland 2000). The design is then engraved using lines and dots and the finished copper plate passed onto a team in the printing shop of a pottery. A mixture of metallic oxides and printing oil is rubbed onto the copper plate before the design is transferred onto sized tissue paper through a printing press. The design is transferred to the ceramic surface by rubbing the paper. The ware is then soaked in water to remove the paper and dried before being fired to 680-700°C to remove the oil prior to glazing (Copeland 2000:11-13). Different individuals were employed for each step of the process, for example, a person was employed to cut the paper designs and another to transfer them to the ceramic ware.

Blue was the first colour used for transfer-printing as the purpose was to create a product to rival the Chinese blue-and-white wares. Many shades of blue transfer-printed wares have been made over the nineteenth century from very-light to very-dark blue and 'almost purple blue' (Little 1969:34). The colour depends on the purity of the cobalt used, the kiln firing-temperature and the craftsman's skill in application (Little 1969). By 1828 other colours were used including black, brown, green, red and yellow. Black is produced using cobalt and iron oxides; brown - a mixture of iron and manganese oxides; purplish-brown - manganese oxide; green – lead and copper oxide; and yellow – tin, lead or antimony (Lockett 1996).

Many transfer-printed designs were applied to ceramic vessels. Coysh & Henrywood (1989:8) recorded hundreds of transfer-printed patterns and concluded that the subject appears 'to be inexhaustible'. Manufacturers originally used book illustrations as their design source but the 1842 Copyright Act brought an end to that (Coysh & Henrywood 1989). Pattern styles were influenced by contemporary fashions and interests. Pattern titles, often named after people or places, were chosen to boost sales rather than relating to the patterns themselves (Coysh & Henrywood 1989). Many British potters adapted their existing transfer patterns to appeal to the American market. Adams, for example, renamed a European romantic scene Columbia, and Enoch Wood & Sons called a design which included an urn, flowers and butterflies, Detroit (Copeland 2000:31). It is not known if patterns titled Sydney, including an urn and shells, or Melbourne, a floral pattern, were aimed at the Australian market, or simply named after prominent British politicians. In the 1870s William Brownfield from Cobridge, Staffordshire, made tableware featuring a kangaroo in a bushland setting but it is not known how many were made (Graham n.d.).

Copper plates for transfer-prints were usually sold along with a pottery (Samford 1997). This means that unmarked vessels are not attributable to a certain maker. Popular patterns, such as Willow and Asiatic Pheasants were produced by many potteries over a long period of time despite copyright acts (Samford 1997). In addition, variations can be found in patterns of the same name. Brooks (2005:44) suggested that patterns may vary more in the centre while the border remains more consistent.

Flown colours

Flow blue was developed in Staffordshire in the 1830s and by the 1840s the wares were available overseas (Samford 1997). Volatile chemicals, such as lime or chloride of ammonia, were added to a firing, causing the colours to flow outside the pattern creating a softer effect (Samford 1997). Flow blue was made principally as an export ware for the American market as it was disliked in Britain (Godden 1999:164-165). Flow mulberry became popular around 1830 but had almost disappeared by 1860. A recipe for mulberry-underglaze consists of manganese and cobalt oxides as the colorants (Kelly 2000 cited in Van Buskirk 2002). Manganese oxide creates a purplish colour which is much cheaper to produce than flown-blue (Stoltzfus & Snyder 1997:15). Godden (1999:166) described mulberry as dark grey, brown or purplish black and this may be the colour that is commonly referred to as 'flow black' (Stoltzfus & Snyder 1997:15).

Hand painting

Many ceramic vessels of this period are hand-painted and are considered relatively cheap tableware (Wakefield 1962:137). This type of decoration is applied under the glaze and can be added to printed or relief-moulded patterns.

Sponge-printing

Sponge-printed vessels are rarely marked as they were aimed at the lower end of the market (Kelly et al. 2001). They originated in Scotland in 1835. Sponges were either cut by potters or bought ready-made from specialist firms. The hardest part of a sponge was fixed to a wooden paddle and used to apply a repeated pattern. Early sponges were soft-edged and of an indeterminate shape resulting in motifs that are dotty, circular and shapeless. In England the earliest colours used were blue and red, and occasionally green. In Scotland purple and brown became popular. By 1870 sponges were hard-edged and were cut into distinct shapes such as flowers or birds and multicoloured prints were made. Kelly, Kowalsky and Kowalsky (2001:10) noted that wares from the six largest firms known to have made sponge-printed items (two firms from England, two from Scotland and two from Holland) are 'virtually indistinguishable from each other'.

Banded, band and line

Banded patterns were very popular in Australia from about 1880 to 1900 (Wilson 1999:321). The most common feature pairs of bands at the rim and another band lower down on hollowware, or towards the centre on flatware. These bands are either the same width or one is wide and the other narrow, a pattern sometimes referred to as 'band and line'.

Glaze

The most common glaze of the nineteenth century is a clear lead glaze. It changes slightly over time. At the start of the nineteenth century the popular tableware of the day, often referred to by archaeologists and collectors (but not manufacturers) as pearlware, contains a small amount of cobalt in the glaze producing a blue tint. Pearlware was produced from about 1780 to 1840 (Lockett 1996). The amount of cobalt added to the glaze decreased over time. By the 1820s whiteware was developing from pearlware (Miller 1991). It has a clear glaze with 'only a minute hint of cobalt' which sometimes produces a bluish-green hue in pooled glaze (Griffin 2001:147). Later nineteenth century clear glaze has no hints of blue.

Rockingham glaze, popular in the nineteenth and early-twentieth centuries, was often put over coloured-body earthenware, usually buff but sometimes yellow. The main flux of this mottled-brown glaze is lead oxide and the colorants, manganese and iron oxides. The most common forms with this glaze are teapots. Other common forms include jugs and spittoons. This glaze was produced by British, Australian and American firms, all being similar. The Rockingham glaze used at the Bendigo Pottery also contains lead as the flux and iron and manganese oxides as the colorants (O'Hoy 1989).

Enamel or over-glaze decoration

In contrast to decoration under the glaze, painting decoration over the glaze requires an additional firing, adding to the cost of a vessel. Enamel is a soft melting glass with oxides used as colorants (Hamer & Hamer 1997:118). It is usually fired about 300°C lower than the original glaze. It fuses itself onto the glaze rather than into the glaze, can be opaque or transparent and has poor resistance to abrasion (Hamer & Hamer 1997:229).

Gilt

This type of decoration refers to the use of gold enamel, particularly for the addition of gold bands. 'Tea leaf' design, introduced mid-1850s, was also common, with 'more than three dozen English potters' producing the pattern (Abrams 1997:13). It was also produced in America in the late-nineteenth century but its popularity waned by 1910 (Abrams 1997). It consists of a gold leaf design in the well of a cup, saucer or plate and gold bands around the rim (Dieringer & Dieringer 2001). Gilt decoration fades over time but leaves a faint grey residue.

Ground-laying and colour-dusting

These are similar processes using oil which is applied to the area needing to be coloured. Powdered oxide colorants and other components are mixed and dabbed onto the vessel. The colour only sticks to the oiled areas. Colour-dusting was used for band and line patterns and ground-laying for larger areas (Graham 1908). Red decoration was used in this manner for Japanese export wares (Stitt 1974:127). Prior to the First World War pigments used for colouring in Japan were imported from Germany, but by 1914 they were mainly sourced from England (NCPI 1922:89).

Decal

Decals are produced by using copper plates, similar to transfer-prints, but are attached over the glaze, requiring an additional firing. They were produced from the end of the nineteenth century, but are more common in the twentieth century.

DATING CERAMIC VESSELS

Dates of manufacture may be suggested by makers' marks, form profiles or through decorative motifs.

Makers' marks

Marks on ceramic items may relate to a pottery, an individual worker at a pottery, a retailer, or a date. They can be incised, impressed, printed or painted. The majority of transfer-printed ceramic vessels were not marked by makers (Samford 1997:1). Many other vessels were also unmarked.

Marks can be applied on the body, either before or after firing, or can be applied over the glaze. Incised marks are those which are scratched into the ceramic body before firing. Impressed marks are also applied before firing by pressing the mark into the clay. Impressed marks are usually smoother than scratched marks which tend to roughen the clay. Other types of marks applied under the glaze are transfer-printed and hand-painted marks. Marks applied over the glaze can be hand-painted or a decal.

Some common marks appear on British-made ceramic vessels in conjunction with the name of the maker or the pattern. Flowery and foliated cartouche marks around pattern names (often without the maker's name included) usually date to about 1830-1840 (Little 1969:31). A garter-shaped mark was used from 1840 onwards, with or without a crown, featuring the maker's name and sometimes the pattern name (Godden 1991:11-13). The Staffordshire Knot was used from 1845, although it was most common in the 1870s and 1880s (Godden 1991:11-12). It features a rope knot which often has the maker's name or initials in the segments (Godden 1991:13). The English Royal Coat of Arms was used by many nineteenth and twentieth century manufacturers, but foreign manufacturers also copied these marks (Godden 1991:552). For example, some Japanese makers use the word 'Royal' in marks.

The British Ornamental Designs Act of 1842 enabled potteries to protect their designs but other people could use them if copyrights lapsed. In 1842 a diamond-shaped mark was introduced to protect the pattern or vessel shape for three years (Godden 1991:526). If the mark is printed, it generally relates to the pattern. If the mark is impressed or moulded, it generally relates to the form of the vessel (Godden 1991). Information regarding the date of manufacture is contained in the corners of the diamond. There are two different configurations of the diamond registration marks, one running from 1842 to 1867 and the other from 1868 to 1883. From 1884, a system of registered numbers was adopted. The numbers run consecutively and appear on vessels with the prefix 'Rd' or 'Rd. No' (Godden 1991:526).

The British Trade-Mark Act was introduced in 1862 but those with the words 'trade mark' usually signify a date after 1875 (Godden 1991:11). 'Limited', 'Ld.' or 'Ltd.' were introduced after 1860, but were not generally used on ceramic vessels before 1880 (Godden 1991:11).

Before 1872 it was customary for a potter to be paid by the 'good from the oven' principle, so it was essential that an individual's work could be identified once it left the kiln (Copeland 2000:37). Impressed or printed single letters or numbers were often used to identify an individual (Little 1969:36).

The American McKinley Act of 1891 required all goods imported into that country to be marked with the country of origin rather than the city. This affected world markets as America was one of the biggest importers of ceramic vessels. The word 'England' was never used before the 1880s (Little 1969:31) but was added to all marks after 1891 to comply with the Act (Godden 1991:11). 'Made in England' signifies a twentieth century date (Godden 1991:11). Some firms have different marks for local and exported wares. For example, Brownfield & Sons, a Staffordshire firm in the second half of the nineteenth century used a printed 'twin globes' mark for export ware and a different mark for local British wares (Peake 2004:49-50).

Much of the porcelain exported from Japan in the period 1868-90 is unmarked (Stitt 1974:143). Japanese export wares from 1891 to 1921 were marked 'Nippon' to comply with the McKinley Act of 1891 even though this act stipulated that marks should be in English. After 1921 this mark became 'Japan' or 'Made in Japan' (Stitt 1974:176).

Marks found on earthenware or stoneware bottles or jars may be those of the maker of their contents rather than the vessel maker's marks. Firm and product labels can be impressed, printed or use a decal.

Vessel profiles

As previously noted, many smaller potteries, which did not have their own resources, used freelance mould makers resulting in uniformity across many forms (Barker 2001) suggesting that profiles of vessels for certain periods will have common characteristics. Although Brooks (2005) cautioned that the use of plate profiles for dating has not been tested in Australia, the majority of tableware found on both Australian and American sites dating from the mid to late-nineteenth century came from the same source – Britain.

In 1997, Randall Moir presented a paper at the Society for Historical Archaeology conference exploring the form-specific attributes of common plates found on American sites. He presented a diagram of forms over time, from about 1750 to 1910, with the sequence based on plates from sites in rural areas. Before 1890 the majority of plates were British-made but after that date more American-made plates were found (personal communication, Dr Randall Moir, 24 July 2007). Figure 5.1 shows the evolution of plate profiles from the early-nineteenth to early-twentieth century. Changes occurred in the foot shape, shoulder angles and rim thicknesses. Salient features of the 1825 plate are the double foot ring and angular shoulder. The rim is slightly thicker than the rest of the marly. By 1845 there is a single foot ring which is flattened on the bottom. The shoulder is still angular but the marly tapers towards the rim. The 1860 plate has a rounded foot ring and a thicker marly than previous plates. The marly is a consistent thickness. Plate thickness increases over the first half of the nineteenth century. The 1875 plate has an interior well but the exterior profile does not have the same shape and is almost straight from the foot to the rim, a feature which persists until the twentieth century. The 1875 plate has a less angular shoulder than previous plates and a rounded foot ring, similar to the 1860 plate. Foot rings get smaller, and plates thinner, from then until the early twentieth century. Jigger marks begin to show from the 1880s onwards. Moir also studied British wares from South African sites dating from 1840s to 1880s and found a similarity between those ceramic plates and the ones he had previously examined in America (personal communication, Dr Randall Moir, 24 July 2007).

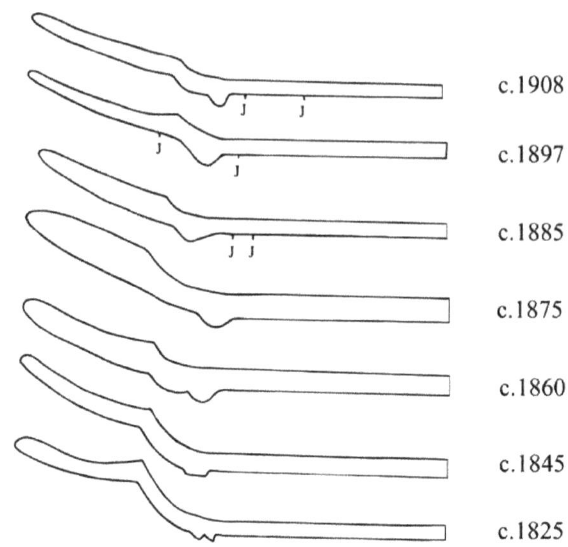

Figure 5.1: Plate profiles 1825-1908 (after Moir 1997; j - jigger marks)

Although no specific research has been done in Australia on plate profiles, it is possible to analyse profiles from dated archaeological sites. A comparison is made between drawn profiles appearing in Allen's 1969 PhD thesis, recording the first professional excavation of a European site in Australia, at Port Essington, Northern Territory, and those in Moir's diagram. Allen (1969:204) noted that ceramics from his site appeared 'to fit into a manufacturing range of 1830-1845' and the drawn plate profiles (Allen 1969, Plate III.17) are similar to Moir's 1825 and 1845 profiles (Figure 5.2). Examination of plates recovered from the Gold Commissioner's residence in Kiandra (KGQ), dating 1860-1862, finds profiles similar to those in Moir's diagram dating 1825 and 1845/1860 (Figure 5.3). None of Moir's later profiles are present at these two sites. Observations of plates recovered from other NSW rural sites with known dates of occupation, has found that generally their profiles are similar to those in Moir's diagram for his figures dated 1825 to 1875.

After 1880 differences appear in plate profiles found on Australian sites to those in Moir's (1997) diagram. There are still some similarities, for example the introduction of jigger marks and the reduction in the thickness of the foot ring. Some later plates found on Australian sites have the same general characteristics but not the exact profiles for the dates outlined by Moir (1997). A Rhine decorated plate, with a Brownfield maker's mark, found at the Old Kinchega homestead is different to any of Moir's later profiles (Figure 5.4). Although the mark on this plate is not completely legible, it appears to have been made in December 1887, and this fits within the period of use for this mark, 1871-1891. This plate has some similarities to the profile to the Rhine plate from KGQ, suggesting that elements of earlier profiles may have been reused.

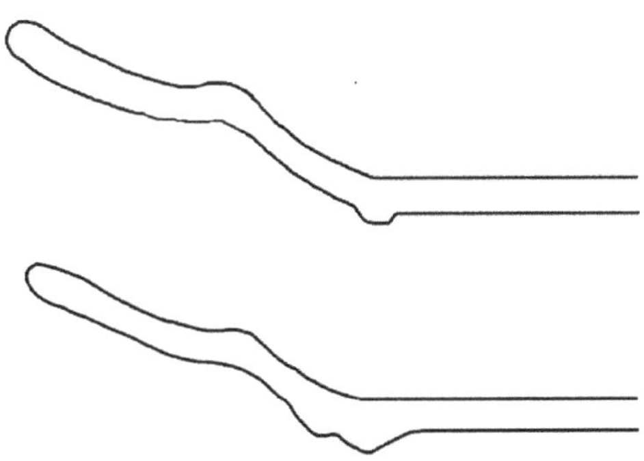

Figure 5.2: Plate profiles from Port Essington (detail after Allan 1969, PlateIII.17, figures b and c)

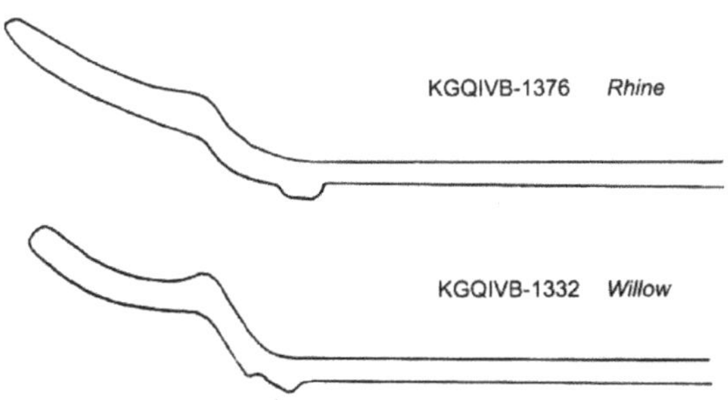

Figure 5.3: Plate profiles from the Gold Commissioner's residence, Kiandra

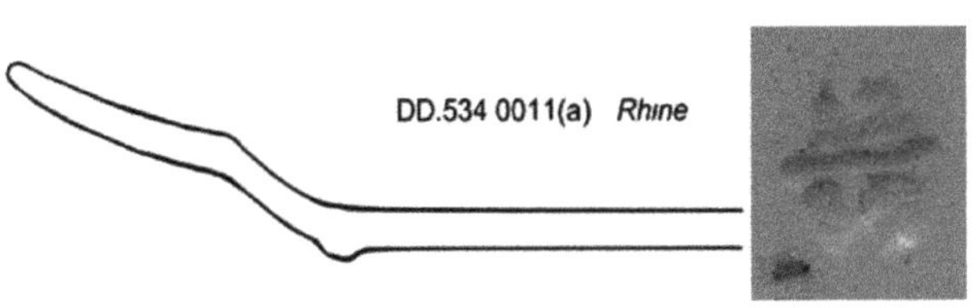

Figure 5.4: Plate profile from Old Kinchega homestead (after Mottram, original drawing courtesy Penelope Allison) and the Brownfield date stamp

Although the sample number of ceramic plates which were compared to Moir's diagram is small, initial investigations suggest that continued research in Australia may find plate profiles to be a useful dating tool for the date of manufacture of ceramic wares but not necessarily their use date. However, this could be important in interpreting how wares were purchased, i.e., new or second-hand, or if they were old stock dumped from Britain.

It may not be possible to date unmarked cups through profiles alone, as many shapes recur over time. Berthoud (1990:ix) detailed pre-1850 examples with reference to 'cup and handle shapes' in his book, A Compendium of British Cups, but after 1850 the shapes became so diverse that most of those he illustrated have dates confirmed by makers' marks. Similarly, Buckrell Pos (2004:63-64) listed cup handle styles for the first half of the nineteenth century but suggested that in the second half of the century earlier styles enjoyed a revival.

Saucers appear to have changing dimensions over time between the exterior foot ring and the interior cup well and this, along with the decoration, may prove to be useful in dating them. To date nothing has been published to confirm this. Examination of a small sample of earthenware saucers with known dates of manufacture suggests that the size of foot and cup well diameters were roughly equal in the mid-nineteenth century. After this, the foot diameter increases until it is about 30-40 mm more than the cup well by the end of the century. The former then reduces in size until the foot and cup well are of a similar size, by about the 1930s. However, more research needs to be done to confirm this.

Decorative motifs

Samford (1997) developed a chronology for nineteenth century transfer-printed ceramics based on their motifs. She examined 3,250 vessels made by 176 different manufacturers which had a known date either by registry mark or by firm name (firm operating less than 40 years). She noted that 'thousands of designs' were produced but the majority of individual vessels were not marked (Samford 1997:1). She identified a series of styles over time. Each was produced over a long period, but had a peak production range of about 20 years (Samford 1997:6). All transfer-printed borders before the influence of Japanese art, except for sheet patterns, appear to have been symmetrical designs. Sheet patterns feature a repeating design which covers the whole vessel. Generally transfer-prints have a central design and a complementary surrounding border design on the marly, with an unpatterned space between the two.

Early transfer-printed designs were copied from Chinese blue-and-white porcelain designs, the most well-known design being Willow which was made by many manufacturers. Central design motifs on these Chinoiserie wares include figures in Eastern garb, junks/sampans, orange trees, pagodas and willow trees (Samford 1997:17). The border designs of these transfers often include dense geometric patterns with key motifs, fish roe, honeycombs, lozenges, latticing and butterflies. They were produced from 1783 to 1834, but their peak period of production was between 1797 and 1814 (Samford 1997:6).

A number of books on British scenery, written and illustrated by William Gilpin (1724-1804), were the stimulus for British view designs (Coysh & Henrywood

1982). These books encouraged other publications in a phase known as 'the cult of the picturesque' and potters copied many of the designs (Coysh & Henrywood 1982:10). Central motifs on vessels include buildings or landscape features (Samford 1997). These central images were often framed with floral designs, some having subsidiary details in medallions (Coysh & Henrywood 1982). They were produced from 1793 to 1868 and their peak period of production was between 1813 and 1839 (Samford 1997:6).

The so-called 'Exotic views' designs were influenced by 'The Grand Tour', a tour of the continent, usually undertaken by wealthy young men in the first half of the nineteenth century (Coysh & Henrywood 1989:8). The opening of the London Zoo in 1828 also created an interest in exotic animals (Coysh & Henrywood 1982), which feature, along with exotic architecture and foreign figures, as central design motifs on vessels (Samford 1997:17). They were made from 1793 to 1868 with a peak period of production between 1820 and 1842 (Samford 1997:6).

In the mid-nineteenth century floral images were naturalistic. Gardens and pressed flowers were very much in vogue and surfaces were covered 'almost entirely with large and coarse flowers' in 'every branch of industrial art' (Evans 1975:152). The popularity of needlework also inspired many blue transfer-printed floral patterns (Coysh & Henrywood 1989). In the second half of the nineteenth century floral images became stylised and included flowers which had been popular earlier rather than those which were popular with gardeners at that time (Evans 1975). Floral motifs were popular for repetitive transfer-printed sheet patterns, particularly between 1826 and 1842 (Samford 1997:18).

Classical designs, produced between 1793 and 1868, were influenced by archaeological excavation reports, particularly from Herculaneum and Pompeii (Samford 1997:12). Central design motifs include acanthus leaves, columned temples, figures in classical garb, Greek key elements and urns (Samford 1997:17). Their peak period of popularity was between 1827 and 1847 (Samford 1997:6).

Romantic and gothic designs followed the major art movements of the time. Romanticism began in opposition to the classical revival and also as a response to industrialisation (Samford 1997). Romantic designs feature small figures engaged in some activity, for example fishing or strolling, in the foreground. The mid-ground design includes a water source such as a pond or river, and the background portrays imaginary buildings (Samford 1997:17). The peak period of production for romantic designs was between 1831 and 1851 although they remained popular throughout the nineteenth century (Samford 1997:6).

The Gothic Revival stemmed from the Romantic Movement and used motifs from medieval illuminated manuscripts. Central designs feature churches, architectural ruins and buildings with arches, battlements, towers and turrets (Samford 1997:17). Border designs are often scrolled motifs. Gothic designs peaked between 1841 and 1852 (Samford 1997:6).

The opening of Japan to Westerners in the mid-nineteenth century led to a strong interest in Japanese arts and literature (Evans 1975). Potters copied elements of Japanese decoration to use for transfer-prints, particularly towards the end of the nineteenth century. The most common motifs were flowers and landscapes and the patterns were asymmetrical (Stitt 1974). Particular motifs include birds, fans, in-filled half circles and prunus branches (Samford 1997:17). These were often produced on ivory bodies over a period from 1864 to 1907 (Samford 1997:6, 17).

Border designs also changed over time, along with the central motifs. For example, in the 1830s and 1840s oval or oblong cartouches enclosed a variety of images often together with floral elements (Samford 1997:21). During the 1840s and 1850s linear border designs were popular. These featured close concentric lines as a background for flowers and scrolls (Samford 1997). However, there were not as many distinctive border motifs as central ones (Samford 1997:21).

Van Buskirk (2002: Chapters 1 & 7) listed changing attributes over time for flown-blue plates. He suggested three distinct periods of decoration: early Victorian 1835-1850s, the initial period of popularity; middle Victorian 1860s-1870s, when not many vessels were produced; and late Victorian 1880s-1920s when there was a revival in popularity. The flown-blue on the early period wares is very heavy with patterns sometimes obscured. The later period flown patterns have more clarity than the earlier ones and much thinner bodies. Plate edges go from panelled in the early period to smooth/rounded in the middle period and embossed/scalloped in the late period. There is no beading on early wares, minimal but large beading on middle period wares, and prolific small beading on later period wares (Van Buskirk 2002:259-266). Popular patterns in the period 1835-1860 feature Oriental designs while from 1860-1880 flowers are the predominant motif (Snyder 2004).

ARCHAEOLOGICAL EVIDENCE FOR AVAILABILITY OF WARES

A comparison of tea and tableware forms recovered from various sites in Australia and New Zealand is shown in Table 5.2 and the range of decorative types in Table 5.3. Data are incorporated from other researchers including Ritchie (1986), Lawrence Cheney (1995), Briggs (2005), Allison and Cremin (2006), Bowen (2007), Brooks (2007), Hayes (2008), Muir (2008) and Dunk (2010).

Table 5.2: Tea and tableware from sites in Australia and New Zealand

Location	Date		FP	SD/L	PL	T	J	D	GB	SP	P-L	P-M	P-S	S	C/M	B	E	R
AUSTRALIA																		
Bungendore1, NSW	1859-1890s	N	x	x	x	x	x				x	x	x	x	x	x	x	10
Bungendore2, NSW	1875-1904	N		x	x	x	x		x		x	x	x	x	x	x	x	10
Kiandra1, NSW	1860s	N		x	x					x	x	x		x	x	x	x	10
Kiandra2, NSW	1860s	N									x			x				10
Kiandra3, NSW	1860-1890	N		x							x	x		x	x	x	x	10
Kiandra4, NSW	1860s-1931	N	x							x	x	x		x	x	x		10
Kiandra5, NSW	1860s-1950s	N	x	x	x	x	x				x	x	x	x	x	x	x	10
Kiandra6, NSW	1880-1930	N		x					x		x			x	x	x		10
Kinchega, NSW	1870s-1950s	N		x	x	x	x			x	x	x	x	x	x	x		4
Lake Innes1, NSW	1830s-1850s	N		x	x	x	x				x	?	?	x	x	x		6
Lake Innes2, NSW	late'50s-1900	N		x	x	x	x	x			x	?	?	x	x	x		6
Narrandera, NSW	1870s-early20th	C	x	x		x	x				x	x		x		x		10
Heidelberg, VIC	1844-1874	N	x	x	x	x	x	x		x	x	x	x	x	x	x	x	7
Dolly's Creek, VIC	1850s-1880s	N				x	x				x	?		x	x	x		2
Melbourne1, VIC	1850s-1910	C		x			x				x	?	?	x	x	x		8
Melbourne2, VIC	1886-1928	C			x		x	x			x	?	?	x	x	x	x	8
Chinaman's Point, VIC	1860s-1900	C										x	x	x	x			5
Port Adelaide, SA	1840-1900	N	x	x	x	x	x			x	x	x	x	x	x	x	x	3
Atherton, QLD	1880-1920	C									x	?	?	x	x		x?	9
NEW ZEALAND																		
Arrowtown	1860s-1920s	C		x		x	x			x	x	?	x	x	x	x	x	1
Cromwell	1860s-1930s	C		x	x	x	x				x	?	x	x	x	x	x	1
Kawarau	1870-1890s	C									x			x	x			1
Luggate	1890-1915	C									x			x	x	x		1
Queensberry	1880-1915	C													x			1

Abbreviations: FP-food preparation vessel; SD/L-serving dish/lid; PL-platter; T-teapot; J-jug; D-drainer; GB-gravy boat; SP-soup plate; P-L-plate-large; P-M-plate-medium; P-S-plate-small; S-saucer; C/M-cup/mug; B-bowl; E-eggcup; R-reference source; N-non-Chinese sites; C-Chinese sites
Key: x = presence; ? = unknown or size unknown for plates; References: 1-Ritchie 1986; 2-Lawrence Cheney 1995; 3-Briggs 2005; 4-Allison & Cremin 2006; 5-Bowen 2007; 6-Brooks 2007; 7-Hayes 2008; 8-Muir 2008; 9-Dunk 2010; 10-Author's catalogues

Table 5.3: Decorative types from sites in Australia and New Zealand

Location	Date		B	SP	HP	TP	F	S	RM	CE	G	CG	D	R
AUSTRALIA														
Bungendore1, NSW	1859-1890s	N	x	x	x	x	x	x	x	x	x	x		9
Bungendore2, NSW	1875-1904	N	x	x		x	x	x	x	x	x	x		9
Kiandra1, NSW	1860s	N		x		x	x	x						9
Kiandra2, NSW	1860s	N		x		x	x							9
Kiandra3, NSW	1860-1890	N	x	x		x	x		x					9
Kiandra4, NSW	1860s-1931	N	x	x		x	x	x	x	x	x		x	9
Kiandra5, NSW	1860s-1950s	N	x	x		x	x	x	x	x	x	x	x	9
Kiandra6, NSW	1880-1930	N				x	x		x	x		x		9
Kinchega, NSW	1870s-1950s	N	x			x	x		x		x	x		4
Lake Innes1, NSW	1830s-1850s	N		x	x	x	x		x	x	x	x		6
Lake Innes2, NSW	late'50s-1900	N	x	x	x	x	x	x	x	x	x	x		6
Narrandera, NSW	1870s-early20th	C	x	x		x			x	x	x	x	x	9
Heidelberg, VIC	1844-1874	N		x	x	x	x		x	x	x	x		7
Dolly's Creek, VIC	1850s-1880s	N	x	x	x	x	x	x	x			x		2
Chinaman's Point, VIC	1860s-1900	C	x			x								5
Port Adelaide, SA	1840-1900	N	x	x	?	x	x	x	?	x	x	x	?	3
Atherton, QLD	1880-1920	C				x	x		x		x			8
NEW ZEALAND														
Arrowtown	1860s-1920s	C	x			x	x		x		x	x	x	1
Cromwell	1860s-1930s	C	x			x			x		x			1
Kawarau	1870-1890s	C				x								1
Luggate	1890-1915	C	x			x			x					1
Queensberry	1880-1915	C				x								1

Abbreviations: B-banded; SP=sponge printed; HP-hand painted; TP-transfer printed; F-flown; S-sprigged; RM-relief moulded; CE-coloured enamel; G-gilt; CG-coloured glaze; D-decal; R-reference source; N-non-Chinese sites; C-Chinese sites
Key: x = presence; Note: creamware and pearlware from other data are not included in this table; References: 1-Ritchie 1986; 2-Lawrence Cheney 1995; 3-Briggs 2005; 4-Allison & Cremin 2006; 5-Bowen 2007; 6-Brooks 2007; 7-Hayes 2008; 8-Dunk 2010; 9- Author's catalogues

The Chinese sites were occupied between the 1850s and 1930s. They include: Arrowtown and Cromwell Chinatowns, Kawarau, Luggate and Queensberry, all in the Otago region of southern New Zealand, occupied 1860s-1930s (Ritchie 1986); a fish curing establishment at Chinaman's Point, near Port Albert, Victoria, 1860s-1900 (Bowen 2007); residences in Melbourne's Chinatown, dating 1850s-1928 (Muir 2008); Atherton Chinatown, about 60 km from Cairns in northern Queensland, dating 1880-1920 (Dunk 2010); and Narrandera Chinatown, NSW, dating from the 1870s to the early-twentieth century (Esposito 2012: Appendix 7).

The non-Chinese sites were occupied from the 1830s to the 1950s. Four sites were examined at Dolly's Creek, Victoria, by Susan Lawrence in 1995. Two were private homes, one a pub and the other a combined pub/store, dating to the period 1857 to 1888 (Lawrence Cheney 1995). These are interpreted as being occupied by Europeans, although there is a possibility of the residents being Chinese as there was a large Chinese presence at Dolly's Creek. Other sites include: cottages on Jane Street, Port Adelaide, dating 1840-1900 (Briggs 2005); Viewbank homestead in Heidelberg, Victoria, occupied by the Martin family, 1844-1874 (Hayes 2008); and the Old Kinchega Homestead, 1870s-1950s (Allison & Cremin 2006). Alasdair Brooks' (2007) data are from the Lake Innes Estate, located near Port Macquarie in New South Wales. There were two distinct periods of occupation, the first, to the early 1850s, when the estate was occupied by the Innes family and the second, after the Innes family left. Although data used came from the main residence and out-buildings, it has been included here only as belonging to the Innes (Lake Innes 1) or post-Innes periods (Lake Innes 2), that is, 1830s-1850s or late 1850s-1900.

Other data are from assemblages examined by the author. A number of the sites are located in Kiandra and these include: 1, the Gold Commissioner's residence, 1860-1862; 2, European miners' huts on Township Hill, 1860s; 3, the Kiandra Hospital, later a private residence, 1860-1890; 4, the Kiandra Hotel, later a private residence, 1860-1931; 5, a Township residence, 1860s-1950s; and 6, a residence on New Chum Hill, 1880-1930. The two sites from Bungendore are; 1, a hotel, 1859-1890s; and 2, a farmstead, 1875-1904 (Esposito 2008).

The early miners' huts, whether Chinese (Ritchie's Kawarau, Queensberry and Luggate) or European (Kiandra's Township Hill), have fewer vessels than other sites which were occupied longer. Most other sites have a wider range of vessels. Similarly, the early miners' huts have fewer decorative types but this is probably results from the presence of fewer vessels. The most common type of decoration, found at all sites, is the transfer-print.

NON-FOOD RELATED VESSELS

This includes vessels used for health and hygiene, writing, cleaning, leisure and ornamental wares.

Earthenware pots and pot lids, once containing ointment, toothpaste or cosmetic preparations are common on nineteenth century Australian sites. They often have transfer-printed designs and were made in a number of places including Britain and Australia. One of the most common ointment pots, or pot lids, is Holloway's ointment. Holloway worked from a variety of premises in London from 1837 until his death in 1883 (Harrison-Barbet 1994); the addresses printed on the pots and lids allow their dating. Other vessels associated with hygiene are basins, ewers, soap dishes and chamber pots. These are often made of earthenware and decorated with transfer prints or simply banded.

Stoneware bottles were used for a variety of products. Small penny inks were durable, cheap to produce and used as ink wells (Vader & Murray 1979). They are usually salt-glazed. Larger ink bottles have a pouring lip and are salt or brown-glazed. Blacking bottles, commonly found on Australian nineteenth century sites in a range of sizes, are usually salt-glazed. Stoneware Bristol-glazed bottles contained oil for boots or saddles. These types of bottles were made locally in Australia as well as being imported from Britain.

Other non-food related items include ornaments, flower pots and tobacco pipes. Ornaments are usually made of bone china or porcelain and often hand-painted. Flower pots are usually made of unglazed red earthenware and tobacco pipes of white earthenware. Other ceramic items made in the nineteenth century, but beyond the scope of this work, are children's toys, industrial pipes and building features.

CHAPTER 6

The assemblage and research methodology

This chapter provides information about the re-examined ceramic collections and outlines the methodology used in cataloguing and technical analysis, and the rationale for this approach. Table 6.1 details the number and weight of recovered sherds, when they were collected and by whom. The ceramic sherds examined are from excavations and surface collections. Although surface sherds are more prone to movement from their original place of discard, previous research has shown that there is a correlation between surface and subsurface artefacts (e.g. Underhill et al. 2008:3).

BRAIDWOOD ASSEMBLAGE

Ceramics from two settlements in the Braidwood region were re-examined: Jembaicumbene Chinese Camp (JCC) and Flanagan's Point Chinese Camp (FPCC). These results are detailed in Chapter 7. The sherds are stored at the Australian National University (ANU). They were previously referenced in Smith (2006: Appendix C3) and were in bags labelled with the square number but no catalogue numbers were present. During re-cataloguing for this research, the square numbers were used to compare the artefacts present with those in Smith's catalogue and where possible, the original number given by Smith was used. This was not always possible as a result of some previous misidentification; therefore there are differences between the catalogues used for this research to those in Smith (2006).

TUMUT ASSEMBLAGE

Two ceramic assemblages from the Tumut region were re-examined: Upper Adelong Chinese Camp (UACC) and Adjungbilly Chinese Camp (ACC). These results are detailed in Chapter 8. The sherds are housed at ANU. Sherds from Adjungbilly Chinese Camp had been catalogued by Smith (2006: Appendix C4) but were found in bags with the square number only. Once again, it was not always possible to match sherds with the number given by Smith. No previous catalogue was found for the sherds collected from Upper Adelong.

KIANDRA ASSEMBLAGE

The Kiandra assemblage is made up of sherds from the Chinese Camp (KCC), Kiandra Valley and the Township stores, all now stored at ANU. The results for KCC are detailed in Chapter 9 and those for Kiandra Valley and the Township stores in Chapter 10. The ANU student catalogues from the 2001 and 2003 field schools feature in Smith (2006: Appendix C5).

Kiandra Chinese Camp

Three separate surveys took place across the camp (Figure 6.1).

Fieldwork 1996-1998

Fieldwork was first carried out on KCC in 1996-1998 by Lindsay Smith as part of his Master's research. He surveyed two areas. These were gridded with one metre squares and originally designated Huts 8 and 13 (Smith 1998) but were later renamed as part of the 2001 field school survey. The grid named Hut 8 covered four hut sites, 2, 25, 26 and 28. Hut 13 became Hut 1. A surface collection was made over the gridded areas and Huts 1 and 2 were excavated.

Table 6.1: Details of ceramic assemblages from Chinese mining settlements in southeast NSW

Settlement	Region	No.	Wt (g)	Collection	Date
Jembaicumbene Chinese Camp	Braidwood	2188	14764.3	LS	2001-2003
Flanagan's Point Chinese Camp	Braidwood	102	1060.5	LS	2001-2003
Adjungbilly Chinese Camp	Tumut	151	421.6	LS	2001-2003
Upper Adelong Chinese Camp	Tumut	119	1863.7	LS	2001-2003
Kiandra Chinese Camp	Kiandra	3393	19088	LS; ANU	1996-1998; 2001; 2003; 2007
Chinese Stores	Kiandra	1005	5931.1	ANU	1996
Kiandra Valley Chinese Camps	Kiandra	371	2853.1	LS, CG-T	2001-2003;2005-2007
Total		7329	45982.3		

Key: ANU - Australian National University field schools; CG-T - Christine Gant-Thompson; LS - Lindsay Smith

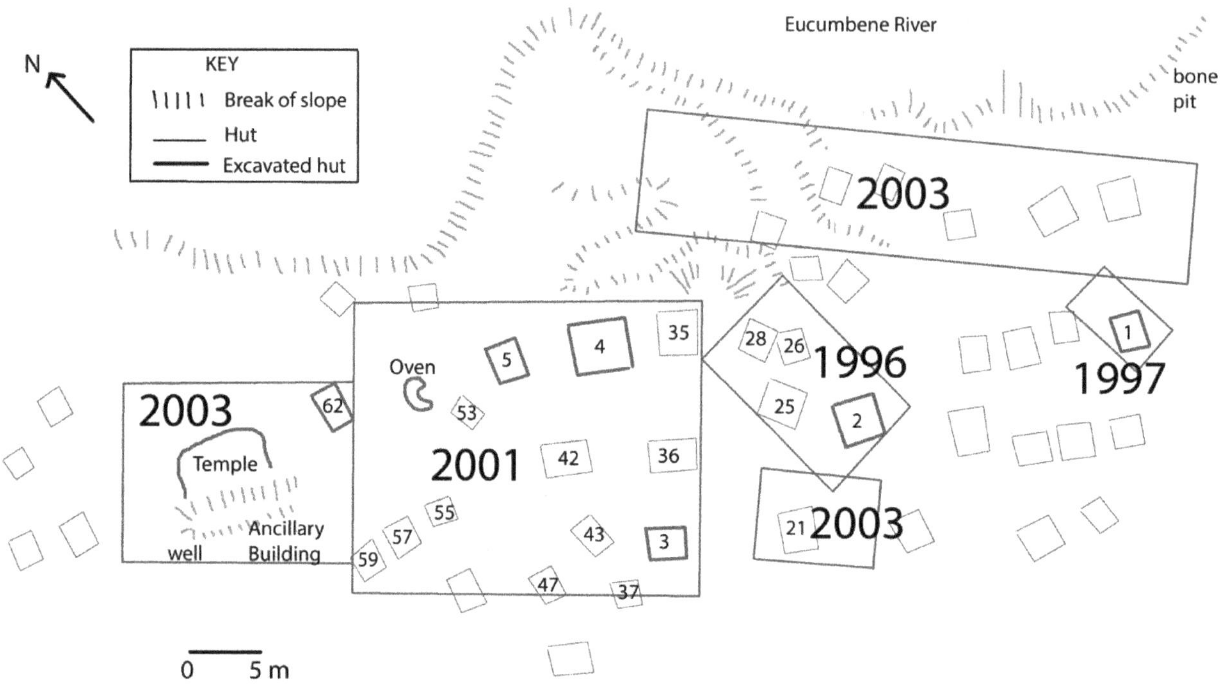

Figure 6.1: Kiandra Chinese Camp showing surveyed areas (after Myles 2004)

Fieldwork 2001

In 2001 three areas were selected for examination at KCC as part of the ANU field school. The main grid was located to the west of the area studied by Smith (1998).

Grid 1, measuring 30 x 25 m, was pegged with one metre squares. Four areas within this grid were excavated including an oven and three hut sites, and a surface collection was carried out over the whole grid. At least 13 huts were identified within this area. For this study these sherds have been considered as being from the nearest hut where they were found. Figure 6.2 shows the areas considered to be related to each hut site. Boundary lines were initially placed half-way between huts and adjusted to take into consideration conjoining sherds.

Grid 2 was located 10 to 12 m from the northwest side of Grid 1. A surface collection was made over 9 m². Two areas were excavated: 4 m² over a depression thought to be a well; and 5 m² over the edge of an area thought to be a terrace retaining wall. The trench dug across the terraced area was re-dug as part of the 2003 Temple Complex.

Grid 3 was located 50 m to the southeast of Grid 1 and designated the Bone Pit because of the concentration of animal bones. It is thought to be a rubbish dump. A surface collection was carried out over 9 m² and 4 m² were excavated.

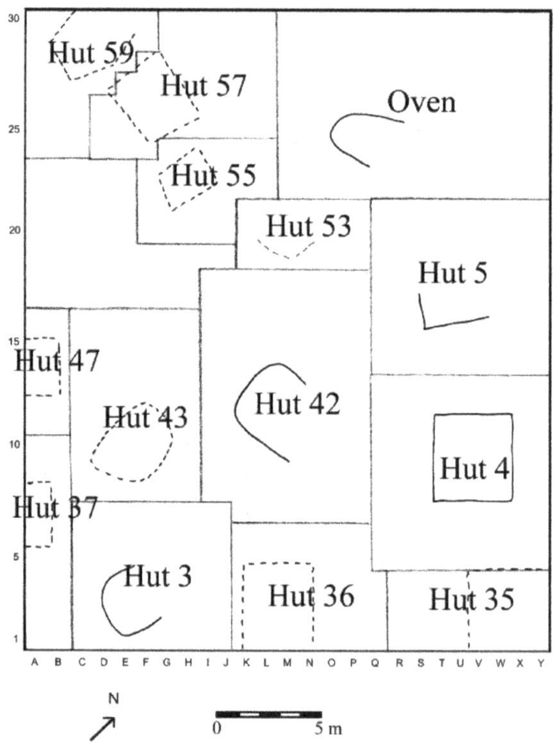

Figure 6.2: Area associated with each hut within the KCC2001 grid

Fieldwork 2003

In 2003 the ANU field school returned to KCC and targeted three areas, designated Grids 1, 2 and 3.

Grid 1, 20 x 15 m, was laid out in one metre squares over a terraced area to the west of the 2001 main grid. Three structures were identified within the grid. The lower terrace is thought to be the site of the Temple and the upper terrace an Ancillary Building. The trench excavated in 2001 ran through part of the Temple building. A small earth platform on the east side of the grid is thought to be the remains of a hut site, designated Hut 62. A surface collection was conducted over the whole grid and parts of each structure were excavated.

Grid 2, KCC2003-2, to the east of the 2001 main grid, covered an area of 45 x 10 m and was pegged with 5 x 5 metre squares. Surface artefacts were collected from this area which covers at least six hut sites.

A surface collection was made from Grid 3, located to the southeast of the 2001 main grid, which covers the remains of a hut site, designated Hut 21.

2007

Fourteen sherds were collected during the 2007 field school, six from Hut 21, three from Hut 43 and five from Hut 46.

Summary

Overall the fieldwork established the presence of a temple, an oven and at least 72 dwellings (Smith 2006:176). Table 6.2 summarises the number and weight of sherds recovered from 1996-1998, 2001, 2003 and 2007, and the sites from which they were collected. Huts 1, 2, 3, 4, 5 and 62 were excavated, along with the Oven, Trench, Well, Bone Pit, Temple, and Temple Ancillary Building.

Kiandra Valley

Five Chinese miners' camps close to Kiandra were examined by Smith (2006) and Gant-Thompson (2008). The ceramics collected are detailed in Table 6.3.

Kiandra Township Stores

A survey of the area was undertaken as part of the 1996 ANU field school, the principal objective of which was to locate the Chinese stores (Heffernan 1996). There were a number of obvious surface features including chimneys, wells, a boundary fence, building platforms and rubbish dumps but parts of the area had been disturbed by bottle hunters and by graders digging modern drainage channels for the road. The latter had disturbed what was probably the front of Ah Yan's store (Heffernan 1996).

A surface collection of artefacts was carried out and small areas were excavated (Figure 6.3). A grid of 2 x 2 m squares covered 40 x 20 m, extending over both allotments, and at the northern end, into that owned by Quinn, a non-Chinese resident. Three areas were excavated: a rubbish dump at the rear of Ah Chee/ Ping Kee's property; part of a twin chimney mound at the front of Ah Yan's property; and part of a stone feature at the back of Ah Yan's property

Table 6.2: Number and weight of sherds from Kiandra Chinese Camp

Collection	No.	Wt (g)	Wt/No.	Sites
1996-1998	356	4641.0	13.0	Huts 1,2,25,26,28
2001	1448	4883.6	3.4	Huts 3,4,5,35,36,37,42,43,47,53,55,57,59 Oven, Bone Pit, Well, Trench
2003	1575	9112.8	5.8	Huts 21,62, Temple, Ancillary Building
2007	14	450.6	32.2	Huts 21,43,46
Total	3393	19088.0		

Note: clay pipes are not included in sherd numbers or weights

Table 6.3: Number and weight of sherds from Kiandra Valley sites

Settlement	Sites	No.	Wt (g)	Wt/No.
Eucumbene Crossing	125, 127, 143, 268	94	901.1	9.6
East Eucumbene	635, 636, 637, 740, 977, 1058	120	701.7	5.8
Southwest Eucumbene	698, 809, 1038	32	254.9	8.0
Giandarra Gully	1079, 1082	6	361.9	60.3
New Chum Hill	485, 742, 1069	119	633.5	5.3
Total		371	2853.1	

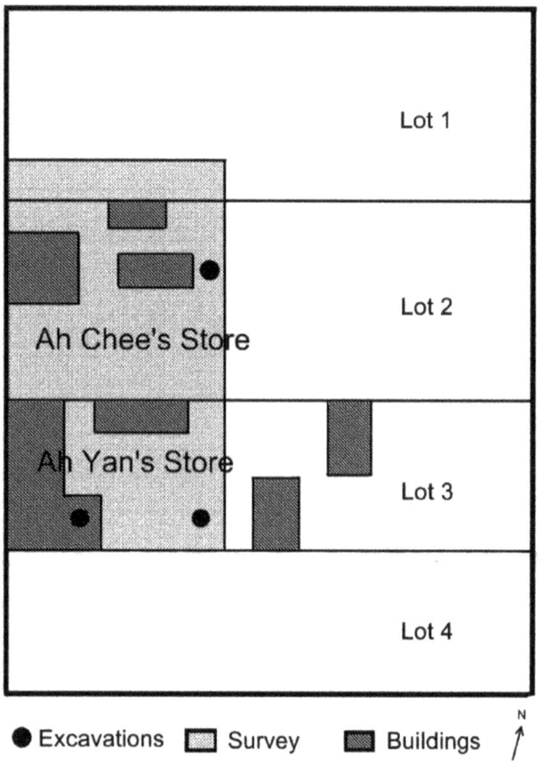

Figure 6.3: Kiandra Town section 14 showing area surveyed in 1996 (after Trueangel 1996:29 – not to scale)

One thousand and five ceramic sherds, weighing 5931.1 g, were recovered. A distribution plot of the sherds revealed that these were concentrated in two main areas: in and around the rubbish dump at the rear of Ah Chee/ Ping Kee's property; and in a deflated area which possibly lay between the residences of Ah Chee/ Ping Kee and Ah Yan. A few sherds were found outside these main areas including: at the front of each allotment; on the northern boundary of the grid; and at the rear of Ah Yan's residence. There were five areas across the grid, one or two metres wide, where no ceramics were found.

For analytical purposes, one of the areas where no ceramics were found, roughly in the middle of the grid, was taken to be the boundary between the artefact assemblages of Ah Chee/ Ping Kee and Ah Yan. Grid rows 19 and 20 may have been located on the Quinn property; as a result these have been excluded from the analysis. These last grid rows contained seven sherds weighing 6.7 g, all of which were British-made. There was a gap of at least four metres between these and other sherds which are considered to be from in and around the rubbish dump area of Ah Chee/ Ping Kee.

Cataloguing of the artefacts confirmed separate assemblages within these areas, that is, sherds from a single vessel were found within one assemblage only. These assemblages have then been further divided into separate areas within each allotment.

Additional Resources

In addition to the sherds detailed previously, over 18,000 sherds weighing more than 50 kg were examined. The majority are from non-Chinese sites in Kiandra (13,283 sherds weighing 27,068.6 g collected by ANU Field Schools 1997-2007) and sites near Bungendore (4,548 sherds weighing 20,927 g collected by Archaeo Analysis 2007-2008). A further 341 sherds weighing 4,203 g were collected by Diana Osborne (2008) from Narrandera Chinatown during her Honours research.

The ANU, in conjunction with the NSW National Parks and Wildlife Service, began investigating sites in Kiandra in 1995, as a long term research project set up within the teaching program. The original aim was to locate buildings within the town by using archaeology to supplement historical records and to examine social and economic aspects of the town. The first four field schools examined the Kiandra Hotel, Chinese stores, Dance Hall, school teachers' residence and another private residence, all located within the township. Field schools in 2001, 2003 and 2004 investigated the Chinese camp and the European miners' village on Township Hill. The field schools in 2006 and 2007 concentrated on Camp Town on the eastern side of Pollock's Creek, which was the area of the initial gold rush in 1859 and 1860; surveying and excavating the Public Hospital, the Government camp and associated buildings and privies.

Cataloguing Methodology

At present there are no standard guidelines for artefact catalogues throughout Australia. All the re-examined ceramic artefacts were re-catalogued by the author, allowing consistency across all catalogues, for comparison purposes. The individual catalogues can be found in Esposito (2012). Various data relating to ceramic sherds were recorded. The most important detail in the catalogues was the place where the artefact was found, as knowing the site and context from which it was recovered allowed the artefact to contribute to the interpretation of that site. Other data recorded for these ceramic catalogues included information about the body, surface decoration, the form or shape of the vessel to which the sherds belonged, the number and weight of sherds, the part of the vessel from which they came and reference to conjoining sherds .

Changes were made to the existing excel spreadsheets rather than using different software because of the time involved in re-cataloguing. Each site had been given a site and year code, for example the Kiandra Chinese Camp, excavated in 2001 was known as KCC2001. Ceramic catalogue numbers began at: 1000 for the ANU 2001 and 2003 field schools and Gant-Thompson; 100 for Smith's catalogues; and 1 for the 1996 field school. Each sherd or group of similar sherds was given a unique number for that site. These numbers were kept during re-cataloguing, although some groups of sherds were separated further as they contained different fabric, decoration or vessels. The catalogue numbers then became 1000.1, 1000.2, 1000.3, and so on or 100.1, 100.2. The ceramic sherds collected by Smith (1998) had not been numbered but appeared in his Appendix 4.1 as a list by square number. These sherds were given numbers from 1000 onwards as part of re-cataloguing. If a vessel contained sherds from both the excavated and surface collections, then the vessel was considered to be excavated rather than from the surface.

There has been much debate over how to group sherds for cataloguing. The majority of researchers sort sherds according to their body-type, i.e. earthenware, stoneware or porcelain, with subcategories of fine and coarse (e.g. Thompson & Wilson 1987). Some have added other bridging categories, such as porcellaneous stoneware (Worthy 1982). Others suggest sherds should be subdivided according to their degree of vitrification, i.e. non-vitreous, semivitreous or vitreous (e.g. Majewski & O'Brien 1987). Although the general categories of ceramic body are earthenware, stoneware and porcelain, bodies which are over- or under-fired may not be correctly assigned; for example low-fired stoneware may look like earthenware. Thick areas of porcelain may be designated as stoneware because they do not fit the usual idea of translucent fabric. To overcome these difficulties, and those associated with classification, sherds were put into the broad categories outlined above. For example, all brown-glazed storage jars were designated stoneware, even though some sherds have been under-fired. Any identification errors are then consistent across all catalogues. The most common earthenware found at the sites in this study is refined white-earthenware. All white-earthenware bodies were catalogued as white-earthenware rather than specifying any particular body-type. Similarly, vessel forms were assigned using the same criteria throughout.

If diagnostic elements were present vessels were assigned forms, for example, cup, bowl, plate or saucer. If the size of a vessel was unknown but the shape appeared to be that of a cup or bowl, it was designated cup/bowl. Likewise, if a vessel appeared to be a plate or saucer but could not be distinguished, it was designated plate/saucer. Vessels which could not be identified were designated unidentifiable hollowware, unidentifiable flatware or unidentifiable. Terminology used for vessel parts is shown in Figures 6.4 and 6.5. Hollowware is defined as a vessel with a concave surface such as teapot, jug, cup or bowl. Flatware, as the name suggests, is defined as a flat vessel such as a platter, shallow dish, plate or saucer.

The Chinese sherds were identified with reference to Chace (1976), Pastron, Gross and Garaventa (1981) Willetts (1981) and Ritchie (1986). The majority of publications covering Chinese-made wares do not include the everyday vessels found at overseas Chinese sites. One exception is an exhibition catalogue of wares found on Malaysian sites, Nonya ware and Kitchen Ch'ing, produced by The Southeast Asian Ceramic Society in 1981, with introductory essays by William Willetts and Lim Suan Poh. The types of Chinese-made vessels found at overseas Chinese sites in America and New Zealand have also been detailed in research by Chace (1976), Pastron, Gross and Garaventa (1981) and Ritchie (1986).

The non-Chinese sherds were identified with reference to Brooks (2005). In 2005 Alasdair Brooks' book, An Archaeological Guide to British Ceramics in Australia 1788-1901, was written as an aid to identifying, processing and cataloguing British ceramics found on Australian sites. Brooks (2005:1) stated that he wrote the book because analyses of ceramics to that point 'was something of a missing link in Australian historical archaeology'. Pattern names for this research were obtained by looking at other collections (including those at Casey & Lowe Pty Ltd, Flinders University, Heritage Victoria and The Modern City Project) and through web sites such as the New Zealand historical ceramics database (NZHCD) and antique sales sites.

Publications regarding non-Chinese made ceramics are numerous, but most containing information about nineteenth century ceramics have been written by ceramic historians or antique dealers targeting collectors. They focus on higher-priced vessels rather than on everyday vessels and mainly cover the pre-1850 time period. Some books are useful for dating through identification of makers' marks (e.g. Godden 1964, 1991; Kovel & Kovel 1986). Some patterns may also be identified (e.g. Snyder 1997, 2004; Dieringer & Dieringer 2001; Neale 2005) however most pattern books concentrate on pre-1850 patterns.

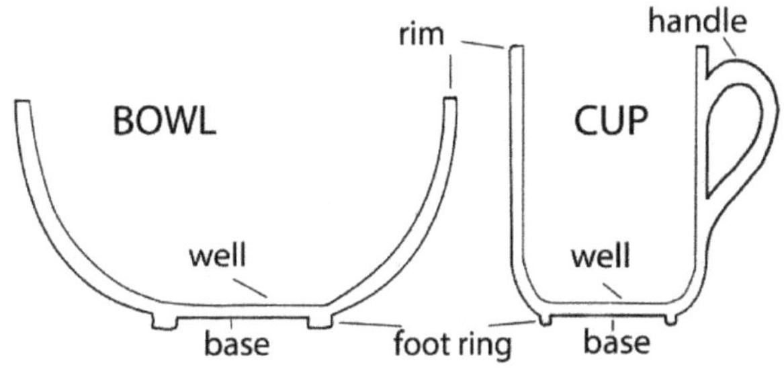

Figure 6.4: Terminology for hollowware

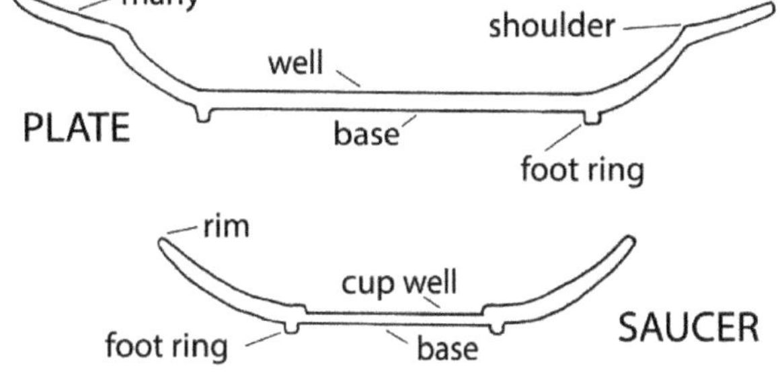

Figure 6.5: Terminology for flatware

After the re-cataloguing was completed, sherds from each site were separated according to body type and decoration to determine conjoins and the minimum number of vessels (MNV). When there was more than one sherd of the same decoration from a hut site and it was not possible to determine whether they were from the same or different vessels, the MNV was counted as one.

Function was assigned to vessels after determination of the MNV. Only four function categories were used: 'food /beverage containers', 'food/beverage consumption', 'smoking' and 'other'. Food/beverage containers includes bottles and jars relating to transport and/or storage of food or drink. The food /beverage consumption category contains vessels used to prepare food, serving vessels and those used for eating and drinking. The smoking category covers opium pipe bowls and European clay tobacco pipes. The 'other' category contains anything not related to food or smoking, for example vessels used for writing and hygiene. The function of vessels forms part of the discussion for each site. Clay tobacco pipes were not part of the re-examined assemblages and therefore have not been included in sherd numbers and weight, but are noted in the MNV. The figures for MNV clay pipes were obtained from previous catalogues prepared by Lindsay Smith (1998, 2006) ANU students (2001, 2003) and Christine Gant-Thompson (2008).

The way sherds are measured for analysis, that is, by number, weight, or MNV, affects the interpretation of a site. During early phases of ceramic study only numbers of sherds were used, but in the 1950s alternatives to this were suggested. Gifford (1951) used weight to describe amounts of sherds and Burgh (1959) counted the MNV. Since then debates about which method best describes a ceramic assemblage have continued. Some researchers favour using weight to describe an assemblage, but this can be biased if comparisons are between different materials. A variety of vessels are found at overseas Chinese sites including heavy stoneware storage jars and lightweight earthenware tea and tableware. If these weights are directly compared there will always appear to be more stoneware as it is heavier and does not weather as easily as the more porous earthenware. Similarly, sherd counts are dependent on fabric type as the ceramic material plays a role in determining how easily a vessel will break. Lower-fired earthenware vessels are not as strong as high-fired porcelain vessels and may break more easily. Furthermore, the frequency of vessel use or how carefully it is treated may determine the likelihood of breakage.

In this research, the numbers, weights and MNV have been used as they can all provide information about the ceramics at a site. The MNV were used to compare the relative proportions of Chinese/non-Chinese vessel types. Although this process has not been in vogue since Lydon's (1999) critique of its narrow focus in determining levels of assimilation, it is useful in highlighting differences between sites as the type of vessels present at any site directly relates to what was available when it was occupied.

Technical analyses

A number of sherds were chosen for further technical analysis. Apparent porosity (water absorption) testing and electron microprobe analysis (EMPA) were used to obtain physical and chemical data to determine the similarities and differences between sherds. These results were interpreted using multivariate statistical analysis and the potters' unity formula.

Apparent porosity or water absorption

Tests for apparent porosity or water absorption, the amount of water absorbed into open pores only, were carried out on 153 samples as a means of comparing the firing temperature for a range of sherds and investigating the uniformity of firing for different vessel types. Porosity decreases as vessels are fired to higher temperatures but all ceramic bodies are porous to some extent.

The water absorption test used in this research was sourced from Singer and Singer (1963:343). It involved heating the sherds in a kiln overnight to 110-115°C and then weighing them, recording this as the dry weight. Groups of sherds were then wrapped in a loosely woven material and suspended in water. They were boiled for one hour and left to soak in that water for a further 23 hours to determine the ability of the clay fabric to absorb water by capillary action. After this time, the sherds were taken out of the water, the surface water was wiped off and they were reweighed. This was recorded as the wet weight. The water absorption or apparent porosity percentage for each sherd was determined by calculating the percentage increase in weight. The test was performed three times and the results averaged for each sherd to obtain the final apparent porosity data (Esposito 2012: Appendix 8).

Compositional analysis

An initial study was carried out on the glazes of 11 sherds to test the viability of using a scanning electron microscope (SEM) to determine the glaze composition. The method used was electron microprobe analysis (EMPA). Consequently, another 98 samples were analysed to determine their body and glaze compositions. These included: 49 Chinese-made food/beverage containers/lids; 28 celadon glazed vessels; 14 non-Chinese samples; and seven samples of unknown origin. The individual samples are listed in Chapter 11, along with their analytical results.

All the Chinese-made samples chosen have diagnostic features, or were the sole sample from a site, to ensure that each was from a different vessel. The highest numbers of diagnostic sherds in this study were from brown-glazed globular jars, wide-mouthed jars and bottles, unglazed wide-mouthed jar lids and celadon-glazed vessels, therefore these forms were chosen to be examined. The samples include: 11 globular jars, 12 wide-mouthed jars, 11 wide-mouthed jar lids, 14 liquor bottles, one unidentifiable brown-glazed vessel (sole sample from ACC), 25 rice bowls, a dish, tea cup and a vessel which is either a cup or bowl. They are from sites at Jembaicumbene, Flanagan's Point, Adjungbilly, Kiandra and Narrandera, NSW.

Fourteen sherds of non-Chinese origin were included for comparison with the Chinese samples. Two are from known Japanese vessels marked 'NIPPON / hand painted'. Other sherds are from vessels, which although unmarked, are more likely to be from either Britain or Australia than China, having decorative types commonly used by those countries in the nineteenth century, including salt glaze, Bristol glaze, Rockingham glaze and transfer-prints. Five bone china samples were chosen to compare the composition of those from Chinese sites with a sample having a known date of pre-1860. Seven samples of unknown origin were examined to determine the probability of them being Chinese.

Small sections, less than 5 mm, were cut from each sherd and embedded in 25 mm resin disks so that the cross-section of each sherd was visible. Ninety-three samples were prepared by staff at the Research School of Earth Sciences, ANU, and five by Australian Petrographic Pty. Ltd. at Queanbeyan, NSW. The disks were sonically cleaned in ethanol and wiped with lint free tissues before carbon coating to enable conductivity during examination.

The physical and mineralogical characteristics of the sherds were examined using a JEOL 6400 SEM, housed at the Centre for Advanced Microscopy in the Research School of Biological Sciences, ANU. The JEOL 6400 SEM was equipped with Oxford ISIS energy dispersive x-ray analysis, sensitive down to boron, and a Robinson backscatter detector. It was optimised for quantitative x-ray analysis. The accelerating voltage was 15 Kv, the beam current 1 nA and the live time for collection of x-rays, 100 seconds. During that time the machine corrected for factors including the atomic number, absorption effects and fluorescence effects (personal communications, Dr Frank Brink, Deputy Director, Centre for Advanced Microscopy, ANU, 21 October 2010). The readings were recorded as combined elements by stoichiometry with oxygen. This means that they were presented as a single oxide, for example FeO, regardless of whether the actual oxide present was FeO or Fe2O3. For any reading, the error rate was given along with the percentage oxide weight. If the error for any element was more than half the element percentage reported, the element was probably not present (personal communications, Dr Frank Brink, 9 February 2010). The standards used for quantitative analysis were constant throughout.

Electron microprobe analysis has been used successfully for provenance determination for over twenty years by researchers on many different ceramic bodies, for example: Chinese porcelain (Tite et al. 1984); Roman / British pottery (Freestone et al. 1985); British / American porcelain (Owen 2001); Korean celadon (Ham et al. 2002); American pottery (Abbott et al. 2008); Indian tiles (Gill & Rehren 2011); and Chinese proto-porcelain (Yin et al. 2011).

The area scanned by the microprobe and the number of readings taken, varies amongst researchers. Tite, Freestone and Bimson (1984) took at least three readings from the body and several readings from the glaze, both of which were averaged. Freestone, Meeks and Middleton (1985:162) scanned an area of about 80 μm in diameter, 'analysing the fine-grained matrix between the coarser temper inclusions' and took an average of 'at least three spots'. Abbott, Lack and Moore (2008:54) took five readings of both the body and temper for each sample and scanned an area of 0.109 mm^2. They chose areas that avoided non-plastic inclusions but found any tiny inclusions to be 'inconsequential' (Abbott et al. 2008:54). Gill and Rehren (2011:23) took an average of three to five readings across a sample avoiding visible quartz and then normalising the total to 100 per cent for analysis 'effectively removing any porosity from the measured areas'. Yin, Rehren and Zheng (2011) took ten readings across body samples at x800 magnification and five readings each of exterior and interior glaze at x2000. When there was insufficient glaze, two to three areas were analysed and if glazes were very thin the magnification was x4000.

In this research, the reported body composition of each sample is the average of ten readings across the sample. No attempts were made to distinguish between the clay and the temper, only the overall composition was considered. On advice from Dr Brink, x200 magnification was consistently used for all body readings, allowing the raster beam to scan an area of about 600 x 400 μm^2. This was the minimum magnification that could be used and largest area scanned to obtain accurate quantitative results on this SEM. This enabled a more precise representation of the overall body composition. One of the problems associated with this process was the presence of pores, resulting in the total of most readings being less than 100 per cent. It was noticeable that the total of any reading directly related to the pore area, with more porous samples having lower totals. Each spectrum reading obtained was checked for peaks to ensure that all elements were accounted for. If unusual readings occurred, as a result of an inclusion, more readings were taken and the anomalies not included in the final average. Although areas with noticeable inclusions were avoided when scanning, some inclusions may have been present immediately below the surface within an area detectable by the microprobe. All totals were normalised to 100 per cent before being averaged to give the body composition used for comparisons between samples.

Readings were taken from both the exterior and interior glaze of each sample, normalised and then averaged for comparisons to be made. The number of readings and magnitude used was dependant on the thickness/visibility of the glaze on each sample. In general, the samples have three to five readings from both the exterior and interior glaze and the magnification ranged from x1500 to x3000.

The number of elements used in provenance determination also varies amongst researchers. The most common major elements in clay, and therefore in ceramic bodies and glazes, are silicon (Si), aluminium (Al), potassium (K), sodium (Na), calcium (Ca), magnesium (Mg) and iron (Fe) and these change in proportion according to the geology of an area. Minor elements that are likely to appear are titanium (Ti), sulphur (S), phosphorus (P) and manganese (Mn). Many trace elements may be present but not always in large enough quantities to be detectable. In this study major and minor elements, rather than trace elements were used for composition comparisons.

Although some researchers have argued that trace elements should be used to determine provenance, the use of major and minor elements alone has been successful. For example, Yap and Hua (1994) used seven elements, Si, Al, K, Na, Ca, Mg and Fe to differentiate between the compositions of ceramic bodies from different regions of China. They found that two groups from Hebei province in northern China were similar but ceramics from Fujian province in southern China were distinctly different (Yap and Hua 1994). Pollard and Hatcher, similarly, grouped 70 North Chinese sherds to their kiln sites using nine elements, Si, Al, K, Na, Ca, Mg, Ti, Mn and Fe, assuming that any other elements present were only in trace amounts (Pollard and Hatcher 1994:42). In another example, Adan-Bayewitz, Asaro, and Giauque (1999:1-2) used eight elements, Si, A, K, Ca, Mg, Ti, Mn and Fe and found that 'the array of elements was sufficient to distinguish between the major production centres' for their samples of Roman terra sigillata, despite having trace element readings to use.

Photomicrographs were taken of all the samples tested. The images in Chapter 11 are back scatter images taken from the JEOL 6400 SEM. The tones in back scatter images relate to the atomic number of the elements present; the lower the number the darker the tone. The tone of any compound is determined by its average atomic number. The lower section of the photomicrographs contain the scale in nanometres, the magnification (e.g. x60), and the two constants for this SEM analysis, the accelerating voltage (15 Kv) and the working distance between the beam and the sample (39 mm).

Analysing the Data

Two methods were used to interpret the microscopic data: statistical analysis and the unity formula used by potters.

Multivariate analysis

The statistical analysis involved the use of principal component analysis (PCA) and canonical discriminant analysis. These are both forms of multivariate analysis which use mathematical and statistical theory to compress the number of variables in order to establish patterns of similarity or difference between examined objects (Shennan 1988). PCA allows the properties of the data to be explored graphically by transforming the original data into a reduced number of uncorrelated variables or principal components, the first of which explains the greatest amount of variance in the data (Shennan 1988). Discriminant analysis can test the probability of group membership suggested by PCA. Canonical roots (the Eigen values associated with canonical functions) are produced, the first function relating to the most, overall, discrimination between groups (StatSoft Inc 2010). Group centroids, the point in space which represents the means of all the variables in a group, are determined once the groups have been calculated. The probability of a sample belonging to a particular group is calculated using the Mahalanobis distance from the group centroid; that is the distance between the two points in space (StatSoft Inc 2010).

Unity formula

Another way of comparing ceramic bodies and glazes is to convert their chemical compositions to comparable ratios. The unity formula, also known as the Seger formula or Empirical formula, is used by potters and commercial manufacturers of clay and glaze materials to compare ceramic bodies or glazes (Hamer & Hamer 1997:146). It requires components to be grouped into columns according to their chemical formulae: monoxides (RO, R_2O); sesquioxides (R_2O_3); and dioxides (RO_2), where R stands for any element in the compound under examination. One column is then adjusted to unity and the other column expressed as a ratio. The monoxide group, also known as the bases or alkalis act as fluxes in a glaze or body. The sesquioxides are amphoteric, having both alkaline and acidic properties. The dioxides are acidic. Traditionally the sesquioxide or R_2O_3 group is brought to unity for ceramic body comparisons and the monoxides or RO group is brought to unity for glaze comparisons. Once each formula is transformed into a unity formula it can be compared directly to any other unity formula. The unity formula of a glaze can also suggest appropriate firing temperature ranges for that glaze, as seen in Table 6.4 (Hamer & Hamer 1997:379).

Table 6.4: Glaze firing temperature ranges (after Hamer & Hamer 1997:379)

Temperature	RO	R_2O_3	RO_2
Leadless glazes			
900-1000	1.00	0.1 to 0.4	1.0 to 3.0
1000-1100	1.00	0.2 to 0.5	2.0 to 3.5
1100-1200	1.00	0.3 to 0.6	2.0 to 4.0
1200-1300	1.00	0.4 to 0.7	2.7 to 7.0
Lead glazes			
900-1000	1.00	0.1 to 0.3	1.5 to 3.0
1000-1100	1.00	0.1 to 0.4	1.7 to 3.2
1100-1200	1.00	0.2 to 0.4	2.0 to 3.2

CHAPTER 7

Braidwood region

Two ceramic collections from the Braidwood region were re-examined. They are from separate Chinese mining settlement systems, one of which was focused at Jembaicumbene and the other at Mongarlowe (Smith 2006). One ceramic assemblage is from the Jembaicumbene Chinese Camp (JCC), designated a central camp in Smith's (2006) Chinese hierarchical settlement pattern, and the other from Flanagan's Point Chinese Camp (FPCC), a second-tier camp in the Mongarlowe system (Figure 7.1). The history of mining in this region is outlined in McGowan's (1996) book, Bungonia to Braidwood.

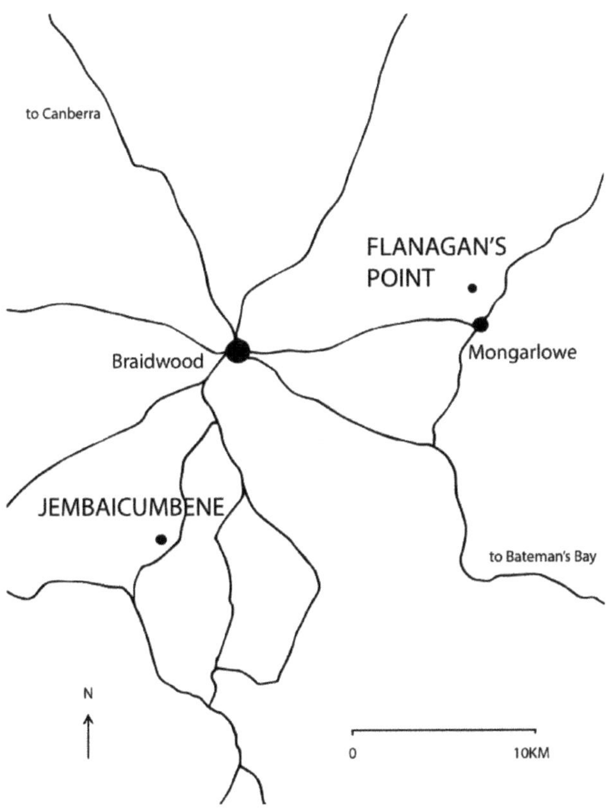

Figure 7.1: Map of Braidwood region showing location of JCC and FPCC (after Smith 2006:79)

Jembaicumbene Chinese Camp

The Chinese settlement was located on the western side of the European village of Jembaicumbene, about 10 km south of Braidwood. It was situated on alluvial flats, an area also known as Jembaicumbene Swamp (Smith 2006:80).

Gold was first found in Jembaicumbene in 1851-1852, but the first mention of Chinese was not until 1859, a time when Jembaicumbene Swamp was the 'favourite goldfield' (McGowan 1996:122). By 1860, the major part of the population at Jembaicumbene was Chinese. A report in the Sydney Morning Herald described the organisation and work ethic of the Chinese:

> People often ask how it is that these Chinese gentry manage to get so much cash together. I believe it is to be attributed solely to their frugal and temperate habits, and to their very constant labour. Many people cry them down as a slovenly, idle, and immoral race, but in regard to the charge of idleness, I believe them to be entirely exempt, for when they do commence operations they stick to their work steadily, and hence are ultimately better rewarded than Europeans...In regard to digging I think if Europeans would adopt their system of unity, and work systematically in companies, as the Chinese do, a great deal more gold would be obtained, and each individual be much better remunerated for his labour -of course these remarks only apply to such diggings as the Swamp, where the claims are deep and wet (The Sydney Morning Herald, 5 November 1860:8).

Other newspaper reports referenced Chinese run facilities in Jembaicumbene, such as stores, and a bank by 1867.

The presence of women at the Jembaicumbene Chinese Camp is confirmed by newspaper articles and by the 1871 census. An article from 1863 mentions 'two European women, the wives of Mongolians' (The Sydney Morning Herald, 7 February1863:7). In 1865, a report stated that although there were 200 European women in the Braidwood region, there was only 'one solitary specimen of female loveliness in Chinese' and she was 'found at Jembaicumbene' (The Sydney Morning Herald, 19 July 1865:5). The following appeared in The Sydney Morning Herald in 1869:

> It has been generally urged as an argument against the Chinese being a suitable class of colonists that they never bring their wives with them like the European races, although for the matter of that those of them who are disposed to marry never appear to be at a loss to find a wife from amongst the people with whom it is held to be impossible for them to amalgamate. One of the Chinese residents of Jembaicumbene (a storekeeper), has determined, however, to set an example to his countrymen in this respect. He has just returned from Melbourne with a Chinese lady whom his parents selected for him in his native land and sent out to him to become his wife. He has married her and taken her to his home at Jembaicumbene. We wish the worthy couple every happiness. We have no doubt that those of the Chinese who obtained their wives from their own country will be far better pleased with them in the end than those who become linked to the most worthless of our own countrywomen (The Sydney Morning Herald, 8 January 1869:5).

The presence of two Chinese women at the Jembaicumbene goldfields was recorded in the 1871 census (HCCDA 2010a:365) in a total camp population of 160 Chinese.

Although there were documented references to a temple and store in the 1860s, maps showing these are from a later date: an 1875 map depicts a block of land containing a store and eight huts (McGowan 1996); and another which probably dates to the 1870s, illustrates a temple and Chinese garden (Smith 2006). Today, little evidence remains of the nineteenth century village as much of the area has been lost to dredging.

In 2003 Smith excavated eight areas thought to contain structures, eight smaller test pits and made a surface collection. Few artefacts were found on the surface as the site was located on an alluvial flood plain (Smith 2006:81).

Sites identified by Smith (2006) include a store, a temple complex, an oven and numerous huts (Figure 7.2). The five hut sites excavated were all typical of those constructed by Chinese miners. These huts along with at least 20 earth platforms suggested 'a temporal sequence' (Smith 2006:91). Three of the excavated huts had stone hearths and may have formed part of the final phase of occupation, while the other more ephemeral platforms were from an earlier phase of occupation. Smith collected numerous artefacts from the sites, which are now housed at ANU.

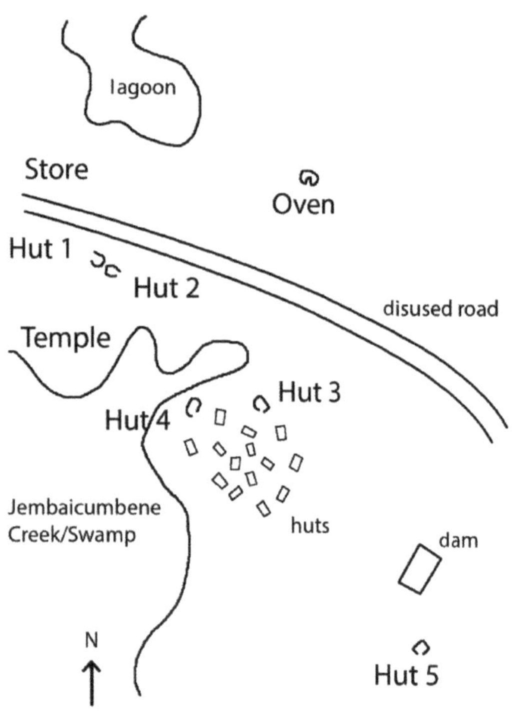

Figure 7.2: Map of Jembaicumbene Chinese Camp (after Smith 2006:84)

The total number of ceramic sherds present in the ANU collection from Jembaicumbene is 2,188; weighing 14,764.3 g. Details of the number and weight of sherds recovered from each site within JCC, along with the minimum number of vessels (MNV), are presented below. Although the results are reported in terms of MNV, there are no whole vessels.

Store

In the early years of the gold rush, merchandise was sold through an overseer, often from his tent, but in later years tents were replaced by more substantial buildings and the store became the social centre of the Chinese community (Wilton 2004:17). Historical records indicate the presence of a Chinese store at Jembaicumbene from at least 1866, when Ah How's store was robbed by bushrangers (The Sydney Morning Herald, 22 November 1866:5). In 1875 You Watt leased or bought that land, which contained a store and other Chinese huts (Smith 2006:85).

Smith surveyed a 10 x 23 m grid on the above mentioned block of land and found a number of postholes, nails and window glass. He stated that the outlined rectangular shape, measuring 6 x 3.8 m, was typical of 'a large weatherboard structure erected on timber posts' and that it was probably a Chinese store (Smith 2006:299). Eight square metres were excavated to a depth of 20 cm.

Ceramic sherds were recovered from all of the eight excavated squares; four squares were at the front of the Store and four at the rear. There were no surface finds. Thirty-seven sherds, weighing 192.5 g, were recovered from the front of the Store (Table 7.1) and 36 sherds, weighing 119.7 g, from the back (Table 7.2). Only one conjoining vessel was found, a celadon-glazed bowl from two squares at the front of the Store. The recovered sherds are not fire damaged. No makers' marks are present.

Store front

A minimum of 14 vessels came from the possible veranda area at the front of the Store. The food storage vessels are all Chinese-made and include: a brown-glazed stoneware globular jar and spouted jar; and a green-glazed ginger jar. The tea and tableware vessels include five Chinese-made vessels: three porcelain Winter Green rice bowls, rim diameters 140 mm, and a liquor cup, rim 50 mm; and a stoneware rice bowl with blue-underglaze decoration, possibly Bamboo. The non-Chinese tableware vessels include: a bone china saucer with a clear glaze, rim 160 mm; a white-earthenware faceted cup with unidentifiable relief-moulding; a cup with a brown floral transfer-printed pattern; and a small plate with an unidentifiable brown transfer-print and possible relief-moulding on the marly. A fragment from a single terracotta opium pipe bowl with light-orange slip was found. The rim diameter is 75 mm, but only the smoking surface is present hence it was not possible to determine the bowl type. A small sherd from a European clay pipe was also recovered.

Table 7.1: Ceramic sherds from JCC Store front

	No.	% No.	Wt (g)	% Wt	Wt/No.	MNV	No./MNV	% MNV
Food/beverage containers								
Chinese	25	67.6	160.3	83.3	6.4	3	8.3	21.4
Non-Chinese	0		0.0			0		0.0
Food/beverage consumption								
Chinese	7	18.9	12.6	6.5	1.8	5	1.4	35.7
Non-Chinese	4	10.8	17.8	9.2	4.5	4	1.0	28.6
Smoking								
Chinese	1	2.7	1.8	0.9	1.8	1	1.0	7.1
Non-Chinese						1		7.1
Total	37		192.5			14		

Note: clay pipe is not in catalogue sherd numbers or weights

Table 7.2: Ceramic sherds from the rear of JCC Store

	No.	% No.	Wt (g)	% Wt	Wt/No.	MNV	No./MNV	% MNV
Food/beverage containers								
Chinese	20	55.6	78.5	65.6	3.9	5	4.0	29.4
Non-Chinese	0		0.0			0		0.0
Food/beverage consumption								
Chinese	2	5.6	2.2	1.8	1.1	2	1.0	11.8
Non-Chinese	13	36.1	38.9	32.5	3.0	8	1.6	47.1
Smoking								
Chinese	1	2.8	0.1	0.1	0.1	1	1.0	5.9
Non-Chinese						1		5.9
Total	36		119.7			17		

Note: clay pipe is not in catalogue sherd numbers or weights

Rear of Store

A minimum number of 17 vessels were excavated at the rear of the Store. The food storage vessels are all Chinese-made and include: a wide-mouthed jar, rim 95 mm, and wide-mouthed jar lid, diameter 90 mm; a barrel jar lid, diameter 330 mm; an unidentified brown-glazed large storage jar; and a green-glazed ginger jar. There are two Chinese-made tableware vessels including a Winter Green cup or bowl, and a porcelain rice bowl with blue-underglaze decoration, but not enough to identify the pattern. The non-Chinese vessels include: a white-earthenware oval serving dish with a gallery for a lid and a thin blue band around the rim; a blue-banded bowl, rim 150 mm; a black transfer-printed bowl; a green band-and-line decorated cup; a blue Fibre transfer-printed cup or bowl; a brown transfer-printed cup or bowl; a clear-glazed straight-sided jug or mug with unidentified relief-moulding; and a bone china saucer with Chelsea Sprig decoration. Fragments from two pipes were recovered: an unidentifiable type of Chinese opium pipe bowl with a light-orange slip and a European clay pipe.

Interpretation of Store site

Origin of ceramics

If the ceramic sherds are considered in total then Smith's (2006:86) statement 'Most of the ceramic artefacts are of Chinese origin' would be correct, 17 Chinese and 14 non-Chinese. However, if the sherds are considered as coming from two locations, the front and the rear of the Store, a different picture emerges. It is conceivable that the function of these two areas was different, perhaps a selling area at the front and a storage or living area at the back. The main difference between the ceramic vessels from the front of the Store to those from the rear is the proportion of Chinese to non-Chinese made vessels. The front has nine Chinese but only five non-Chinese vessels, while the back of the Store has a roughly equal number of Chinese/non-Chinese vessels, eight/nine. The majority of the non-Chinese made ceramics are from Britain.

Dating ceramic sherds

There are no datable makers' marks present. Both the non-Chinese nameable patterns, blue Fibre transfer-print and blue-sprigged Chelsea Sprig can only be dated to the nineteenth century. Fibre, with stylised fern-like tree branches, was a very common pattern in mid to late-nineteenth century Sydney (Casey & Lowe 2009) and in southeast New South Wales, blue being its most common colour. Chelsea Sprig, also commonly found on Australian sites, possibly dates from 1820 to the late-nineteenth century (Brooks 2005:42-43). None of the other transfer-prints have enough of the pattern present to speculate on the dates of manufacture. Banded patterns were popular in the last quarter of the nineteenth century, particularly 1880-1900 (Wilson 1999:321).

Function of ceramic vessels

All the food/beverage container sherds are Chinese in origin. Green-glazed ginger jars were found in the front and rear. Two brown-glazed jars, globular and spouted, were from the front of the store. Two jars, wide-mouthed and an unidentified large jar, and two lids, wide-mouthed and barrel jar were from the rear of the store.

The proportion of Chinese to non-Chinese food/beverage consumption vessels found at the front of the store was 5:4, while at the rear it was 2:8. The vessels in both locations are predominantly hollowware.

Opium pipe bowls and European clay pipes were found at both the front and back of the store. The opium pipe bowls cannot be identified to type, as only small sherds are present. Opium pipe smoking in Chinese camps was often centred on areas where socialisation occurred, such as stores (Wylie & Fike 1993:290).

Decorative types

Winter Green is the most common Chinese decoration occurring on four vessels from the front of the Store and one vessel from the rear of the store. Blue-underglaze is found on two vessels, a stoneware rice bowl, possibly Bamboo, from the front and a porcelain rice bowl from the back.

Two decorative types are on the non-Chinese vessels from the front of the Store: relief-moulding and brown transfer-prints. Seven decorative types are found on the non-Chinese vessels from the rear of the Store: blue sprigs, blue and green bands, black, blue and brown transfer-prints and a relief-moulded pattern. No sets are present but blue-banding is present on a serving dish and a cup/bowl and green-banding on a cup. The Chinese-made rice bowl also has a blue band at the rim.

Summary

The two areas that were excavated at the Store may have been used for different purposes. In particular, the tea and tableware vessels reveal the main difference between the assemblages. The rear of the Store has far more variety in the non-Chinese vessels. There is a serving vessel and decorative types which are complementary. The Chinese-made blue-underglaze bowl is of a better quality than that at the front. No obvious faults can be seen in the transfer-prints from the back, although the sherds are small. The brown transfer-printed cup at the front has an obvious fault with the edge of the transfer missing. If the front was the shop and the back a residence, the occupants were using a better quality and wider range of ceramics than they were selling; the majority of tea and tableware for sale was Chinese while the majority being used by the occupants was British.

Temple Complex

Larger Chinese settlements often had a temple in which mixtures of religions were practised. Confucianism, Taoism and Buddhism have coexisted as part of the national Chinese ideology since the Tang dynasty, 618-907 (TOCAOSC 2006:13). The site identified by Smith (2006:88) as the Temple Complex was located about 50 m south of the Store. A grid of 13 x 15 m was pegged out in one metre squares. Earth mounds and postholes suggested a building measuring 6 x 3 m, alongside with a number of smaller structures. The presence of nails and glass, together with the postholes, inferred the larger building was weatherboard, while the smaller structures may have been constructed differently. Smith (2006) excavated 17 m² to depths of 20 to 40 cm. The excavations were in four distinct areas across the Temple Complex and for this research they have been labelled as Areas 1 to 4 (Figure 7.3). There were 72 sherds, weighing 466.7 g, from the site as a whole. There were no surface finds.

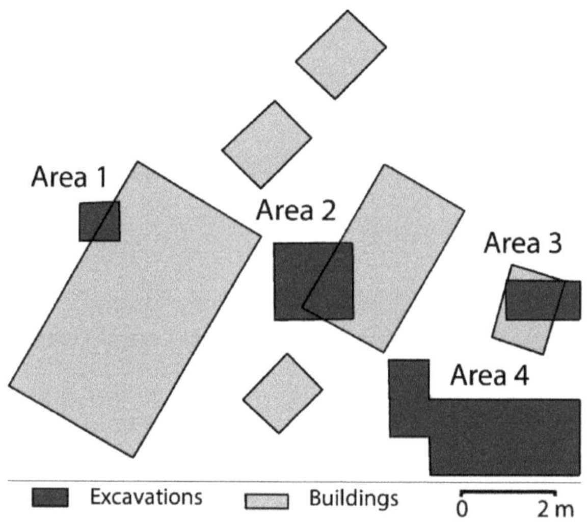

Figure 7.3: Map of Temple Complex areas excavated (after Smith 2006:311)

Area 1

Area 1 was located along the western wall of the Temple building where 1 m² was excavated. Four sherds, weighing 28.2 g, account for a MNV of three (Table 7.3). These include a large Chinese-made brown-glazed storage jar, and two non-Chinese made vessels, both transfer-printed. A cup or bowl has a black romantic-type pattern and an unidentifiable vessel, possibly tableware, of which the base only is present, has a green foliate pattern.

Area 2

Area 2 was located immediately to the northeast of the Temple taking in the southwest corner of another rectangular structure measuring about 3.5 x 2 m.

Seventeen sherds, weighing 84.8 g, represent a MNV of 13 (Table 7.4). No maker's marks were present.

The storage vessels are both Chinese brown-glazed stoneware: a barrel jar, rim 330 mm; and an unidentifiable small-medium storage jar. The tea and tableware vessels include: a Chinese porcelain rice bowl with a blue-underglaze decoration but not enough present to identify the pattern; and seven vessels of British origin including two bone china and five white-earthenware vessels. The bone china consists of two saucers, one with an unidentifiable blue sprig and the other, with a worn enamel band, possibly pink, at the rim and other gold decoration. The white-earthenware includes: a faceted cup with a clear glaze; a cup with a blue Willow transfer-print; a cup or bowl with an unidentified blue transfer-print; a plate, rim 160 mm, with a floral relief-moulded pattern around the marly and the edge only of a brown transfer-print below; and a saucer with a clear glaze, rim 175 mm. The Chinese-made vessel with an unknown function is possibly a miniature straight-sided porcelain pot with a diameter of 30 mm, having a clear glaze and enamel floral pattern in pink, green and brown, possibly Four Seasons. Ritchie (1986:224) described a similar pot from Cromwell's former Chinatown, in New Zealand. The two vessels of unknown origin are both porcelain, and represented by small sherds only. One is a very fine white rim sherd which is too small to determine if it is flat or hollowware or its size. The other thin sherd belongs to a hollowware vessel and has a whitish-grey body with faded pink enamel on the exterior. Its decoration does not appear to be applied by brush but rather as a stencil or transfer, therefore it may have been Japanese-made.

Area 3

Area 3 was to the east of Areas 1 and 2, across a small structure, possibly a hut. Twenty-four sherds, weighing 98.2 g, represent a MNV of 14 (Table 7.5). No maker's marks were present.

The containers are both Chinese-made: a brown-glazed spouted or wide-mouthed jar and a liquor bottle. There is one Chinese-made tea/tableware vessel: a porcelain hollowware vessel having a clear glaze with a green cast, possibly a teapot. The sherd is from the shoulder of the vessel which has a body diameter of 80-100 mm. Although the sherd has no glaze on the interior, the vessel may have been glazed unevenly leaving some areas without glaze. Porcelain has a porosity of close to zero so glazing the inside is not necessary for holding liquid. The non-Chinese made tea and tableware vessels include a bone china saucer with a green enamel pattern and eight white-earthenware vessels: a plate in blue Albion, rim 240 mm; a plate with a brown floral transfer-print and possible relief-moulding on the marly; a plate/saucer with a flown-black floral pattern; a saucer with a blue sponged-printed pattern, rim 175 mm; a saucer with a green geometric transfer-print, rim 175 mm; a cup or bowl with an unidentifiable blue transfer-print; a cup or bowl with an unidentifiable flown-blue pattern; and a possible mixing bowl with a thick body, foot diameter 90 mm. Other vessels include: a Chinese light-orange slipped terracotta opium pipe bowl, with a circular rim, 70 mm, and ten faceted sides; and a white-earthenware paste/cosmetic/ointment pot, rim 80 mm.

Table 7.3: Ceramic sherds from JCC Temple Complex, Area 1

	No.	% No.	Wt (g)	% Wt	Wt/No.	MNV	No./MNV	% MNV
Food/beverage containers								
Chinese	2	50.0	12.3	43.6	6.2	1	2.0	33.3
Non-Chinese	0		0.0			0		0.0
Food/beverage consumption								
Chinese	0		0.0			0		0.0
Non-Chinese	2	50.0	15.9	56.4	8.0	2	1.0	66.7
Total	4		28.2			3		

Table 7.4: Ceramic sherds from JCC Temple Complex, Area 2

	No.	% No.	Wt (g)	% Wt	Wt/No.	MNV	No./MNV	% MNV
Food/beverage containers								
Chinese	6	35.3	55.0	64.9	9.2	2	3.0	15.4
Non-Chinese	0		0.0			0		0.0
Food/beverage consumption								
Chinese	1	5.9	1.1	1.3	1.1	1	1.0	7.7
Non-Chinese	7	41.2	24.4	28.8	3.5	7	1.0	53.8
Other								
Chinese	1	5.9	4.0	4.7	4.0	1	1.0	7.7
Non-Chinese	0		0.0			0		0.0
Unknown	2	11.76	0.3	0.4	0.2	2	1.0	15.4
Total	17		84.8			13		

Table 7.5: Ceramic sherds from JCC Temple Complex, Area 3

	No.	% No.	Wt (g)	% Wt	Wt/No.	MNV	No./MNV	% MNV
Food/beverage containers								
Chinese	11	45.8	55.5	56.5	5.0	2	5.5	14.3
Non-Chinese	0		0.0			0		0.0
Food/beverage consumption								
Chinese	1	4.2	1.0	1.0	1.0	1	1.0	7.1
Non-Chinese	10	41.7	27.5	28.0	2.8	9	1.1	64.3
Smoking								
Chinese	1	4.2	4.7	4.8	4.7	1	1.0	7.1
Non-Chinese						0		0.0
Other								
Chinese	0		0.0			0		0.0
Non-Chinese	1	4.2	9.5	9.7	9.5	1	1.0	7.1
Total	24		98.2			14		

Area 4

Area 4 covered 10 m² and was located south of Area 3. Although no structures are drawn on Smith's map, earthworks depicted indicate the possibility of a structure/s. Twenty-seven sherds, weighing 255.5 g, represent a MNV of 15. A clay pipe brings the total MNV to 16 (Table 7.6). A partial maker's mark, 'hine' (Rhine) printed in grey within a cartouche, is present on the underside of a white-earthenware base but there is no maker's name. One fire damaged sherd was also found in this area.

The storage vessels are both Chinese-made: a barrel jar lid, rim 310 mm and a large brown-glazed storage jar. Tea and tableware includes three Chinese-made and nine non-Chinese made vessels. There are two Chinese porcelain rice bowls with blue-underglaze decoration: one with Bamboo pattern, rim 140 mm; and the other with a rim of 130 mm, but not having enough of the pattern present to identify it. A shallow dish, rim 130 mm, with enamel decoration, possibly Four Seasons is included as being Chinese-made even though the body was whiter than most other Chinese porcelain found at JCC. The body is a similar colour to the miniature pot found in Area 2 of the Temple Complex. The non-Chinese made vessels include a British bone china saucer with a clear glaze and eight white-earthenware vessels: a cup or bowl in grey Fibre; a relief-moulded mug or jug, body diameter 100 mm; a plate or saucer in blue Albion; a small plate or saucer with a grey floral pattern, possibly Asiatic Pheasants; a plate or saucer in green Rhine; and a plate with a brown foliate transfer-printed pattern, rim 200 mm. There are two unidentifiable vessels, possibly tableware but only the bases are present, one in grey Rhine and the other a blue transfer-print. The vessel of unknown function is a British bone china slip-cast hollowware vessel with a textured pineapple-like pattern and a cream glaze. It may have been a vase or a jug. A European clay pipe is also present.

Table 7.6: Ceramic sherds from JCC Temple Complex, Area 4

	No.	% No.	Wt (g)	% Wt	Wt/No.	MNV	No./MNV	% MNV
Food/beverage containers								
Chinese	10	37.0	197.1	77.1	19.7	2	5.0	12.5
Non-Chinese	0		0.0			0		0.0
Food/beverage consumption								
Chinese	3	11.1	4.5	1.8	1.5	3	1.0	18.8
Non-Chinese	13	48.1	34.5	13.5	2.7	9	1.4	56.3
Smoking								
Chinese	0		0.0			0		0.0
Non-Chinese						1		6.3
Other								
Chinese	0		0.0			0		0.0
Non-Chinese	1	3.7	19.4	7.6	19.4	1	1.0	6.3
Total	27		255.5			16		

Note: clay pipe is not in catalogue sherd numbers or weights

Interpretation of Temple Complex

Historical reports place at least two Chinese temples in Jembaicumbene, one of which was opened in 1861 (Smith 2006:80). The Temple Complex identified by Smith consists of a number of structures, of which three (or four) were partially excavated. The main structure is the Temple itself. Other ancillary buildings are of various sizes. A minimum number of 46 vessels in total were recovered from the Temple Complex.

Origin of ceramics

Smith (2006:88) stated that 'Most of the ceramic artefacts' from the Temple Complex are 'of Chinese origin'; however this is only correct if interpreting by the weight (Table 7.7). The sherd numbers are almost equal and the minimum number of Chinese-made vessels is half that of the non-Chinese made vessels.

Table 7.7: Origin of ceramic sherds from Temple Complex

	Area 1	Area 2	Area 3	Area 4	Total
Number					
Chinese	2	8	13	13	36
Non-Chinese	2	7	11	14	34
Unknown	0	2	0	0	2
Weight					
Chinese	12.3	60.1	61.2	201.6	335.2
Non-Chinese	15.9	24.4	37.0	53.9	131.2
Unknown	0.0	0.3	0.0	0.0	0.3
MNV					
Chinese	1	4	4	5	14
Non-Chinese	2	7	10	11	30
Unknown	0	2	0	0	2

The majority of non-Chinese vessels are British-made. The sponge-printed saucer may have originated elsewhere in Europe, for example Holland. If the site was occupied into the twentieth century, the white-earthenware relief-moulded mug/jug may have been Japanese. The unidentifiable porcelain may also be of Japanese origin.

Dating ceramic sherds

No datable makers' marks were found. There are four nameable transfer-printed patterns, Albion, Willow, Rhine and Fibre, recovered along with Chelsea Sprig. All these patterns are common at Australian nineteenth century archaeological sites, all produced from at least 1830 and into the twentieth century. A black transfer-printed romantic pattern was recovered from Area 1. Although the peak production period for these types of patterns was 1831-1851 they remained popular throughout the nineteenth century (Samford 1997:6). Two vessels have sponge-printed patterns which are of indeterminate shape, suggesting a date before 1870 (Kelly et al. 2001:9).

Function of ceramic vessels

In 1867, the Sydney Morning Herald featured an article entitled 'A Day in Canton' written by a correspondent describing his visit to a temple:

> ...we went to the Temple of the five Hundred Gods, or Virtues, the approaches to which are mean and shabby looking; nor is the principal building itself handsome... The principle (sic) god, before whom joss-sticks were burning, in vases filled with the ashes of former joss-sticks, was really a majestic figure in bronze, of enormous size, and represented an emperor of the second dynasty, who is venerated as one of the great benefactors of China. Behind this statue was a painting which contained three figures, typifying the Past, the Present, and the Future. Within the walls of this vast temple was a pagoda and a number of other buildings, and we were taken to visit the high-priest in a very comfortable apartment, where he received us with many "chin-chins," begged us to sit down, and ordered tea and sweetmeats for us. The Chinese drink tea almost scalding hot, without milk or sugar, and we followed the custom as far as we could (The Sydney Morning Herald, 26 April, 1867:6).

This description of a temple suggests that a number of buildings comprised a temple complex and that visitors were often served refreshments. Food and drink were also associated with temples in Australia. In addition to being a place of worship, they were meeting places and a focus of traditional festivals (Wilton 2004:85). The majority of the recovered artefacts from JCC Temple are tea and tableware (Table 7.8). It is possible that some of the ceramics may be associated with a domestic assemblage for a caretaker. However, saucers and small plates dominate the assemblage. These non-Chinese vessels are a similar size and shape to traditional Chinese shallow serving dishes.

Table 7.8: Tea and tableware excavated from the Temple Complex

Vessel		Area 1	Area 2	Area 3	Area 4	Total
Rice bowl	C		1		2	3
Shallow dish	C				1	1
Teapot?	C			1		1
Cup	N		2			2
Cup/bowl	N	1	1	2	1	5
Saucer	N		3	3	1	7
Plate/saucer	N			1	3	4
Small plate	N		1	1		2
Medium plate	N			1	1	2
Jug/mug?	N				1	1
Total		1	8	9	10	28

Key: C-Chinese; N-non-Chinese

Both the small plates may have been children's plates which, in the Victorian era, often featured relief-moulding on the marly and a transfer-printed pattern in the well. These plates were often teaching tools, featuring the alphabet or moralistic script. A writing slate was also found from the same area as one of these plates (Smith

2006: Appendix C3). It is possible that children's china may have been used as a tool, either by adults or children, to learn English.

Other vessels include a possible Chinese-made miniature pot, as described by Ritchie (1986:224), from Area 2 and a bone china vase or jug from Area 4. Vases were used in temples for displaying flowers and as incense containers. The small-medium brown-glazed storage jars may also have been reused for this purpose. In 1859, a newspaper reporter described the Chinese Joss-House at Strike-a-Light Flat, near Jembaicumbene. The interior featured an altar on which was placed 'a round earthenware vessel with the handle broken off, filled with earth and flags'; other 'tubs' placed around the room were also filled with flags (The Sydney Morning Herald, 25 October 1859:2).

Communal Oven

Communal ovens, colloquially referred to in Australia as 'pig ovens', were used for feasts of both 'spiritual and secular' natures in Chinese communities (Smith 2006:50). They were often associated with a group of dwellings and were built into the ground using stone and earth and had a circular interior. To 2006, there were a total of 83 Chinese ovens known in Australia (Smith 2006:53).

A circular stone arrangement, about 0.8 m in diameter, supported by an earth mound, indicated the presence of a Chinese oven at Jembaicumbene. It was located about 65 m east of the Store. Smith (2006:298) pegged out a grid measuring 12 x 10 m and excavated 17 m² to depths of 10 to 40 cm.

Ceramic sherds were found in only two of the 17 squares excavated. Six, weighing 173.8 g, represent a MNV of four (Table 7.9). Only one of the sherds recovered, a brown-glazed fragment, is fire damaged.

The two Chinese-made vessels are brown-glazed stoneware storage jars, one large, possibly a barrel jar, and the other a smaller type of jar, but no diagnostic sherds are present.

One of the non-Chinese sherds is Australian-made lead-glazed terracotta (personal communication, Dr Mary Casey, 27 June 2009). It may have been part of a pan or dish. The other vessel is clear-glazed white-earthenware, rim diameter 210 mm. Only the rim and part of the side are present. The flat rim, width 17 mm, is perpendicular to the slightly curving side. It is not known how deep the vessel was. The shape appears similar to pictures seen of a pie dish or wash basin. It is included in the food/beverage consumption category, assuming that it was used for food at the Oven site.

Interpretation of Oven site

Smith (2006:89) reported that there were 569 artefacts recovered from the Oven site. These mostly comprise unidentifiable pieces of metal, charcoal and animal bone. There were 46 sherds of glass, mainly alcohol bottles.

Origin of ceramics

In comparison to the other artefacts, only six ceramic sherds were recovered. These represent a minimum number of four vessels, two Chinese-made, one Australian-made and one possibly of British origin.

Dating ceramic sherds

None of the ceramic vessels from the Oven can be dated.

Function of ceramic vessels

The Oven was somewhat detached from the older part of the camp and closer to the Store. The low number of ceramic vessels recovered suggests that the Oven may only have been used for festive occasions, rather than for communal eating on a daily basis, at least in the latter period of occupation. The large brown-glazed vessel, possibly a barrel jar, may have been re-used to store water. The terracotta vessel may have been a pan or dish related to food preparation or serving while the white-earthenware vessel may have been used to prepare/serve food.

Table 7.9: Ceramic sherds from JCC Oven

	No.	% No.	Wt (g)	% Wt	Wt/No.	MNV	No./MNV	% MNV
Food/beverage containers								
Chinese	4	66.67	122.7	70.6	30.7	2	2.0	50.0
Non-Chinese	0		0.0			0		0.0
Food/beverage consumption								
Chinese	0		0.0			0		0.0
Non-Chinese	2	33.33	51.1	29.4	25.6	2	1.0	50.0
Total	6		173.8			4		

Hut sites

The Chinese initially set up their camps with tents, but calico was replaced with more durable material as the population became more settled. The archaeological survey of JCC found that 'the settlement extended along the northern bank of Jembaicumbene Creek for a distance of about 250 m with a concentration of individual dwellings at its centre' (Smith 2006:91). Table 7.10 details the grid size set out by Smith, the internal dimensions of each hut, the excavated area and a brief description.

Hut 1

This hut was located between the Store and Temple. Ceramic sherds were recovered from five of the excavated squares. There were no surface finds. Conjoining sherds were only found within one square, belonging to one brown-glazed wide-mouthed jar. Ninety-six sherds, weighing 604.0 g, represent a MNV of 15 (Table 7.11). There are no makers' marks present on any sherds. Several of the recovered sherds are fire damaged.

The storage jars include two Chinese brown-glazed wide-mouthed jars, rims 80 mm and 90 mm, and another large Chinese brown-glazed jar (possibly a barrel or globular jar); and a large Bristol-glazed stoneware storage jar from either Britain or Australia.

Two Chinese-made tableware vessels were recovered: a Chinese Winter Green porcelain rice bowl, rim 140 mm, and a Four Seasons decorated spoon. The seven non-Chinese tea/tableware vessels include six of white-earthenware and one clear-glazed bone china cup. The earthenware vessels are: a bowl with a blue transfer-printed floral pattern; a cup or bowl with a flown-black floral/foliate transfer pattern; a cup or bowl with unidentifiable decoration resulting from fire damage; an eggcup with a clear glaze; a dinner plate with blue transfer-printed Gem pattern; and a children's plate which has relief-moulded daisies around the marly and black transfer-printed script below the shoulder, but not enough to be legible. A foot ring sherd with a clear glaze and a diameter of 125 mm is also present but this has not been included in the MNV as it may have belonged with the blue transfer-printed bowl.

Fragments from two Chinese opium pipe bowls were recovered, both with light-orange slip: one with a circular rim and ten faceted sides, rim diameter 70 mm; and the other with a circular rim and smooth sides with a single raised ridge.

Hut 2

This hut was located between the Store and Temple. Ceramic sherds were recovered from nine of the excavated squares. There were no surface finds. One thousand, eight hundred and two ceramic sherds, weighing 11,588.6 g, represent a MNV of 121. A European clay pipe brings the MNV to 122 (Table 7.12). The majority of the sherds recovered from this hut site are fire damaged.

A total of 18 makers' marks are present, 11 Chinese (Table 7.13) and seven non-Chinese (Table 7.14).

Table 7.10: JCC hut dimensions, area excavated and description (after Smith 2006:92, 298)

Hut No.	Grid size	Int dimen m²	Exc m²	Description
1	6 x 6	7.5	12.0	Free-standing tented structure with tamped earth floor and hearth
2	6 x 6	7.5	10.0	Free-standing tented structure with tamped earth floor and hearth
3	6 x 6	6.0	7.0	Free-standing tented structure with tamped earth floor and hearth
4	6 x 6	6.0	5.0	Built into slope with tamped earth floor
5	6 x 6	Unknown	2.0	Free-standing tented structure with tamped earth floor

Table 7.11: Ceramic sherds from JCC Hut 1

	No.	% No.	Wt (g)	% Wt	Wt/No.	MNV	No./MNV	% MNV
Food/beverage containers								
Chinese	45	46.9	457.5	75.7	10.2	3	15.0	20.0
Non-Chinese	1	1.0	9.5	1.6	9.5	1	1.0	6.7
Food/beverage consumption								
Chinese	2	2.1	11.8	2.0	5.9	2	1.0	13.3
Non-Chinese	37	38.5	107.4	17.8	2.9	7	5.3	46.7
Smoking								
Chinese	11	11.5	17.8	2.9	1.6	2	5.5	13.3
Non-Chinese						0		0.0
Total	96		604.0			15		

Table 7.12: Ceramic sherds from JCC Hut 2

	No.	% No.	Wt (g)	% Wt	Wt/No.	MNV	No./MNV	% MNV
Food/beverage containers								
Chinese	970	53.8	5206.5	44.9	5.4	39	24.9	32.0
Non-Chinese	93	5.2	1986.1	17.1	21.4	4	23.3	3.3
Unknown	2	0.1	18.6	0.2	9.3	2	1.0	1.6
Food/beverage consumption				0.0				
Chinese	92	5.1	426.8	3.7	4.6	17	5.4	13.9
Non-Chinese	501	27.8	3144.7	27.1	6.3	38	13.2	31.1
Smoking				0.0				
Chinese	94	5.2	245.9	2.1	2.6	9	10.4	7.4
Non-Chinese						1	0.0	0.8
Other								
Chinese	0		0.0			0		0.0
Non-Chinese	28	1.6	375.1	3.2	13.4	11	2.5	9.0
Unknown	22	1.2	184.9	1.6	8.4	1	22.0	0.8
Total	1802		11588.6			122		

Note: clay pipe is not in catalogue sherd numbers or weights

Table 7.13: Chinese makers' marks from JCC Hut 2

Cat.No.	Mark	Vessel	Details
715	hand painted	rice bowl	Reign type mark - complete
731	hand painted	rice bowl	Reign type mark - edge of mark only
773.1	hand painted	tea cup	Reign type mark - edge of mark only
190	hand painted	liquor cup	Small rectangular mark - complete
175	hand painted	liquor cup	S shaped mark - complete
734.1	hand painted	liquor cup	Small blue mark - edge of mark only
713	embossed	ginger jar	Three Chinese characters inside oval cartouche
106.1	scratched	opium pipe bowl	Chinese characters scratched into lower part of bowl before firing, only one character *hao* could be identified
176	impressed	opium pipe bowl	Two impressed Chinese stamps on lower part of bowl, one square and one rectangular, right side of square stamp has Chinese characters for water and east
723.2	scratched	opium pipe bowl	Chinese characters scratched into lower part of bowl before firing, only one character *er* could be identified

Table 7.14: British makers' marks from JCC Hut 2

Cat.No.	Mark	Vessel	Details
146.2	transfer print	plate	Black printed cartouche with crown and script 'NSI...GTO' / 'H'. Possible pattern name Kensington, unidentifiable maker
198.5	transfer print	plate	Partial blue printed maker's mark 'R' in cartouche (Rhine)
198.6	transfer print	plate	Blue printed fire damaged pattern garter on back of marly, Cable pattern, possibly Pinder, Bourne & Co. (1862-1882)
744	transfer print	saucer	Purple printed maker's mark 'H. N. & Co.' below crown; unidentifiable maker
701.7	impressed	bowl	Impressed star under base
708	impressed	pot	'B.E.& Co / 1oz': Bates Elliot & Co., Burslem, 1870-1875
729, 764 & 762.7	transfer print	toothpaste pot lid	Black transfer: 'WHITE ROS/ AN ELEGANT/ PREPARATION FOR/ PRESERVING & WHITENIN/ THE TEETH/ PURIFYING & PERFUMING THE/ BREATH/ PREPARED BY/ S. MAW, SON & SONS/ ALDERSGATE ST/ LONDON/ ENGLAND' scroll pattern in corners of lid, after 1901

The blue-underglaze painted marks on the rice bowls and liquor cups do not include any Chinese characters (Figure 7.4). The marks on the ginger jar lid relate to the county or city name but cannot be translated (personal communication, Po-yi Chiang, 2 November 2010). Both scratched and impressed marks are present on the opium pipe bowl sherds. Catalogue numbers JCC-106.1 and JCC-723.2 had characters scratched into the surface when they were made (Figure 7.5). The Chinese characters are identified as *hao* and *er* respectively, but neither of these have any meaning by themselves. Catalogue number JCC-176 has two impressed stamps, one of which contains the Chinese characters for water and east, which may relate to the workshop. The rest of these stamps are illegible (translation, Po-yi Chiang, 2 November 2010).

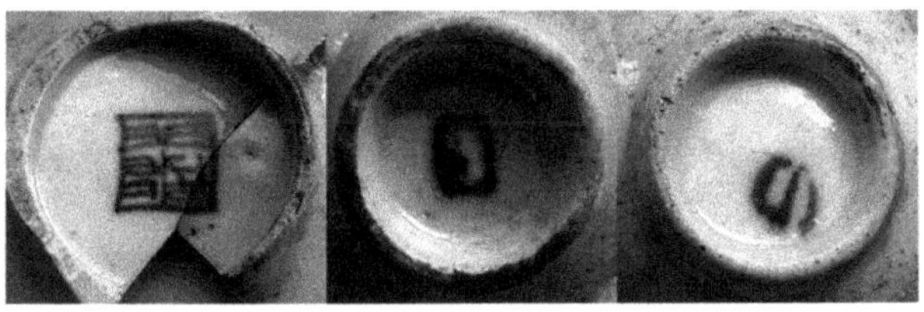

Figure 7.4: Marks on celadon-glazed vessels from JCC Hut 2 (from left: Cat nos. JCC-715, JCC-190, JCC-175; foot diameters 53 mm, 19, mm and 18 mm)

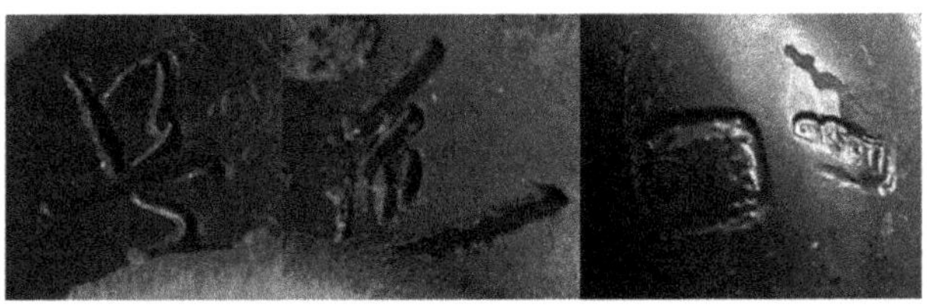

Figure 7.5: Marks on opium pipe bowls from JCC Hut 2 (from left: Cat nos. JCC-106.1, JCC-723.2, JCC-176; mark widths c.5 mm)

Three of the seven British marks are shown in Figure 7.6. The first mark has an unidentifiable maker, but the pattern name is possibly Kensington. The Cable patterned plate has a fire damaged mark on the underneath of the marly, similar to that of Pinder Bourne & Co. (1862-1882, Godden 1991:495) and Pinder, Bourne & Hope (1851-1862, Godden 1991:495). H. N & Co. is unidentifiable. This mark was also found in the Lake Innes assemblage from the post-Innes period, after 1860 (Brooks 2007). An ointment/paste pot, made by Bates Elliot & Co., Burslem, dates 1870-1875 (Godden 1991:59). The White Rose toothpaste lid dates to the beginning of the twentieth century. The family pharmaceutical business became S. Maw, Son & Sons in 1901 when it was incorporated as a limited liability company (Matlach 2010).

Conjoined sherds were found within and between squares. Twelve vessels have conjoining sherds excavated from different squares, some up to three metres apart. At least another 20 vessels have conjoining sherds recovered from within the same square. Conjoined vessels account for at least a quarter of the total MNV.

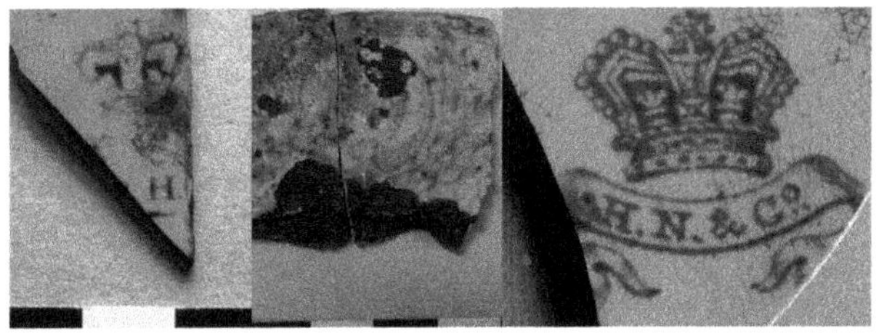

Figure 7.6: British makers' marks on tableware from JCC Hut 2 (from left: Cat nos. JCC-146.2, JCC-198.6 and JCC-744)

There are 39 Chinese-made food/beverage containers, 4 non-Chinese made vessels and two of unknown origin used for food/drink storage.

The Chinese storage vessels comprise: a barrel jar lid, two globular jars, 11 wide-mouthed jars, nine wide-mouthed jar lids, five spouted jars, six liquor bottles, one storage jar lid, two ginger jars and two ginger jar lids. The barrel jar lid has a diameter of about 370 mm, a height of 53 mm and is thinly glazed on the exterior and interior. The globular jars have rims of 85 mm and 90 mm. There are five sizes of wide-mouthed jars, 70 mm (2), 80 mm (3), 90 mm (2), 100 mm (3) and 110 mm (1) but only three sizes of lids. The lids have diameters of 70 mm (1), 90 mm (6) and 100 mm (2). One of the wide-mouthed jars, rim 70 mm, is the shape described by Hellmann and Yang (1997:186) as a 'subtle form variation'. The other wide-mouthed jars appear to be the shape described by Ritchie, with the slightly splayed foot (1986:239). There are two sizes of spouted jars, one size having a rim diameter of 35 mm (2) and the other 40 mm (3). The six liquor bottles all have rim diameters of 55-60 mm. The storage jar lid has a rim diameter of 75 mm and a flange diameter of 55 mm; only the exterior top of the lid is glazed in dark-brown and the rest unglazed. Both ginger jars are round and have green-glazed exteriors and brown-glazed interiors. The ginger jar lids also have a brown-glazed interior. One of the lids is complete and has a band of green glaze measuring about 10 mm wide around the exterior rim; the height of the lid is 44 mm (Figure 7.7). The exterior glaze appears red as a result of fire damage. Oxygen has been withdrawn from the copper oxide in the glaze by carbon monoxide during the fire, and as a result the green copper glaze has changed to red. The other lid, of which only part is present, is a similar size.

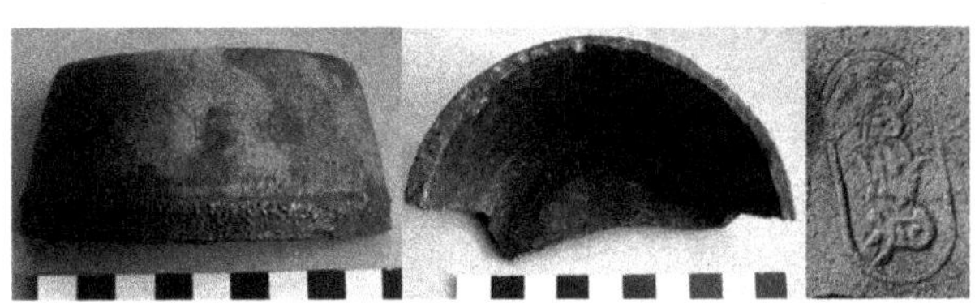

Figure 7.7: Fire damaged ginger jar lid from JCC Hut 2: exterior, interior and embossed mark (Cat no. JCC-713)

The non-Chinese storage vessels include two two-tone Bristol-glazed stoneware jars, shoulder diameters 220 mm and 230 mm; a white-earthenware food jar, rim 67 mm, having a thickened rim with indented ridges below the rim and above the foot; and a salt-glazed stoneware blob-top ginger beer bottle, body diameter 80 mm. These vessels may have been made in Britain or Australia.

Two storage items of unknown origin are both lids. One is an unglazed stoneware storage jar lid, diameter 110 mm, but only a small sherd is present. It has a height of about 10 mm; one side of the rim is flat and the other curved. The shape of the rim edge matches that in an illustration by Ritchie (1986:255) of a possible globular jar lid. Ritchie described his lid as unglazed earthenware, whereas the JCC sherd is unglazed stoneware. Alternatively, the sherd could be the edge of a lid for a European style stoneware bung jar. The other lid fragment is part of a brown-glazed knob that again could be from either a Chinese or European-style crock lid. It has a diameter of 32 mm and height of 10 mm.

Tea and tableware vessels include 17 of Chinese origin, 37 non-Chinese and one of unknown origin.

The Chinese-made tea and tableware comprises: a blue-underglaze shallow dish, rim 145 mm, with Rocks and Orchid pattern; three porcelain rice bowls in Winter Green, rims 140 mm; a Bamboo decorated porcelain rice bowl, rim 130 mm; a stoneware Double Happiness rice bowl, rim 140 mm; two porcelain Winter Green tea cups, feet 30 mm; a small porcelain Winter Green tea cup measuring 25 mm at the foot and having a height of 40mm; two porcelain spoons in Four Seasons pattern; a porcelain blue-underglaze teapot lid, rim 85 mm; and five Winter Green liquor cups with varying measurements. One cup has a rim diameter of 50 mm, foot 19 mm, height 25 mm and another a rim of 50 mm, foot 18 mm, and height 24 mm. Only the rims are measurable on the other three cups, one being 50 mm and the other two 45-50 mm.

The 38 non-Chinese food preparation/consumption vessels include 13 bowls, four cups, one mug, three cup/bowls, an eggcup, a tureen/serving dish lid, seven plates, two soup plates, four saucers and two plate/saucers. One of these vessels is a bone china saucer, rim 160 mm, with partial blue sprigs which may have been grapes or flowers. A porcelain eggcup has a clear glaze. All the rest are made of white-earthenware in a variety of decorative types.

Six vessels have nameable transfer-printed patterns. These include: a grey Rhine bowl, rim 140 mm, foot 70-75 mm and height 80 mm; a blue Albion plate, rim 240 mm; a blue Cable plate, rim 265 mm; a blue Rhine plate, rim 270 mm; a green Rhine plate, rim 250 mm; and a blue Willow soup plate, rim 260 mm. A child's plate has a purple transfer-print in the well with the title, The Poor Boy and The Loaf. The plate has red bands at rim and shoulder, relief-moulded flowers between the bands and a rim diameter of 170 mm. Other vessels with transfer-printed decoration are: three bowls with geometric patterns, one in blue, rim 140 mm, one in green, rim 130 mm, and another in green, rim 140 mm, foot 60 mm and height of 80 mm; a cup, rim 85-90 mm, in a blue floral pattern; a cup, rim 100 mm, in a blue geometric pattern; a child's alphabet mug, YZ, which features a black transfer-print with hand-painted areas of blue and yellow; two cup/bowls in blue geometric patterns; a cup/bowl in a purple floral pattern; a plate, foot diameter 85 mm, with a black floral pattern, possibly Kensington; a plate, foot 140 mm, in a green geometric pattern; a fire damaged soup plate, foot 130 mm, in a floral pattern of unknown colour; a saucer, rim 160 mm, in a blue geometric pattern; and a saucer, rim 175 mm, in a purple geometric pattern.

Vessels with other decorative types include: a blue-banded bowl, rim 160 mm, which has three thin bands at rim and one thick band below; two bowls with rims of 120 mm, feet 57 mm and height 65 mm, which feature three thin blue bands at the rim; a saucer, rim 165 mm, has red band-and-line decoration; a bowl with a very-flown blue pattern, rim 150 mm; a cup, rim 90 mm, having a flown-black scenic pattern on the exterior and a floral pattern on the interior; a sponge-printed bowl, rim 150 mm, with a repeated purple leaf motif and bands; a bowl with a scalloped rim of 160 mm and a repeated purple sponge-printed motif; a sponge-printed bowl, rim 130 mm, with a purple band and possible repeated flower motif; a plate/saucer with a purple band and an unidentifiable repeated sponge-printed motif; a small plate or saucer with a relief-moulded possible floral pattern around the rim; a cup with foliate relief-moulding and a green glaze; two bowls with rim diameters of 140 mm and unknown fire damaged decoration. The tureen/serving dish lid with a rim of 175 mm has unidentifiable decoration resulting from fire damage.

Sherds from a minimum of nine Chinese terracotta opium pipe bowls and a European 'White's Rifle' clay pipe were recovered from Hut 2. There are four types of opium pipe bowls present. The first type is C1, one with light and one with dark-orange slip, both having rims of 70 mm. The second type is C3, two with light-orange slip, rim diameters 80 mm, one with light-orange slip, rim 70 mm, and the other with dark-orange slip, rim 80 mm. The third type of pipe bowl is octagonal; two have light-orange slip, one of which has an impressed geometric pattern around the rim. The final type of pipe bowl has a circular rim and body but only small sherds were found. This opium pipe bowl was thrown on a potter's wheel, whereas all the other pipe bowls from Hut 2 were pressed into a mould.

Non-food related items include paste/ointment pots, blacking bottles and an unidentifiable item. Five paste/ointment pots and four pot lids were recovered from Hut 2. These include: a pot with a body diameter of 60 mm, made by Bates Elliot & Co., Burslem, 1870-1875 (Figure 7.8, left); two pots with diameter of 60 mm, one of 70 mm and one of 40 mm; two pot lids with diameters of 65 mm and one of 85 mm, all with a clear glaze only;

and a square S. Maw, Son & Sons White Rose toothpaste lid measuring 65 x 65 mm (Figure 7.8, right). Two salt-glazed stoneware blacking bottles from either Britain or Australia were recovered from Hut 2. One has a rim of 45 mm and a body diameter of about 55 mm, and the other a rim of 50 mm and a body diameter of about 70 mm. Many sherds from an unidentifiable vessel, of unknown origin, were found (Figure 7.9).

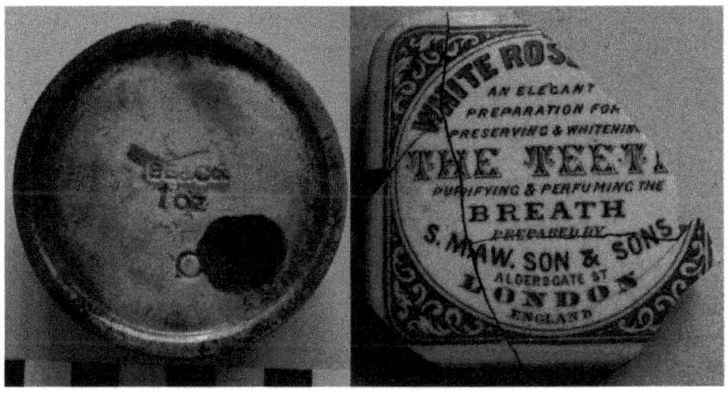

Figure 7.8: Paste pot - Bates, Elliot & Co, (Cat no. JCC-708) and White Rose toothpaste lid (Cat nos. JCC-729, JCC-762.7, JCC-764)

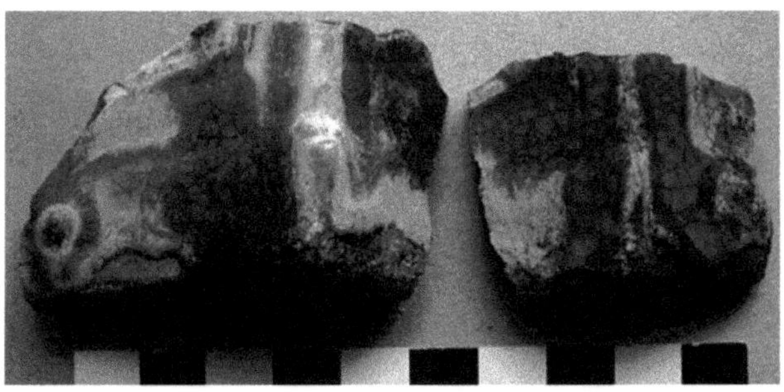

Figure 7.9: Sherds from an unidentifiable vessel, JCC Hut 2 (Cat no. JCC 762.3)

The stoneware vessel was a hexagonal shape with cut-away sections in the sides. Each side is about 75-80 mm wide, with protruding bosses near the foot, and there is a hole in base with a diameter of about 80 mm. The cut-away sections and the hole in the base are part of the vessel's design, not later additions. Although there are many sherds they do not conjoin to give an idea of what the sides look like or what the height of the object is. The interior of the vessel is unglazed and water staining from burial is evident. The exterior features a cream, green, blue and brown fluid glaze. The clay body is similar to that of the Chinese brown-glazed storage jars and the underneath of the base resembles the underneath of a spouted or wide-mouthed jar. Although the body looks similar to Chinese-made brown ware, no vessels of this description have been found in the literature. No object of this nature is present in the database of the Asian American Comparative Collection; although there is a hexagonal pot with a 'gray-green to blue glaze', and protruding bosses have been noted on green-glazed barrel jars (personal communication, Dr Priscilla Wegars, 16 February 2010).

Hut 3

This hut was located to the southeast of the Temple Complex. Ceramic sherds were recovered from four of the excavated squares. Seven excavated sherds, weighing 54.0 g, represent a MNV of five (Table 7.15). No conjoining vessels were found. Only one sherd is fire damaged.

A minimum of one medium-large Chinese-made brown-glazed stoneware storage jar is present but there are no diagnostic sherds to determine the type.

Three tableware vessels are of British origin: two are white-earthenware and one bone china. They are: a bowl of unknown size with a flown-blue decoration; a plate of unknown size with a double foot ring and a clear glaze only present on the sherd; and a bone china saucer with a clear glaze. The remaining vessel is a porcelain saucer, probably of Japanese origin, with gilt-banded decoration and jigger marks on the base. A white-earthenware foot ring, diameter 115 mm, was recovered but it has not been included in the MNV because it only has clear glaze and may have belonged with the flown-blue sherd.

Hut 4

This hut was located to the southeast of the Temple Complex. Ceramic sherds were recovered from all of the excavated squares. There were no surface finds. Forty sherds, weighing 432.4 g, represent a MNV of 11. A European clay pipe brings the total MNV to 12 (Table 7.16). No conjoining vessels were found.

Table 7.15: Ceramic sherds from JCC Hut 3

	No.	% No.	Wt (g)	% Wt	Wt/No.	MNV	No./MNV	% MNV
Food/beverage containers								
Chinese	2	28.6	19.1	35.4	9.6	1	2.0	20.0
Non-Chinese	0		0.0			0		0.0
Food/beverage consumption								
Chinese	0		0.0			0		0.0
Non-Chinese	4	57.1	33.5	62.0	8.4	3	1.3	60.0
Unknown	1	14.3	1.4	2.6		1		20.0
Total	7		54.0			5		

Table 7.16: Ceramic sherds from JCC Hut 4

	No.	% No.	Wt (g)	% Wt	Wt/No.	MNV	No./MNV	% MNV
Food/beverage containers								
Chinese	30	75.0	327.3	75.7	10.9	3	10.0	25.0
Non-Chinese	1	2.5	84.0	19.4	84.0	1	1.0	8.3
Food/beverage consumption								
Chinese	1	2.5	0.5	0.1	0.5	1	1.0	8.3
Non-Chinese	8	20.0	20.6	4.8	2.6	6	1.3	50.0
Smoking								
Chinese	0		0.0			0		0.0
Non-Chinese						1		8.3
Total	40		432.4			12		

Note: clay pipe is not in catalogue sherd numbers or weights

The three Chinese-made containers include: a brown-glazed stoneware spouted or wide-mouthed jar, foot 140 mm and shoulder diameter 130 mm; a large brown-glazed stoneware storage jar; and a blue-underglaze decorated ginger jar. The non-Chinese item is a buff-earthenware lid with a screw thread on the flange. It has a rim diameter of 67 mm, a flange diameter of 46 mm, a thread depth of 23 mm and a clear glaze on the top only.

The tea and tableware consists of a Chinese Winter Green porcelain liquor cup, rim 45-50 mm; an English bone china saucer with blue sprigged decoration, rim 170 mm, and five white-earthenware vessels: a cup in blue Willow, rim 90 mm; a cup/bowl in a blue floral transfer-printed pattern; a saucer in a blue sponge-printed pattern; a plate/saucer in blue Willow; and an unidentifiable vessel with a foot diameter of 30 mm. It has a dark brown opaque glaze on the exterior and a cream glaze on the foot and interior.

Only a small sherd is present from the unidentifiable European clay pipe.

Hut 5

This hut was located at the southern end of the site. Ceramic sherds were recovered from both the excavated squares in Hut 5; there were no surface finds. Seventeen sherds, weighing 439.2 g, represent a MNV of eight. A European clay pipe brings the total MNV to nine (Table 7.17). Only one vessel, an opium pipe bowl, has conjoining sherds and they were excavated from the same square. There are no makers' marks present.

The containers are both Chinese-made brown-glazed stoneware: a globular jar, foot 110 mm, and a liquor bottle.

One of the tea and tableware vessels is Chinese-made, a Winter Green porcelain cup or rice bowl. The non-Chinese vessels are both transfer-printed white-earthenware but only small sherds are present: a plate/saucer had a black floral/foliate pattern; and a cup/bowl, a red geometric/floral pattern.

Fragments from three opium pipe bowls, all of different types, were recovered from Hut 5: one C1 with dark-orange slip; one C3 with light-orange slip, rim 80 mm; and an octagonal bowl with light-orange slip and an impressed dotted band at the rim. A small sherd was found from a European clay pipe.

Surface collection

A surface collection was made from an area to the south of Huts 3 and 4. This artefact scatter, which was in a region heavily disturbed by wombats, was found in and around huts (Smith 2006:299). Fifty-seven sherds, weighing 655.4 g, represent a MNV of 24 (Table 7.18).

Table 7.17: Ceramic sherds from JCC Hut 5

	No.	% No.	Wt (g)	% Wt	Wt/No.	MNV	No./MNV	% MNV
Food/beverage containers								
Chinese	3	17.6	422.6	96.2	140.9	2	1.5	22.2
Non-Chinese	0		0.0			0		0.0
Food/beverage consumption								
Chinese	1	5.9	0.8	0.2	0.8	1	1.0	11.1
Non-Chinese	3	17.6	2.7	0.6	0.9	2	1.5	22.2
Smoking								
Chinese	10	58.8	13.1	3.0	1.3	3	3.3	33.3
Non-Chinese						1		11.1
Total	17		439.2			9		

Note: clay pipe is not in catalogue sherd numbers or weights

Table 7.18: Ceramic sherds from JCC surface collection

	No.	% No.	Wt (g)	% Wt	Wt/No.	MNV	No./MNV	% MNV
Food/beverage containers								
Chinese	16	28.1	241.3	36.8	15.1	2	8.0	8.3
Non-Chinese	0		0.0			0		0.0
Food/beverage consumption								
Chinese	3	5.3	18.1	2.8	6.0	2	1.5	8.3
Non-Chinese	37	64.9	389.2	59.4	10.5	19	1.9	79.2
Other								
Chinese	0		0.0			0		0.0
Non-Chinese	1	1.8	6.8	1.0	6.8	1	1.0	4.2
Total	57		655.4			24		

The containers are both Chinese-made: a brown-glazed stoneware globular jar, foot diameter 120 mm; and a blue-underglaze decorated ginger jar with a clear glaze.

The two Chinese-made tableware vessels are a porcelain liquor pot or sauce serving bottle, foot 75 mm, with blue-underglaze Sweet Pea decoration and a Four Seasons decorated porcelain spoon.

The 19 non-Chinese tea/tableware vessels are all white-earthenware. These include: a pudding bowl with a clear glaze, rim 230 mm, foot 120 mm and height 130 mm; a large plate/platter or dish with green key fret and scrolls and a scalloped edge; four bowls, two cups, two cup/bowls, five plates, two plate/saucers and two vessels which are possibly tableware, one a small sherd only and the other a base only. There are three vessels with nameable transfer-prints: a purple Antique cup; a green Rhine dinner plate; and a blue Willow plate of unknown size. A cup or bowl has a purple transfer-print, possibly Orange Blossom. Seven other vessels have transfer-printed patterns: a bowl, rim 140 mm, with a blue geometric pattern; a bowl, foot 75 mm, with a blue scenic pattern; a bowl, rim 155 mm, with a green floral pattern; a cup, rim 85-90 mm, with a purple foliate pattern; a plate or saucer with a blue geometric pattern; and the two unidentifiable vessels, one in purple and the other in blue, possibly Asiatic Pheasants. Vessels with other decorative types include: a bowl, rim 140 mm, with an unidentifiable, very-flown blue pattern; a cup or bowl with a flown-green foliate pattern; a saucer with a purple sponge-printed motif at the rim and shoulder; a child's alphabet plate with a relief-moulded pattern at rim and embossed script below; a child's alphabet plate with a moulded wave pattern on the rim edge, a blue band and embossed script below, rim 160 mm; and a small plate, possibly a child's, with a thin red band at the shoulder. The foot of a child's plate, diameter 120 mm, was also found but not included in the MNV as it could belong with one of the rims. It has transfer-printed script 'Ma(y)' in brown in the well.

One Holloway's ointment pot lid, rim 85 mm, was recovered but no features are present to allow dating.

Test pits

Eight test pits were excavated to depths of 10 to 20 cm, each pit measuring 1 m². Ceramic sherds were found in five of these test pits, designated Grids 9, 10, 11, 12 and 14 (Smith 2006).

Grid 9
Nine sherds weighing 25.5 g were excavated from Grid 9. A small sherd is from a Chinese light-orange slipped type C3 opium pipe bowl. The other sherds are all from a non-Chinese made brown transfer-printed plate with an ivory body, rim 230 mm.

Grid 10
A single sherd weighing 0.9 g was excavated from Grid 10. It is from a Chinese wide-mouthed jar lid, rim 90 mm.

Grid 11
A single sherd weighing 1.1 g was excavated from Grid 11. The stoneware sherd is from a Chinese small-medium brown-glazed storage jar.

Grid 12
Four sherds weighing 5.1 g were excavated from Grid 12. One sherd belongs to a Chinese buff-earthenware wide-mouthed jar lid, rim 110 mm. A bone china sherd is from a saucer with a clear glaze, rim 165 mm. The other small sherds are from an unidentifiable white-earthenware blue transfer-printed vessel, which is possibly tableware.

Grid 14
Three sherds weighing 5.4 g were excavated from Grid 14. Two of the sherds are from an unidentifiable small-medium Chinese brown-glazed storage jar. The other sherd is from a Winter Green rice bowl, rim 140 mm.

Interpretation of hut sites

Huts 1 and 2, with internal areas of 7.5 m², are larger than Huts 3 and 4, which have internal areas of 6.0 m², while the platform of Hut 5 could not be measured. Nevertheless, all huts within the camp are within the size range characteristic of the Chinese (Smith 2006:90). Over half, in terms of MNV, of the ceramic vessels recovered from hut sites are from Hut 2, a hut which may have burnt down while it was occupied. The majority of sherds in Hut 2 are fire damaged, whilst those from other huts at JCC are not. Scattered sherds, which conjoin, suggest that people sifted through the debris and smashed burnt vessels, as burnt areas and melted glass do not extend over sherd edges. The sherds are water stained, signifying their burial in soil after breakage. If Hut 2 did burn down while it was still occupied, it would be more representative of the total vessels used than other hut sites.

Origin of ceramics

Table 7.19 details a comparison of the MNV recovered from each hut site and the surface, in terms of their origin. Huts 1 and 2 have roughly equal amounts of Chinese and non-Chinese made vessels, while Huts 3 and 4, and the surface collection, have predominantly non-Chinese made vessels. The surface collection came from an area which was disturbed by wombats, thus it could be expected that what was brought to the surface is representative of what is underneath. In contrast, Hut 5 is dominated by Chinese-made vessels.

Table 7.19: Summary of MNV percentages per site in terms of origin

Origin	Hut 1	Hut 2	Hut 3	Hut 4	Hut 5	Surface
Chinese	46.7	53.3	20.0	33.4	66.7	16.7
Non-Chinese	53.3	43.4	60.0	58.3	33.3	83.3
Unknown	0.0	3.3	20.0	8.3	0.0	0.0

Note: this table includes European clay pipes

Dating ceramic sherds

A number of British makers' marks were found in Hut 2 and the dates range from the middle of the nineteenth century into the early-twentieth century. No datable marks were found from the other hut sites.

The examination of plate profiles shows that a number of plates from Huts 1, 2, 3 and the surface collection have a double foot ring, possibly dating to the first half of the nineteenth century; or a flat or slightly rounded foot and angular shoulders dating around the middle of the nineteenth century. Profiles of plates from Huts 1 (Gem) and 2 (Albion, Willow) and a saucer from Hut 4 (sponge-printed) are similar to those excavated from the Gold Commissioner's Quarters in Kiandra, occupied 1860-1862. Blue tints could be observed in the pooled glaze of some vessels from Huts 1, 3 and the surface collection. This also suggests an early to mid-nineteenth century date as later nineteenth century clear glaze contains no hints of blue. The vessels from Hut 2 are fire damaged so blue tints are not evident. The Cable plate from Hut 2 has a rim/shoulder profile which is closer to the examples from the second half of the century. The ivory-bodied plate from Grid 9 and the porcelain saucer from Hut 3 with jigger marks on the base indicate a date of manufacture closer to the end of the nineteenth century, possibly into the early twentieth century.

Decorative motifs and patterns suggest date ranges. The Antique pattern, recovered from the surface near Huts 3 and 4, was identified through the NZHCD (2012). Examples of this pattern from New Zealand were all manufactured by Pinder, Bourne & Co (1862-1882). This pattern was also found at Lake Innes, NSW, with Pinder, Bourne & Hope (1851-1862) marks (Brooks 2007). Other nameable patterns found include Albion, Chelsea Sprig, Rhine and Willow. These patterns were manufactured throughout the second half of the nineteenth century and are commonly found at nineteenth century sites in Australia. A vessel, recovered from the surface, features a Greek key fret, a typical classical motif, found in patterns manufactured to 1868 (Samford 1997:6,17). The flow blue vessels in Huts 2 and 3, and from the surface, suggest a date before 1880 (Van Buskirk 2002:12). Sponge-printed patterns on vessels from Hut 2 and the surface have defined motifs, suggesting a date after 1870, while the saucer from Hut 4 has an indeterminate pattern indicating a date before the 1870s (Kelly et al. 2001:9). Banded patterns are only found in Hut 2. These were very common in Australia about 1880-1900 (Wilson 1999:321).

In summary, the date range for the non-Chinese made ceramics found at the hut sites is from the first half of the nineteenth century to the early twentieth century. The early date of some of the ceramic vessels may indicate that they were purchased by the suppliers through an auction of second-hand goods rather than as new stock.

Function of ceramic vessels

Table 7.20 details the function of recovered vessels per site as a percentage MNV. Food/beverage consumption vessels are dominant in four of the five hut sites, followed by food/beverage containers. Only Hut 5 has more smoking-related than storage or consumption vessels.

Table 7.20: Summary of MNV percentages per site in terms of function

Function	Hut 1	Hut 2	Hut 3	Hut 4	Hut 5	Surface
Storage	26.7	36.9	20.0	33.3	22.2	8.3
Consumption	60.0	45.1	80.0	58.3	33.3	79.2
Smoking	13.3	8.2	0.0	8.3	44.4	0.0
Other	0.0	9.8	0.0	0.0	0.0	12.5

Note: this table includes European clay pipes

All the hut sites have at least one food/beverage container or lid, as do four of the five excavated test pits and the surface collection. Table 7.21 details the numbers of small-medium and medium-large food/beverage containers. Small-medium containers include wide-mouthed jars and lids, spouted jars, ginger jars, food jars and bottles. Medium-large containers include barrel jars and lids, and globular jars. Figure 7.10 shows a range of container sherds. A small brown-glazed sherd, catalogue number JCC-756.3, from an unknown vessel has a partial pattern which may be part of a maker's mark or lid decoration.

Table 7.21: Number of food and beverage containers in terms of size

	Hut 1	Hut 2	Hut 3	Hut 4	Hut 5	Test pits	Surface	Total
Medium-large: Chinese	1	3	1	1	1		1	8
Medium-large: Non-Chinese	1	2		1				4
Small-medium: Chinese	2	36		2	1	4	1	46
Small-medium: Non-Chinese		2						2
Total	4	43	1	4	2	4	2	60

Note: two unidentifiable lids from Hut 2 not included

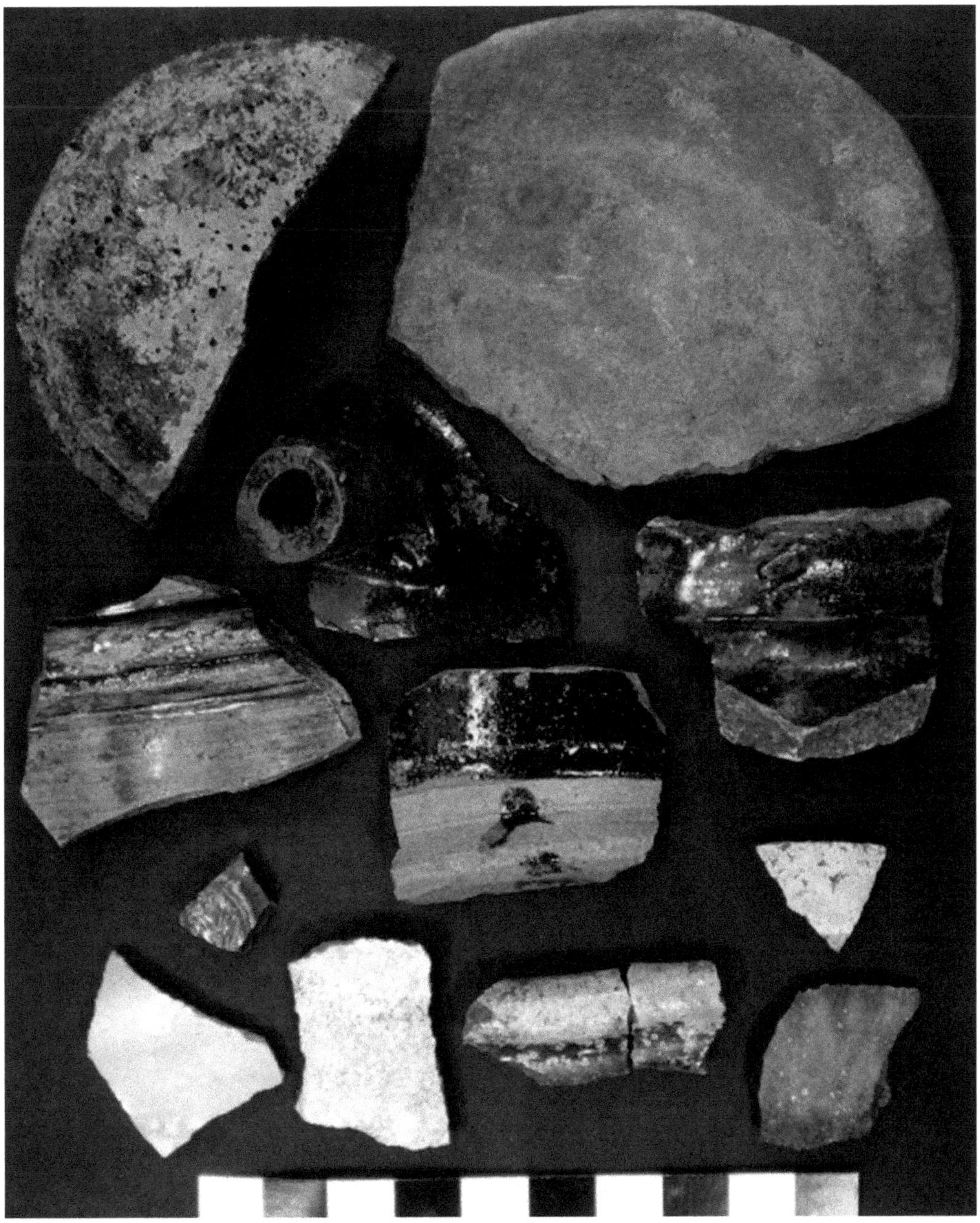

Figure 7.10: Chinese container sherds (Cat nos. from top left: 709.3 storage jar lid, 739 spout, 186 wide-mouthed jar lid; 2nd row: 128 interior bottle join seam, 792.2 foot of wide-mouthed or spouted jar, 700.18 globular jar rim; 3rd row: 756.3 brown-glazed patterned sherd, 1206.2 green-glazed ginger jar; 4th: 1260.2 and 788.4 blue-underglaze ginger jars, 748.5 and 725.1 green-glazed ginger jars)

Chinese-made jars and lids dominate the assemblage. Only Huts 1 and 2 have jars or bottles that were made in either Britain or Australia. Hut 2, with its large number of vessels, has far fewer large than small jars. All hut sites have at least one medium-large jar. Hut 3 is the only site not to have any small-medium jars.

While ginger jars and unglazed lids are more easily identified, there are numerous brown-glazed sherds without any diagnostic features. Table 7.22 shows the number and weight of brown-glazed sherds per site, the average weight per sherd and the average number of sherds per vessel.

Table 7.22: Average weight and number of sherds per vessel of brown-glazed wares

	No.	Wt (g)	MNV	Wt/No.	No./MNV
Hut 1	45	457.5	2	10.2	22.5
Hut 2	867	4564.0	26	5.3	33.3
Hut 3	2	19.1	1	9.6	2.0
Hut 4	28	321.1	2	11.5	14.0
Hut 5	3	422.6	2	140.9	1.5
Test pits	3	4.8	2	1.6	1.5
Surface	5	222.4	1	44.5	5.0

Huts 1 and 2 have a higher number of sherds per MNV. Hut 1 has seven sherds with diagnostic features, all from two identified jars. Hut 2 has only 75 diagnostic sherds from a total of 867. There may have been other jars or bottles present in the assemblage but it is not possible to determine this from the fragments.

The most common containers found are wide-mouthed jars (13 jars and 11 lids), liquor bottles (7) and spouted jars (5). No barrel jars have been positively identified from the hut sites but some of the unidentifiable medium-large jars may have been barrel jars. Hut 2 has a barrel jar lid. Other types of containers found include globular jars and ginger jars. All of these vessels are common at nineteenth century overseas Chinese sites. One of the wide-mouthed jars, having a subtle variation in the shape of its foot, is the only one of its kind to be identified at JCC.

Each hut site has more non-Chinese than Chinese-made food/beverage consumption vessels (Table 7.23). The two vessels of unknown origin are porcelain, possibly Japanese-made.

Table 7.23: Number of food/beverage consumption vessels by origin

	Hut 1	Hut 2	Hut 3	Hut 4	Hut 5	Surface	Total
Chinese	2	17	0	1	1	2	23
Non-Chinese	7	37	3	6	2	17	72
Unknown	0	1	1	0	0	0	2
Total	9	55	4	7	3	19	97

Most of the Chinese-made tea and tableware vessels found across the hut sites (Table 7.24) are common to contemporary overseas Chinese sites. The exception is a blue-underglaze liquor warmer or sauce serving bottle, which is part of the surface collection near Huts 3 and 4. The difference between a warmer and a sauce bottle is that the latter has no handle or lid (Wegars 2012:6) To date, these have not been commonly found in Australia (Muir 2008:60). Sweet Pea is now the standard term for the decoration on these vessels, although it has previously been known as Plant with Central Flower, Simple Flower or Shanghai Ware (Wegars 2012:6). An illustration in Nonya ware and Kitchen Ch'ing (The Southeast Asian Ceramic Society 1981:78) described the vessel as a sauce pot with a pattern of 'sweet pea blossoms and foliage below (the shoulder), and at the bottom a border based on parallel oblique strokes'. The most common vessels found from JCC huts are rice bowls and liquor cups.

Seventy-six food/beverage consumption vessels are of non-Chinese origin (Table 7.25). Most of these vessels are British-made; two saucers may have been Japanese-made. The most common vessels of non-Chinese origin are bowls (19) followed by plates (17). Thirteen of the 17 measurable bowls have rim diameters of 130, 140 or 150 mm, similar to those of the Chinese rice bowls. Four bowls are slightly smaller (120 mm) or slightly larger (160 mm). Huts 3 and 4 have no identifiable plates but have two saucers and one vessel each which may have been a plate or saucer. Hut 5 has only one flatware vessel, possibly a plate or saucer. Five plates and one mug designed for use by children were found across the hut sites, one vessel from Hut 1, two from Hut 2 and three from the surface collection between Huts 3 and 4. Three vessels feature the alphabet, two feature relief-moulded flowers on the marly and script in the well, while one does not have enough decoration present to identify. The decoration on two fire-damaged vessels can be identified to their source, a plate and mug from Hut 2 (Figures 7.11 and 7.12).

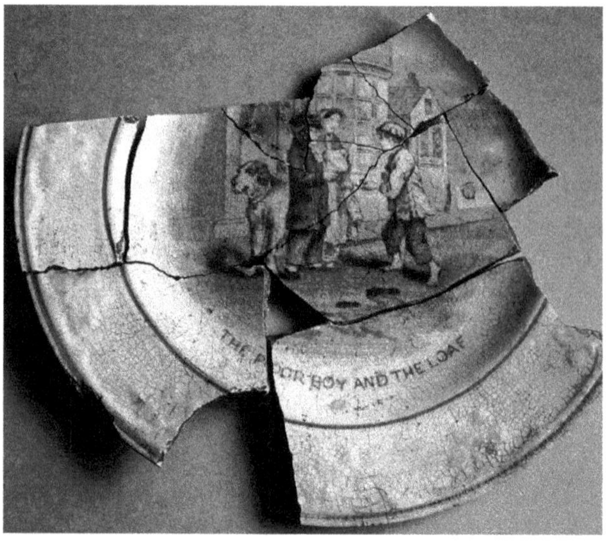

Figure 7.11: The Poor Boy and The Loaf (Cat nos. JCC-720.4, JCC-757.3; rim diameter = 170 mm)

Table 7.24: Chinese tea and tableware from JCC hut sites

Vessel	Hut 1	Hut 2	Hut 3	Hut 4	Hut 5	Test pits	Surface	Total
Shallow dish - rim = 145 mm		1						1
Bowl - rim = 130 mm		1						1
Bowl - rim = 140 mm	1	4				1		6
Tea cup		2						2
Small tea cup		1						1
Liquor cup		5		1				6
Spoon	1	2					1	4
Teapot lid		1						1
Liquor warmer or sauce serving bottle							1	1
Cup/bowl					1			1
Total	2	17	0	1	1	1	2	24

Table 7.25: Food/beverage consumption vessels of non-Chinese origin from JCC huts

Vessel	Hut 1	Hut 2	Hut 3	Hut 4	Hut 5	Test pits	Surface	Total
Pudding bowl							1	1
Platter/dish							1	1
Serving dish lid		1						1
Bowl - rim = 120 mm		2						2
Bowl - rim = 130 mm		2						2
Bowl - rim = 140 mm		5					2	7
Bowl - rim = 150 mm		2					2	4
Bowl - rim = 160 mm		2						2
Bowl - unknown size	1		1					2
Plate - small (150-190)		2						2
Plate - medium (200-240)		1				1		2
Plate - dinner (250+)	1	3					1	5
Soup plate		2						2
Child's plate	1	1					3	5
Plate - unknown size							1	1
Tea cup	1	4		1			2	8
Child's mug		1						1
Saucer		4	2	2		1		9
Eggcup	1	1						2
Plate/saucer		2	1	1	1		2	7
Cup/bowl	2	3		1	1		1	8
Hollowware				1			1	2
Total	7	38	4	6	2	2	17	76

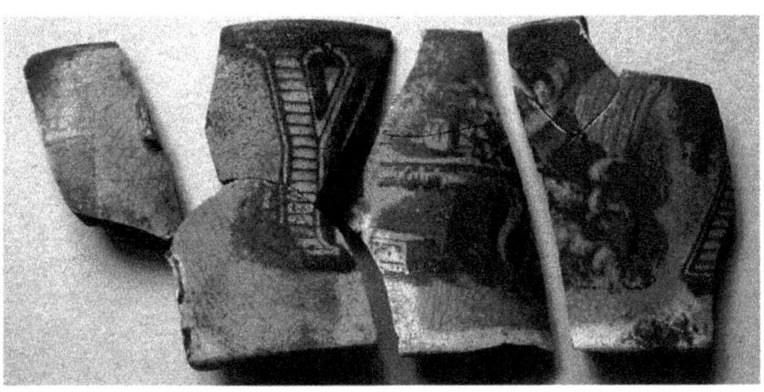

Figure 7.12: Y Z alphabet mug (Cat nos. JCC-179, JCC-180.5, JCC-195.3, JCC-762.2; height = 65 mm)

The plate, titled The Poor Boy and The Loaf, features a transfer-print copied from The Child's Companion, published in October 1854 (Shipkowitz & Shipkowitz 2002). This plate demonstrates how ceramic vessels of the Victorian era were intended to teach children 'the ethos of loving their labors' and the need for charity since 'not everyone could help themselves' (Shipkowitz & Shipkowitz 2002:105).

The child's mug from Hut 2 features the letters Y and Z and is from the series Alphabet of Virtues, based on a book published in June 1856 (Shipkowitz & Shipkowitz 2002). Y stands for yearn and Z for zeal. The following verses are from the book:

> Y is to teach us our bosom should Yearn
> With tender affection and love, in return
> To our father and mother, and teachers and friends,
> And we ought to thank God who such comforters sends!
>
> Z stands for Zeal, which should fire our breasts
> When the cause we're engaged in on equity rests;
> What our hands find to do must be done with good will,
> In old age it will be time enough to stand still!
> (Shipkowitz & Shipkowitz 2002:214).

It is questionable whether the user of the mug would have been familiar with the verse or even if they knew what the letters stood for. Although these vessels were intended for children's use they may also have been used by adults as an aid to learn the alphabet or read English. McGowan (2004a:329) cited an example of a family of European descent, on the Mongarlowe goldfields, who taught Chinese the alphabet and read them the bible.

There are at least four types of opium pipe bowls from the hut sites, decorated in either light or dark orange slip (Table 7.26).

Table 7.26: Opium pipe bowl types from JCC hut sites

Pipe bowl	Hut 1	Hut 2	Hut 5	Test pits	Total
C1: dark-orange		1	1		2
C1: light-orange	1	1			2
C3: dark-orange		1			1
C3: light-orange	1	3	1	1	6
O1: light-orange		2	1		3
Thrown		1			1

Following Ritchie's typology (1986:373-375), C1 has a circular smoking surface, flared rim and a smooth side with a single ridge; C3 has a circular smoking surface, flared rim and ten faceted sides; and O1 has an octagonal smoking surface with eight faceted sides. The octagonal pipe bowls are the same shape but one has a geometric stamped pattern at the rim, one a dotted line at the rim and the last no further decoration. The thrown bowl is circular in shape and may be similar to Ritchie's C2, but not enough is present to verify this. Ritchie described the bowl decoration as 'slipped' and 'unslipped' and this may correspond to the dark and light orange slip. The most common opium pipe bowl type found at JCC hut sites is C3, followed by C1. No opium pipe bowls were from Huts 3 or 4. A minimum number of three European clay pipes were recovered for the hut sites, one each from Huts 2, 4 and 5. Figure 7.13 shows some of the opium pipe bowl sherds from JCC.

Only two paste/ointment pots lids were able to be identified: the White Rose toothpaste lid and the Holloway's ointment. The function of the other pots and pot lids is unknown. They may have contained foodstuffs such as meat or fish paste, or may have been toothpaste or ointment. Although most of the pots and pot lids are unmarked, they are all made of white-earthenware; therefore their origin was probably Britain or Australia. No references have been found to earthenware Chinese pot lids, only porcelain ones (e.g. Muir 2008:62 - Chinese toothpowder pot lid with label).

Two salt-glazed blacking bottles were found in Hut 2. These bottles originally contained blacking used for cleaning boots, saddles or stoves but it is not known whether the Chinese would have been using the blacking for this purpose or were reusing the bottles for other purposes.

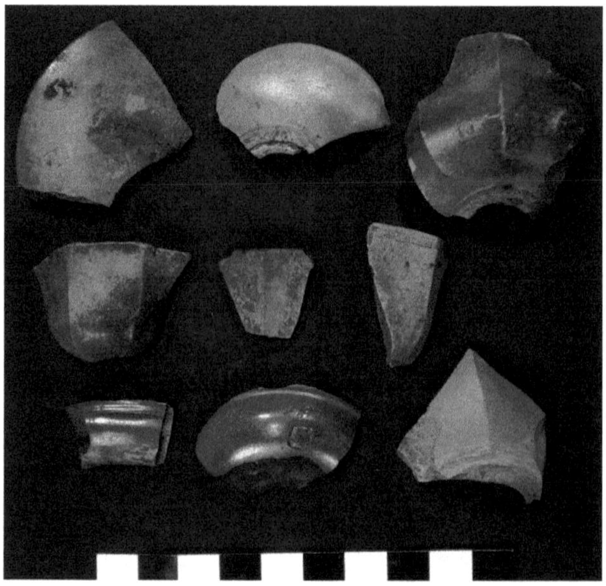

Figure 7.13: Opium pipe bowl sherds from JCC (Cat nos. from top left: 770.1, 102, 191.1, 10004, 754.2, 719, 711.1, 176, and 1253.1)

The stoneware hexagonal vessel, recovered from Hut 2, has not been identified. It may have been a planter but the hole in the base is quite large, or it may have been used for bulbs, sitting on a saucer of water (personal communication, Dr Priscilla Wegars, 16 February 2010). Other suggestions are that it may be a lantern as there are cut-away sides, or a receptacle for incense.

Decorative types

Six nameable Chinese patterns were identified: celadon-glazed Winter Green; enamelled Four Seasons; and four blue-underglaze patterns, Bamboo, Double Happiness, Rocks and Orchid, and Sweet Pea. The teapot lid has an unidentifiable blue-underglaze pattern. The most common decoration is Winter Green occurring on 15 vessels, followed by Four Seasons found on the four spoons. The other patterns are found on one vessel each. The Four Seasons wares all exhibit in-filled brush strokes over outlines. The spoon from Hut 1 has a calligraphic dark-green outline. Figure 7.14 shows a selection of Chinese sherds from JCC.

The most common type of non-Chinese decoration is a transfer-print; blue is the most common colour (20 vessels), followed by green (7), then purple (6). Hut 2 has the widest range of colours (Table 7.27). Other decorative types include: flown patterns in black, blue and green; sponge-prints in blue and purple; banding in blue, green and red; and relief-moulded patterns. Figure 7.15 shows a selection of non-Chinese sherds from JCC.

There are no tea or tableware sets present at any of the hut sites. Hut 4 is the only one to have a matching cup and plate/saucer and that is a Willow transfer-print, while Hut 2 has many vessels which could have been complementary, in terms of pattern or colour.

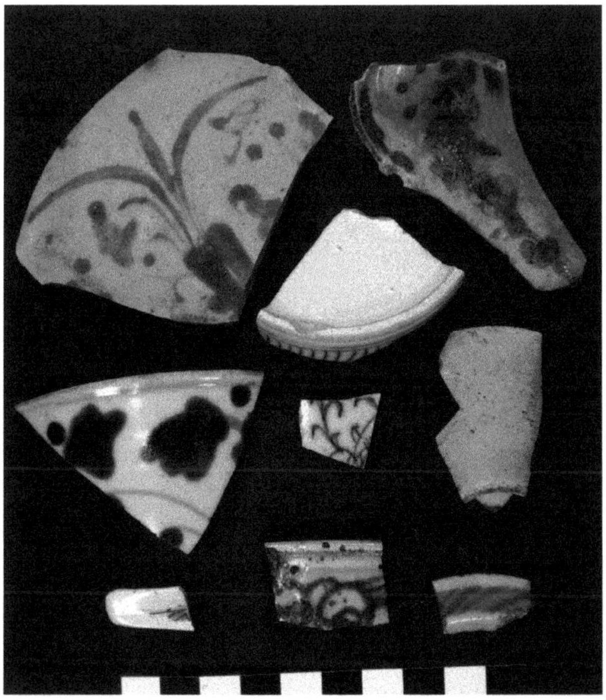

Figure 7.14: Chinese tableware sherds from JCC (Cat nos. from top left: 199 Rocks & Orchid dish, 1260.1 Sweet Pea liquor pot, 745 Four Seasons spoon, 705 Bamboo bowl, 1260.1 part of liquor warmer or sauce serving bottle, 773.1 Winter Green small tea cup, 114 Four Seasons spoon, 194.2 Double Happiness bowl, 189.3 teapot lid)

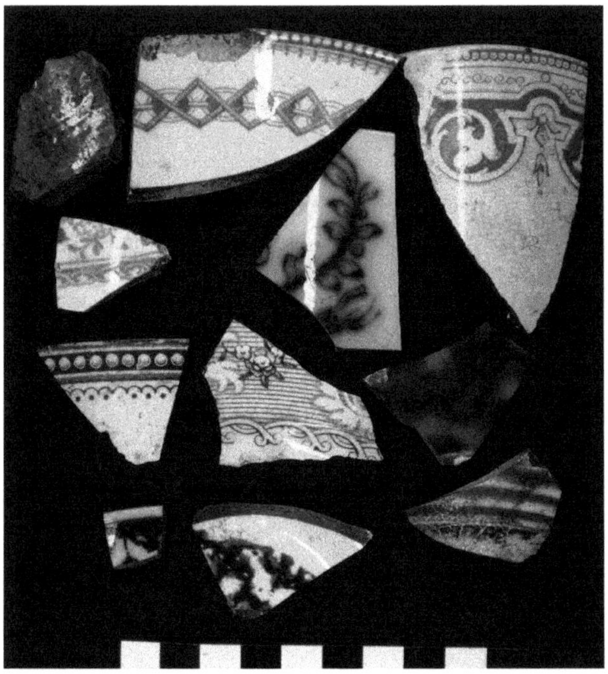

Figure 7.15: Non-Chinese sherds from JCC (Cat nos. from top left: 101 Australian-made terracotta pan/dish, 772.1 purple transfer-print - maker H.N. & Co., 762.1 blue transfer-print, 133 blue transfer-print, 180.3 flow blue, 749.5 green transfer-print, 757.12 green Rhine, 1216 flow blue, 156.4 purple sponge-print, 788.1 blue sponge-print, 757.7 blue bands)

Table 7.27: Transfer-print colours from JCC hut sites

	Black	Blue	Brown	Green	Grey	Purple	Red
Hut 1	x	x					
Hut 2	x	x		x	x	x	
Hut 3		x					
Hut 4		x					
Hut 5	x						x
Test pits		x	x				
Surface		x		x		x	

Identifying hut occupants

Huts 1 and 2

Although Hut 2 has a far greater number of vessels than Hut 1, there are some similarities in the ceramic assemblages. A re-examination of the ceramic assemblage concludes that Smith's (2006:346) statement that all the ceramic artefacts from Huts 1 and 2 are 'Chinese in origin' is incorrect. Non-Chinese made ceramics were recovered from both huts, for example, British-made transfer-printed white-earthenware.

Some of the ceramics found in these huts date to the mid-nineteenth century. Other ceramics from Hut 2 indicate that the site was still occupied into the twentieth century. These huts are the only ones to have eggcups and children's ceramics. Although Hut 1's ceramics do not suggest an end date for the site, fragments of Bakelite were found dating to the early-twentieth century (Smith 2006: Appendix C3). Hut 1 also has a small metal decorative drawer handle and Hut 2, stemmed wine glasses (Smith 2006: Appendix C3).

Both these huts appear to have had long-term occupants possibly from the mid-nineteenth century until the early-twentieth century. It is feasible that the occupants of these huts were families. Although children's vessels do not necessarily mean that there were children at the site, the length of occupancy, the documented presence of women at the camp, and the importance of family to the Chinese, makes it is reasonable to conclude that children were present.

Huts 3 and 4

Huts 3 and 4 both have predominantly non-Chinese made ceramics. In contrast to the other huts, there are no opium pipe bowls from these sites. The only Chinese-made vessel in Hut 3 is an unidentified large storage jar. This hut is the only one to have more flatware than hollowware. Two of the five vessels recovered are saucers, one bone china and the other, gilt-banded porcelain. Other artefacts include nails, tin, oyster shell, bone, charcoal, and a 'small blue glass bead' (Smith 2006:346). The storage jars in Hut 4 are Chinese-made, although a screw-thread lid may have belonged to a British or Australian-made demijohn. Most of the tea and tableware is probably British-made. A matching cup and plate/saucer in Willow pattern is present. The sole Chinese tableware vessel is a liquor cup.

The glass bead, from Hut 3, may have belonged to a female. Although there are very few vessels from Hut 3, there is more flatware than hollowware, including two saucers which would have been considered genteel. It is suggested that one of the occupants of Hut 3 was a female of European descent. European women on the goldfields were looked down upon by society in general, described by a reporter in The Sydney Morning Herald, 8 January 1869, as the 'most worthless of our own countrywomen'. While Hut 4's assemblage does not give any clues about the gender or number of occupants, it is comparable to Hut 3 rather than those of the other huts. As the camp contracted these huts would have been on the outskirts of the settlement.

Hut 5

Hut 5 has a different assemblage to the other huts, in that smoking-related vessels account for almost half of the total MNV. Opium tins and a Chinese tincture vial were also recovered from this hut. Along with three ceramic opium pipe bowls and a European clay pipe, there is a globular jar and liquor bottle, a rice bowl, a plate/saucer and a cup/bowl, suggesting single miners occupied this hut. The hut platform is ephemeral in nature, signifying a shorter occupation or earlier time period.

Summary - JCC

Table 7.28 summarises the total number and weight of sherds, and MNV for each site. It also shows the average weight of sherds collected and the average number of sherds per MNV. The final column records the MNV per square metre excavated.

The larger average weights per sherd in Hut 5 and the Oven are both accounted for by Chinese brown-glazed stoneware. In Hut 5, two sherds from a globular jar weigh 414.9 g. In the Oven, two sherds from an unidentified storage jar weigh 111.1 g. Similarly, the surface collection and Hut 4, with average weights per sherd of over 10.0 g, both have larger fragments of brown-glazed stoneware.

Clay tobacco pipes were not included in the MNV figures but only six were found, one each from Huts 2, 4, 5 and the Temple Complex, and two from the Store. The addition of the clay pipes to the MNV figures increases the MNV per excavated area to 12.2, 2.4, 4.5, 3.9 and 2.7 respectively.

Hut 2 stands out, not only for having the largest number, weight and MNV, but also for having the largest MNV per square metre excavated. The majority of sherds in Hut 2 are fire damaged and the large number of sherds per vessel supports the theory that the hut burnt down while it was still inhabited. Table 7.29 summarises the vessel types found at each site.

Re-examination of ceramic artefacts questions the assumption made by Smith (2006:91) that the artefacts point to an all-male population and that there 'was no evidence to indicate occupation by females or children'. Documents also confirm the presence of both Chinese and European women at the Jembaicumbene Chinese Camp. However, these results do confirm the occupation dates suggested by Smith (2006); the site was intensively occupied from the 1860s but few people remained by the early 1900s.

Table 7.28: Summary of number and weight of sherds, and MNV from JCC

Site	No.	Wt(g)	Wt/No.	MNV	No./MNV	Exc. Area(m^2)	MNV/Exc Area
Hut 1	96	604	6.3	15	6.4	12	1.3
Hut 2	1802	11588.6	6.4	121	14.9	10	12.1
Hut 3	7	54	7.7	5	1.4	7	0.7
Hut 4	40	432.4	10.8	11	3.6	5	2.2
Hut 5	17	439.2	25.8	8	2.1	2	4.0
Oven	6	173.8	29.0	4	1.5	17	0.2
Store	73	312.2	4.3	29	2.5	8	3.6
Temple	72	466.7	6.5	45	1.6	17	2.6
Surface	57	655.4	11.5	24	2.4		
Test pits	18	38	2.1	9	2.0	8	1.1
Total	2188	14764.3		271			

Note: clay pipes are not in catalogue numbers and weights, therefore not included in MNV numbers for this table

Table 7.29: MNV from the Jembaicumbene Chinese Camp

Jembaicumbene Chinese Camp - MNV													
Function	Origin	Vessel	Store	Temple	Oven	Hut 1	Hut 2	Hut 3	Hut 4	Hut 5	Surface	Test pits	Total
Food / beverage container													
	Chinese	barrel jar		1									1
	Chinese	barrel jar lid	1	1			1						3
	Chinese	ginger jar	2				2		1		1		6
	Chinese	ginger jar lid					2						2
	Chinese	globular jar	1				2			1	1		5
	Chinese	wide-mouthed jar	1			2	11						14
	Chinese	wide-mouthed jar lid	1				9					2	12
	Chinese	spouted jar	1				5						6
	Chinese	storage jar - small-medium		2	1				1			2	6
	Chinese	storage jar - medium-large	1	2	1	1		1	1				7
	Chinese	storage jar lid					1						1
	Chinese	liquor bottle		1			6			1			8
	Non-Chinese	beer/stout bottle					1						1
	Non-Chinese	food jar					1						1
	Non-Chinese	storage jar-large				1	2						3
	Non-Chinese	storage jar lid							1				1
	Unknown	storage jar lid					2						2
Food / beverage consumption													
	Chinese	cup/bowl	1							1			2
	Chinese	rice bowl	5	3		1	5					1	15
	Chinese	shallow dish - medium		1			1						2
	Chinese	spoon				1	2				1		4
	Chinese	tea cup					2						2
	Chinese	tea cup - small					1						1
	Chinese	teapot?		1									1
	Chinese	teapot lid					1						1
	Chinese	liquor cup	1				5		1				7
	Chinese	liquor pot									1		1
	Non-Chinese	bowl	2			1	13	1			4		21
	Non-Chinese	cup	3	2		1	4		1		2		13
	Non-Chinese	cup/bowl	2	5		2	3		1	1	2		16
	Non-Chinese	eggcup				1	1						2
	Non-Chinese	hollow ware	1	1	1				1				4
	Non-Chinese	pudding/mix. bowl		1							1		2
	Non-Chinese	mug					1						1
	Non-Chinese	pan/dish			1								1
	Non-Chinese	plate	1	4		2	7	1			5	1	21
	Non-Chinese	plate/saucer		4			2		1	1	1		9
	Non-Chinese	platter/dish									1		1
	Non-Chinese	saucer	2	7			4	1	2		1	1	18
	Non-Chinese	serving dish/tureen	1										1
	Non-Chinese	serving dish/tureen lid					1						1
	Non-Chinese	soup plate					2						2
	Non-Chinese	possible tea/table		3							2	1	6
	Unknown	saucer						1					1
Smoking													
	Chinese	opium pipe bowl	2	1		2	9			3		1	18
	Non-Chinese	clay pipe	2	1			1		1	1			6
Other													
	Chinese	miniature pot		1									1
	Non-Chinese	blacking bottle					2						2
	Non-Chinese	paste/ointment pot		1			5						6
	Non-Chinese	paste/ointment pot lid					4				1		5
	Non-Chinese	hollow		1									1
	Unknown	unknown		2			1						3
Total MNV			31	46	4	15	122	5	12	9	24	9	277

Flanagan's Point Chinese Camp

Flanagan's Point is located about 5 km north of the Chinese camp and cemetery at Mongarlowe, on the western bank of the Mongarlowe River (Smith 2006:113). It is situated in a relatively flat area, measuring 100 x 140 m, comprising open forest and short grasses and the remains of alluvial mining.

The first recorded evidence of the presence of Chinese at Flanagan's Point was in 1862 when some bought a claim from Owen Flanagan, a European miner (Smith 2006). By the end of that year, 17 Chinese miners lived there. The last historical record of Chinese at Flanagan's Point is from 1895 when a resident, Quong Chong, died from bronchitis (Smith 2006:116). There is a scarcity of documentary evidence for the intervening years.

Smith's survey identified 'at least 13 hut sites, a communal oven and several other cultural features' (Smith 2006:113). He suggested that a temporal sequence could be seen in the huts, some of which had earth platforms and stone structures, while others had ephemeral platforms which could have been used for temporary tent structures. Smith made a surface collection across the whole site and excavated five areas including an oven and four huts (Figure 7.16).

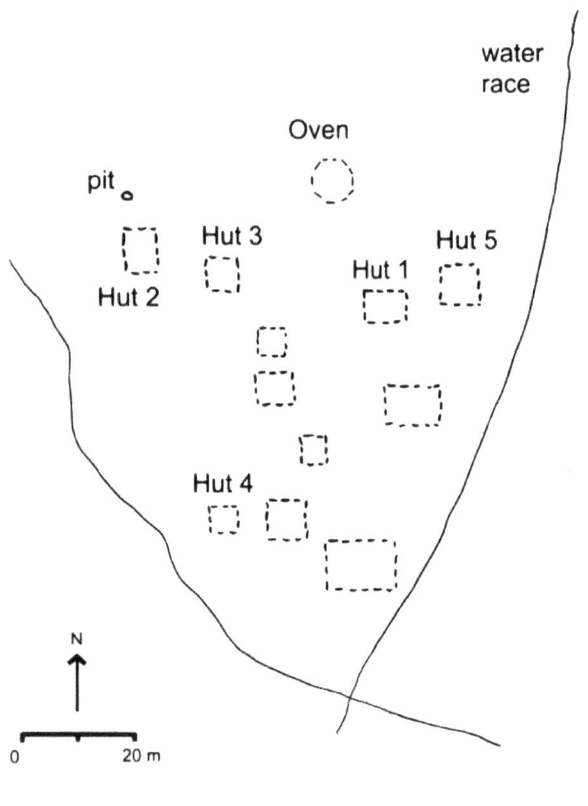

Figure 7.16: Map of Flanagan's Point Chinese Camp (after Smith 2006:434)

One hundred and two sherds, weighing 1060.5 g, are present in the collection housed at ANU. The details of ceramics sherds found at each site are presented below. Although the results are reported in terms of MNV, there are no whole vessels.

Communal Oven

A random rubble structure of about a metre in diameter indicated the presence of a Chinese oven. A grid, measuring 8 x 6 m, was pegged out in one metre squares by Smith who excavated a total of 8 m² (Smith 2006). Ceramics were found in five of the excavated squares. In total, 24 sherds, weighing 86.4 g, represent a MNV of seven (Table 7.30).

The container is a medium-large Chinese brown-glazed storage jar.

The tea and tableware vessels are all transfer-printed white-earthenware: an oval platter and two plates in blue Willow; a plate with a flown-blue floral pattern; a cup with a flown-blue floral pattern; and a cup with a black or very-dark blue grape pattern.

Interpretation of Oven site

Origin of ceramics

All excavated vessels are British-made. The Chinese-made storage jar sherd was recovered from the surface.

Dating ceramic sherds

No makers' marks were recovered. The heavily-flown blue patterned cup suggests a date before 1880 (Van Buskirk 2002). The only transfer-print to be identified is Willow, which was manufactured by numerous potteries throughout the nineteenth and twentieth centuries.

Function of ceramic vessels

All ceramic vessels from the Oven relate to food; the majority are for serving or consuming food.

Hut sites

Huts 1 to 4 were excavated and a surface collection was made around Hut 5. No ceramics were recovered from Hut 4. Table 7.31 details the grid size set out by Smith, the internal dimensions of each hut, excavated area and a brief description.

Hut 1

Hut 1 was located southeast of the Oven. Ceramic sherds were recovered from only two squares. Three sherds in total, weighing 2.5 g, were excavated and these represent a MNV of two.

Table 7.30: Ceramic sherds from FPCC Oven

	No.	% No.	Wt(g)	% Wt	Wt/No.	MNV	No./MNV	% MNV
Food/beverage containers								
Chinese	1	4.2	31.1	36.0	31.1	1	1.0	14.3
Non-Chinese	0		0.0			0		0.0
Food/beverage consumption								
Chinese	0		0.0			0		0.0
Non-Chinese	23	95.8	55.3	64.0	2.4	6	3.8	85.7
Total	24		86.4			7		

Table 7.31: FPCC hut dimensions, area excavated and description (after Smith 2006:116, 436)

Hut No.	Grid size	Int dimen m²	Exc m²	Description
1	6 x 8	19.3	15.5	Free-standing structure with random rubble base, tamped earth floor and stone fireplace
2	6 x 8	17.5	2.5	Free-standing structure with random rubble base, tamped earth floor and stone fireplace
3	6 x 6	12.3	5.0	Free-standing structure with random rubble base, tamped earth floor and internal hearth
5	9 x 8	Unknown	0.0	Free-standing structure with tamped earth floor

One sherd belongs to a small Chinese brown-glazed storage jar. The other two sherds have flown-blue decoration and may have belonged to the same cup or bowl. There is not enough of this decoration to identify or describe it.

Hut 2

Hut 2 was located southwest of the Oven. Ceramics were found in all the excavated squares. There are 16 sherds, weighing 73.7 g, representing a MNV of four. Two European clay pipes bring the total MNV to six (Table 7.32).

Nine sherds are from the same vessel, a bowl, rim diameter 140 mm, with a fluted body and relief-moulding around the rim, possibly a wheat pattern. The remaining sherds represent three vessels: a bowl, having a possible thick blue band on the exterior only; a blue transfer-printed plate or saucer, with a floral or foliate pattern; and a green transfer-printed cup with a vegetal pattern. There are at least two European clay pipes present, one having a maker's mark 'SAYWELL / SYDNEY'. Saywell was a Sydney tobacconist from 1863.

Hut 3

Hut 3 was located southwest of the Oven, to the east of Hut 2. Only one sherd weighing 2.5g was excavated from this hut. It has a sponge-printed decoration combining a green leaf-motif with red dots. The vessel may have been a cup or bowl.

Hut 5

Hut 5 was located to the southeast of the Oven and east of Hut 1. There are 58 sherds weighing 895.4g representing a MNV of six (Table 7.33).

The storage vessels are all Chinese-made. These include a barrel jar lid, diameter 340 mm; a globular jar, foot 120 mm; an unidentified small-medium storage jar; and a straight-sided jar lid with a dark brown-glazed exterior, unglazed interior and a diameter of about 240 mm.

Table 7.32: Ceramic sherds from FPCC Hut 2

	No.	% No.	Wt(g)	% Wt	Wt/No.	MNV	No./MNV	% MNV
Food/beverage containers								
Chinese	0		0.0			0		0.0
Non-Chinese	0		0.0			0		0.0
Food/beverage consumption								
Chinese	0		0.0			0		0.0
Non-Chinese	16	100.0	73.7	100.0	4.6	4	4.0	66.7
Smoking								
Chinese	0		0.0			0		0.0
Non-Chinese						2		33.3
Total	16		73.7			6		

Note: clay pipes are not in catalogue sherd numbers or weights

Table 7.33: Ceramic sherds from a surface collection around FPCC Hut 5

	No.	% No.	Wt(g)	% Wt	Wt./No.	MNV	No./MNV	% MNV
Food/beverage containers								
Chinese	55	94.8	884.2	98.7	16.1	4	13.8	66.7
Non-Chinese	0		0.0			0		0.0
Food/beverage consumption								
Chinese	0		0.0			0		0.0
Non-Chinese	3	5.2	11.2	1.3	3.7	2	1.5	33.3
Total	58		895.4			6		

Two vessels of tea and tableware were found: a purple transfer-printed cup, rim 120 mm, with Alhambra pattern; and a blue transfer-printed bowl with an exterior foliate pattern and an interior scrolled pattern.

Interpretation of hut sites

Origin of ceramics

Chinese-made ceramics dominate by number and particularly by weight of sherds. However, when considering the MNV, those of non-Chinese origin are in the majority. The MNV of the European clay tobacco pipes is shown separately (2) as sherd numbers and weights were not in the re-catalogued ceramics (Table 7.34).

Dating ceramic sherds

The non-Chinese ceramics in this assemblage are all typical of the second half of the nineteenth century. The only nameable pattern is Alhambra, found in purple. Alhambra has been found at a number of other sites in NSW and in New Zealand. It is a Moorish design with ornate foliated scroll motifs, probably referring to the palace of the same name built in the thirteenth and fourteenth centuries in Granada, Spain (Coysh & Henrywood 1989). Other Alhambra patterned vessels have been found at Kiandra, from a hotel site which was later a residence, occupied 1860s-1931, and from the Hibernian Hotel near Bungendore, NSW, occupied from the mid to late-nineteenth century. All of these were purple transfer-prints, as were those found to date in New Zealand (NZHCD 2012). Two green-printed Alhambra vessels were recovered from a Sydney site, 19-41 Reservoir Street, dating from 1860 onwards (Casey & Lowe 2009).

The flow blue vessel found at Hut 1 suggests a date before 1880. Heavily-flown vessels were fashionable in the early to mid-Victorian periods, 1835-1870s (Van Buskirk 2002:12). By the 1880s, the extent to which the patterns flowed had lessened considerably.

The sponge-printed cup/bowl from Hut 3 has a distinct shape to the pattern, perhaps using a harder sponge, suggesting a date after 1870 (Kelly et al. 2001:9).

The relief-moulded wheat pattern, found in many versions, was made by many manufacturers over a long period (Sussman 1985). The pattern features a 'raised design combining heads of grain and grass-like leaves' and occurs on 'ironstone' or 'white granite' (Sussman 1985:7). It was first made in the 1840s and was still produced into the twentieth century.

Function of ceramic vessels

All the ceramic food/beverage containers are Chinese-made while all the food/beverage consumption vessels are of non-Chinese origin, probably from Britain. There are no children's ceramics present.

No storage jars or lids were found in Huts 2 or 3. A single small-medium storage jar was recovered from Hut 1. In contrast, there were four vessels found in Hut 5. At least three of them are medium-large storage jars.

There are no Chinese-made tea or tableware vessels. The eight non-Chinese made vessels from the hut sites are mainly hollowware; bowls and cups being the most common forms (Table 7.35). No plates could be positively identified, but one vessel from Hut 2 may have been a plate or saucer.

Table 7.34: Origin of ceramic vessels from FPCC hut sites

	No.	% No.	Wt(g)	% Wt	Wt./No.	MNV	No./MNV	% MNV
Chinese	56	71.8	886.4	91.0	15.8	5	11.2	33.3
Non-Chinese	22	28.2	87.7	9.0	4.0	8	2.8	53.3
Non-Chinese-clay pipes						2		13.3
Total	78		974.1			15		

Table 7.35: Food/beverage consumption vessels per hut site at FPCC

Vessel	Hut 1	Hut 2	Hut 3	Hut 5	Total
Bowl - rim = 140 mm		1			1
Bowl - unknown size		1		1	2
Cup		1		1	2
Plate/saucer		1			1
Cup/bowl	1		1		2
Total	1	4	1	2	8

Decorative types

Decorative types from the hut sites include: transfer-prints in blue (2 vessels), green (1) and purple (1); and one vessel each of blue-banding, flown-blue, sponge-printing and relief-moulding. Figure 7.17 shows a selection of sherds from FPCC.

Summary - FPCC

Table 7.36 presents a summary of the number, weight and MNV recovered from FPCC. Over half of the sherds are from a surface collection around Hut 5; the larger than average weight of sherds accounted for by the number of heavier Chinese brown-glazed stoneware sherds. The number of sherds per MNV is also higher in Hut 5. There are 43 non-diagnostic brown-glazed fragments in the surface collection which cannot be positively assigned to a type of storage jar. In contrast, the white-earthenware fragments are easier to assign to vessels because of the different decorative types. A summary of vessel types found at each site is shown in Table 7.37.

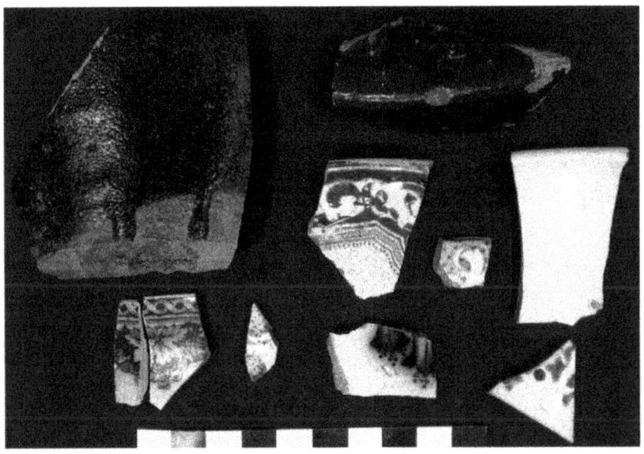

Figure 7.17: Selection of sherds from FPCC (top: Cat nos. 122.3 Chinese brown-glazed globular jar, 122.7 straight-sided jar lid; non-Chinese earthenware, middle row: Cat nos. 125 Alhambra, 121 blue transfer-print, 112.3 relief-moulded wheat pattern; bottom row: Cat nos. 105 black transfer-print, 113 blue transfer-print, 102 flow blue, 123 sponge-print)

Table 7.36: Summary of number and weight of sherds, and MNV from FPCC

Site	No.	Wt(g)	Wt/No.	MNV	No./MNV	Exc. Area (m^2)	MNV/Exc Area
Hut 1	3	2.5	0.8	2	1.5	15.5	0.1
Hut 2	16	73.7	4.6	4	4.0	2.5	1.6
Hut 3	1	2.5	2.5	1	1.0	5	0.2
Hut 5	58	895.4	15.4	6	9.7		
Oven	24	86.4	3.6	7	3.4	8	0.9
Total	102	1061		20			

Note: clay pipes are not in catalogue sherd numbers or weights, therefore not included in MNV numbers for this table

Table 7.37: MNV from Flanagan's Point Chinese Camp

Flanagan's Point Chinese Camp - MNV									
Function	Origin	Vessel	Oven	Hut 1	Hut 2	Hut 3	Hut 4	Hut 5	Total
Food / beverage container									
	Chinese	barrel jar lid						1	1
	Chinese	globular jar						1	1
	Chinese	straight-sided jar lid						1	1
	Chinese	storage jar - small-medium		1				1	2
	Chinese	storage jar - medium-large	1						1
Food / beverage consumption									
	Non-Chinese	bowl			2			1	3
	Non-Chinese	cup	2		1			1	4
	Non-Chinese	cup/bowl		1		1			2
	Non-Chinese	plate	3						3
	Non-Chinese	platter	1						1
	Non-Chinese	plate/saucer			1				1
Smoking									
	Chinese	opium pipe bowl							0
	Non-Chinese	clay pipe			2				2
Total MNV			7	2	6	1	0	6	22

Origin of ceramics

The ceramic vessels from the Oven and Huts 2 and 3 are predominantly of British origin. Only Hut 5 has a greater number of Chinese-made vessels. All the storage vessels are Chinese-made and the food consumption vessels are of non-Chinese origin.

Dates of occupation

Smith (2006) stated that the glass and metal artefacts suggested an occupation period from the 1860s to the late 1880s or early 1890s. The ceramic artefacts all fit within this timeframe.

Function of ceramic vessels

The majority of the vessels from FPCC are related to food consumption (Table 7.38). Most of the food storage vessels are from Hut 5, the only site to have large food containers. A platter and three plates were found at the Oven site while no plates could be positively identified from the hut sites. This may be indicative of communal eating; a number of vessels to serve the cooked meats at the Oven and the bowls and cups used for consumption found in the hut sites. While there are at least two European clay pipes, there are no Chinese opium pipe bowls. An opium tin was recovered from Hut 5 (Smith 2006: Appendix C3).

Table 7.38: Function of vessels from FPCC

	Hut 1	Hut 2	Hut 3	Hut 5	Oven	Total
Food containers	1	0	0	4	1	6
Food consumption	1	4	1	2	6	14
Smoking	0	2	0	0	0	2
Total	2	6	1	6	7	22

Note: this table includes European clay pipes

Overall, very few ceramic vessels were recovered from FPCC. Fossickers' holes were noted by Smith (2006), therefore it is possible that some vessels have been removed from the site. All the ceramic artefacts were recovered from hut sites within 40 m of the Oven. The other excavated hut was located about 60 m from the Oven and no ceramics were found there, perhaps indicating the shorter use of that site. The Oven appears to have been used for communal eating. Re-examination of the ceramics has agreed with Smith's (2006) statement that the camp was occupied by single male miners.

Discussion - Braidwood Region

Analysis of the ceramics from two important camps in the Braidwood region indicates a series of possible conclusions.

A re-examination of the ceramic artefacts from Jembaicumbene Temple establishes the dominance of tea and tableware and the presence of possible ornamental vessels, indicating that this was a place of socialisation and perhaps education as vessels with script and slates were present. Tea and tableware are also present in the excavated, but unanalysed collection of ceramics from the temple at Mongarlowe Chinese Camp (Smith 2006:413, Figure 3.3.23).

A comparison of artefacts from JCC and FPCC ovens suggest that they were used differently. There are few ceramic artefacts from JCC Oven; while in contrast, FPCC Oven has the highest MNV of any site at that camp. The latter appears to have been occupied for a shorter period by single male miners who may have used the Oven for communal eating. The camp demographic at Jembaicumbene changed over time from single males to families. The low number of artefacts at JCC Oven suggests that it was only used for festivities in the latter period of occupation.

Overall, JCC has more ceramic artefacts than FPCC, however, a mix of Chinese and non-Chinese, mainly British, vessels were found at both settlements. A Chinese brown-glazed globular jar, a barrel jar lid, straight-sided jar lid and three unidentifiable small-medium jars were from FPCC. In comparison, wide-mouthed jars, spouted jars, ginger jars and liquor bottles were from JCC in addition to the types found at FPCC. While JCC has some food containers manufactured in either Britain or Australia, FPCC does not. No Chinese-made tea or tableware vessels were recovered from FPCC, although fossickers may have removed all traces of these sherds. A number of non-Chinese made forms were at JCC but not FPCC: eggcups, pudding/mixing bowls, pan/dish, serving dish tureen lid, soup plates and children's ceramics. The most common colours at both camps were blue and purple. No opium pipe bowls were at FPCC but at least four different types were at JCC. European clay pipes were found at both settlements.

In summary, differences are seen between the main and peripheral camp, particularly in the use of their oven sites. The low number of artefacts retrieved from the huts at FPCC, which are predominantly cups or bowls, suggests that the camp was occupied for a relatively short time, by single male miners. In contrast, artefacts from JCC suggest that while some sites may have been more ephemeral, others were occupied from the mid-nineteenth century into the early-twentieth century. Smith (2006) stated that the Chinese camps contracted over time as the population declined and that the remaining residents lived in close proximity to the store, temple and oven. A re-examination of the ceramic artefacts supports this statement but adds that the sites which remained until the end of the century were occupied by families rather than single miners.

CHAPTER 8

Tumut region

Ceramic artefacts were re-analysed from two Chinese settlements in the Tumut region, Upper Adelong and Adjungbilly Chinese Camps (Figure 8.1). Although the results are discussed in terms of MNV, there are no whole vessels.

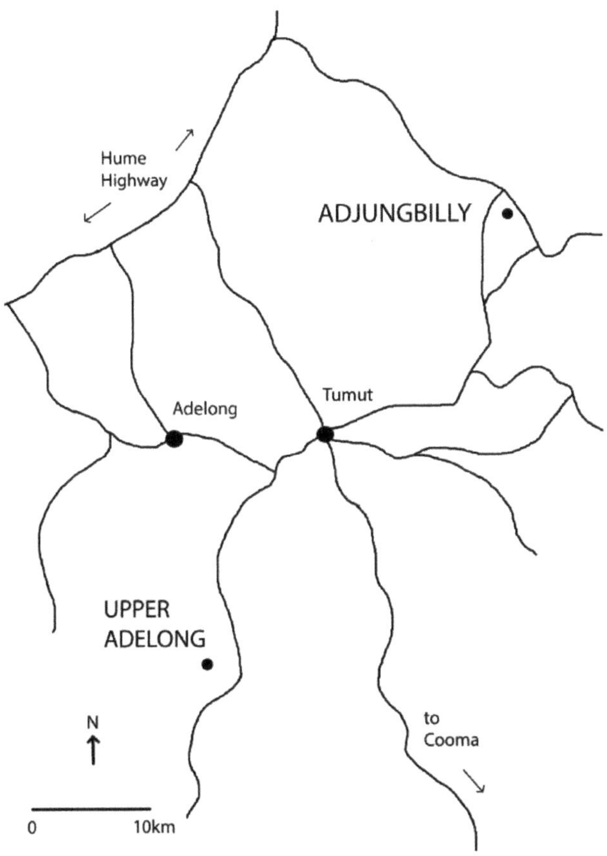

Figure 8.1: Map of Tumut region (after Smith 2006:513)

Upper Adelong Chinese Camp

The Chinese settlement of Upper Adelong (UACC) was located 20 km southeast of Adelong Township in a wooded area which slopes down to Hindmarsh Creek, a tributary of Adelong Creek (Smith 2006:142). The first major gold was found at Adelong in 1852 and by 1858 there were 250 Chinese on the goldfield at Upper Adelong (Smith 2006:138). A newspaper report from 1861 describes a number of Chinese stores, restaurants and gambling houses at Upper Adelong and notes that 'we understand that the place on Saturdays and Sundays presents a very lively picture of Chinese interior life, without the trouble of visiting China' (Maitland Mercury & Hunter River General Advertiser, 27 April 1861:6). Historical records for the cemetery, located about 100 m to the northeast of the Chinese settlement, confirm Chinese were buried there during the period 1860-1901 and Smith (2006:148) proposed that the settlement may have been contemporary to the earlier burials, from around the 1860s-1870s. His survey confirmed the presence of a store and at least four huts, as well as races, pits, which were possibly wells, and an old roadway. Smith (2006:144) 'observed' 134 ceramic artefacts on the surface at the site, along with '33 glass artefacts, 3 pieces of metal, and 2 miscellaneous artefacts'. Most of these could not be associated with a particular structure as they were found in a deflated area to the north of the Huts 2 to 4 (Figure 8.2).

The artefact assemblage from Upper Adelong, housed at ANU, comprises 119 ceramic sherds, weighing 1863.7 g, which represent a minimum number of vessels (MNV) of 42. Nine ceramic sherds are featured in Smith's Figure 4.2.26 (2006:534), eight of which are missing from the collection catalogued for this research. The photographed sherds include fragments of a Bamboo decorated rice bowl, the base of a 'blue/grey-ware bowl' with maker's mark, a celadon-glazed bowl and spoon, and a globular jar. There is a small fragment of a Bamboo decorated rice bowl present in the ANU collection and part of a celadon-glazed spoon which may belong with Smith's photographed ceramics. There are no other celadon-glazed sherds. The globular jar sherd in the photograph is also catalogued. As a result, there are 15 unknown sherds which may have been observed but not collected by Smith. None of the collected sherds are fire damaged.

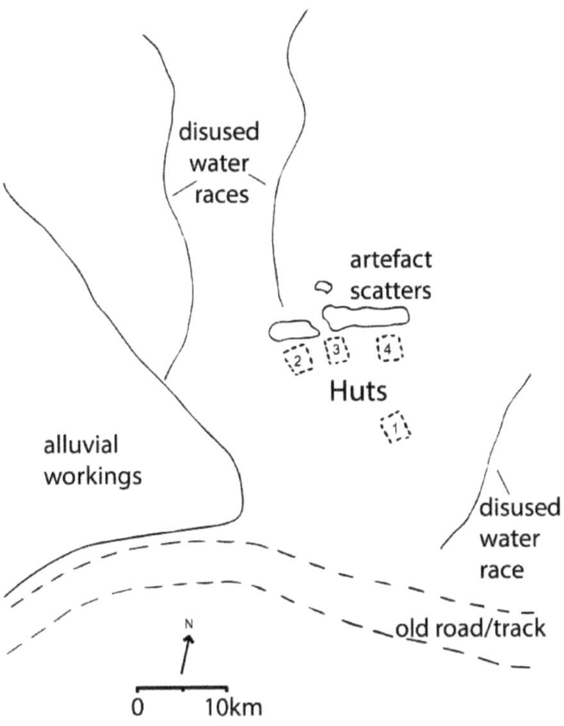

Figure 8.2: Map of UACC (after Smith 2006:530)

Hut 1

Hut 1 was a tented structure with a tamped earth floor and an internal area of 6.4 m² (Smith 2006:146).

Two sherds, weighing 4.7 g, represent a MNV of two. Both sherds are of Chinese origin. One is from a small-medium brown-glazed stoneware storage jar and the other from a rice bowl with Bamboo decoration.

Hut 2

Hut 2 was a tented structure with a tamped earth floor and an internal area of 7.2 m² (Smith 2006:146).

Twenty four sherds, weighing 418.4, represent a MNV of two. Twenty three of the sherds are from a Chinese-made brown-glazed globular jar, foot 120 mm. The other sherd is from a British-made cup or bowl with a black transfer-printed pattern featuring a floral interior and a scenic exterior.

Road

The road was located about 20 m south of Huts 1 and 2.

Two sherds, weighing 73.3 g, were collected from the road alongside the huts. These are from a small-medium Chinese-made brown-glazed storage jar and a non-Chinese made salt-glazed blacking or ink bottle, body diameter 70 mm.

Artefact scatter

The majority of artefacts from UACC were found scattered to the north of the Huts 2, 3 and 4.

There were ninety-one sherds, weighing 1367.3 g, and representing a MNV of 36 (Table 8.1).

The food containers are all Chinese-made brown-glazed storage jars. Three are globular jars, one with a rim diameter of 90 mm; the other two have foot diameters of 120 mm and 130 mm. The rim does not appear to belong to either of these jars. The other jar is a barrel jar with a rim diameter of about 350 mm.

One Chinese-made tableware vessel is present, a porcelain celadon-glazed spoon. The non-Chinese made vessels include: six bowls; two cups; six vessels which are either cups or bowls; two unidentified hollowware vessels; seven plates; four saucers; and three unidentified flatware vessels. All vessels, apart from a grey dyed-body saucer, are made of white-earthenware.

The six bowls include: a rimmed bowl, diameter 180 mm, with a blue floral transfer-print; two bowls with purple transfer-prints, one with rim 140 mm and geometric pattern, the other a floral pattern; a bowl of unknown size with a red transfer-printed pattern featuring foliated scrolls and bands; a bowl, rim 160 mm, with an 18 mm blue band at the rim with thinner bands below; and a blue-banded bowl, rim 150 mm, with at least three 4.5 mm bands at the rim.

Only two cups were identified. One has a clear glaze, and the other has flown-blue short vertical lines on the interior but no decoration is present on the exterior of the sherd or on the handle. Six other vessels may have been cups or bowls and these are all decorated differently. Three nameable patterns are present, blue Kulat, blue Rhine and purple Crystal. The other patterns include a black transfer-printed foliate pattern, a purple transfer-printed floral or foliate pattern and a flown-purple bead and foliate pattern.

Six of the seven plates are dinner size and one is small. The decoration on the dinner plates includes: transfer-prints in blue Albion, purple Gem, green Rhine, and blue, possibly Rhine; a flown-blue floral pattern; and a blue-banded pattern with two thin bands at the rim and shoulder. The small plate has a blue Willow pattern.

Two of the four saucers have transfer-printed patterns: a purple Crystal pattern with a foot and cup well diameter of 55 mm; and a smudgy black pattern, possibly scenic. Another saucer has a blue sponge-printed repeated leaf motif. The remaining saucer features a grey dyed-body with scrolled relief-moulding that extends from the rim to the cup well. The rim diameter is 160 mm, the foot 65 mm and the cup well 40 mm; the clear glaze is worn and partially crazed.

Table 8.1: Number, weight and MNV of ceramic sherds from UACC surface scatter

	No.	% No.	Wt (g)	% Wt	Wt/No.	MNV	No./MNV	% MNV
Food/beverage storage containers								
Chinese	38	41.8	1088.3	79.6	28.6	4	9.5	11.1
Non-Chinese	0		0.0			0		0.0
Food/beverage consumption								
Chinese	1	1.1	2.9	0.2	2.9	1	1.0	2.8
Non-Chinese	51	56.0	269.1	19.7	5.3	30	1.7	83.3
Other								
Chinese	0		0.0			0		0.0
Non-Chinese	1	1.1	7.0	0.5	7.0	1	1.0	2.8
Total	91		1367.3			36		

One of the unidentified flatware vessels, with a blue floral transfer-print, may have been a large plate or platter, the rim being 6.5 mm thick. Another vessel, which features embossed dots all over the marly, is possibly a plate. The final flatware vessel features a purple transfer-print and may have been a saucer.

One of the hollowware vessels may have been a bowl or jug, with a 5.5 mm blue band at the shoulder. The other vessel is an oval dish or bowl with a green Rhine transfer-print.

A single sherd was found from a salt-glazed blacking or ink bottle with a body diameter of 60 mm.

Interpretation of UACC

The majority of sherds recovered from UACC came from the artefact scatter north of Huts 2 to 4. Table 8.2 presents a summary of the MNV from UACC and Figure 8.3, a selection of sherds.

Origin of ceramic sherds

The sherds of Chinese origin are in the majority by number and weight; however non-Chinese made vessels dominate the MNV (Table 8.3).

The average weight of the Chinese-made sherds is over four times that of the non-Chinese sherds as most are larger stoneware fragments in comparison to the lighter earthenware. The number of sherds per vessel is also higher; many of the brown-glazed stoneware sherds have no diagnostic features whereas the earthenware can be separated by decoration.

All the food/beverage containers found at UACC are of Chinese origin while the tea and tableware is predominantly British. The transfer-printed vessels are probably all from Britain. The sponge-printed earthenware may have been manufactured in Britain or elsewhere in Europe and the salt-glazed stoneware, in Britain or Australia.

Dating ceramic sherds

No makers' marks were found.

Profiles of recovered plates, Willow (double foot ring) and possible Rhine plate (flat foot), correspond to Moir's (1997) 1825 and 1845 diagrams respectively. The shoulders of the Albion and Gem plates are angular, similar to the pre-1850 plates in Moir's diagram.

Six nameable transfer-printed patterns were recovered: Albion, Crystal, Gem, Kulat, Rhine and Willow. Most of these patterns were made from the 1830s and into the twentieth century (Coysh & Henrywood 1982, 1989; Kowalsky & Kowalsky 1999). No information has been found in reference books regarding Kulat, a pattern which features a border with foliated scrolls around cartouches, containing two fishermen on the banks of a river and a turreted building in the background. The central design features a figure in Oriental dress, standing in the archway of a columned building, with further buildings in the background. A Kulat patterned plate was found at the Kiandra Gold Commissioner's Quarters with the maker's mark of Pinder, Bourne and Hope, 1851-1862. A number of vessels with this design, found on Sydney sites, have been attributed to Pinder, Bourne & Co., dating 1862-1882 (Casey & Lowe 2009).

Table 8.2: MNV from Upper Adelong Chinese Camp

Upper Adelong Chinese Camp - MNV							
Function	Origin	Vessel	Hut 1	Hut 2	Road	Scatter	Total
Food/beverage container							
	Chinese	barrel jar				1	1
	Chinese	globular jar		1		3	4
	Chinese	unid.st.jar-small	1		1		2
Food/beverage consumption							
	Chinese	rice bowl	1				1
	Chinese	spoon				1	1
	Non-Chinese	bowl				6	6
	Non-Chinese	cup				2	2
	Non-Chinese	cup/bowl		1		6	7
	Non-Chinese	flatware				3	3
	Non-Chinese	hollowware				2	2
	Non-Chinese	plate				7	7
	Non-Chinese	saucer				4	4
Smoking							
	Chinese	opium pipe bowl					0
	Non-Chinese	clay pipe					0
Other							
	Non-Chinese	blacking/ink bottle			1	1	2
Total MNV			2	2	2	36	42

Table 8.3: Number, weight and MNV of ceramic sherds from UACC by origin

Origin	No.	% No.	Wt (g)	% Wt	Wt/No.	MNV	No./MNV	% MNV
Chinese	65	54.6	1518.3	81.5	23.4	9	7.2	21.4
Non-Chinese	54	45.4	345.4	18.5	6.4	33	1.6	78.6
Total	119		1863.7			42		

The decoration on the blue sponge-printed saucer from the artefact scatter appears to be hard-edged, indicating a date after 1870 (Kelly et al. 2001:9).

Smith (2006:148) stated that the glass and nails from the settlement proposed an occupation date from the 1860s to the late 1870s. The ceramic artefacts would fit with this date; however it would also suggest that the blue-banded vessels, and to a lesser extent the sponge-printed saucer, would have been fashionable vessels at that time. In summary, the ceramic artefacts confirm that the settlement was contemporary with the nearby cemetery, which contained Chinese graves dating 1860-1901.

Function of the ceramic vessels

The majority of vessels recovered from UACC are related to food preparation and consumption but overall, the highest weight is in the food/beverage container sherds (Table 8.4).

Two globular jars, both with chipped rims, and a barrel jar are the only identifiable types of jar, however there are at least two unidentifiable smaller jars. Two Chinese-made vessels, a Bamboo decorated rice bowl and a celadon-glazed spoon, were recovered from the site. Although a number of non-Chinese made bowls, cups, saucers and plates can be identified, many sherds are small and can only be described as cup/bowl, hollowware or flatware. Two vessels are either blacking or ink bottles. There were no opium pipe bowls.

Form and decoration

Over half of the food consumption vessels are hollowware (Table 8.5). Some bowl sizes, 140-150 mm, are similar to those of Chinese rice bowls. Two bowls, rim diameters 160 mm and 180 mm, are a similar size to Chinese serving bowls.

Chinese decorative types include Winter Green and Bamboo. The most common non-Chinese tea and tableware decorative types are transfer-prints in blue (7 vessels), purple (7), black (3 vessels), green (2), and red (1). Other types of decoration are: banding in blue (4); flow blue (2) and flow purple (1); relief-moulding (2); and sponge-printing in blue (1).

Table 8.4: Function of vessels recovered from UACC

Function	No.	% No.	Wt (g)	% Wt	Wt/No.	MNV	No./MNV	% MNV
Storage	63	52.9	1512.4	81.2	24.0	7	9.0	16.7
Consumption	54	45.4	276.7	14.8	5.1	33	1.6	78.6
Other	2	1.7	74.6	4.0	37.3	2	1.0	4.8
Total	119		1863.7			42		

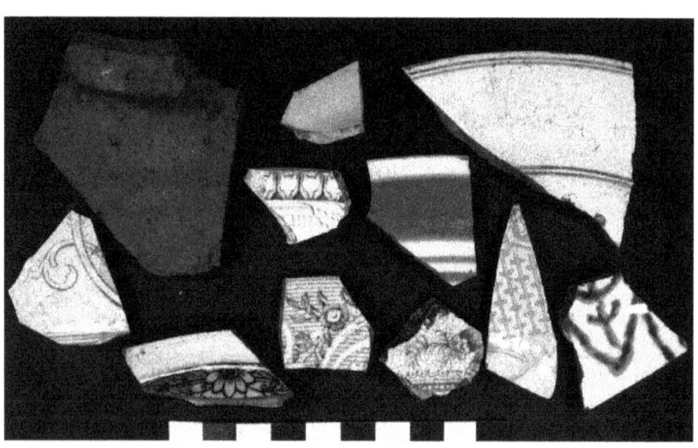

Figure 8.3: Sherds from UACC (top – Cat nos. 128 globular jar, 110 celadon spoon, 140 banded; middle – Cat nos. 117 red transfer, 116 blue transfer, 127 banded, bottom row – Cat nos. 157 flow blue, 143 Rhine, 156 Gem, 146 Crystal, 144 sponge-print)

Table 8.5: Tea and tableware from UACC

Form	Hut 1	Hut 2	Scatter	Total
Flatware			12	12
Hollowware	1	1	16	18
Total	1	1	28	30

ADJUNGBILLY CHINESE CAMP

Adjungbilly Chinese Camp (ACC) was located on the northern side of the Adjungbilly Creek, about 2.5 km to the southeast of the present village of Adjungbilly. It was situated on a spur overlooking the creek which is permanently spring fed and covers an area of about 130 x 120 m (Smith 2006:154). Historical documents detail the European presence in the area, but scant documentary evidence exists about the presence of Chinese. Smith (2006:154) found only one reference to the Chinese. In May, 1862, The Sydney Morning Herald, reported that about 50 Chinese were at Adjungbilly and that they were 'apparently doing well'. Another newspaper reported 'Several Chinese have passed through (Adjungbilly) en route to Lambing Flat, from the Snowy River' and that the 'Adjungbilly diggings are slowly going along' (The Empire, 23 November 1861:3).

An archaeological survey revealed a Chinese settlement of at least 19 huts and a communal oven (Smith 2006). Seven areas were excavated including an oven, hut sites and water race, and artefacts were collected from the surface (Figure 8.4).

The total number of ceramic sherds recovered from ACC was 151, weighing 421.6 g. Only two sherds were surface finds, the rest were excavated. No fire damaged sherds were recovered.

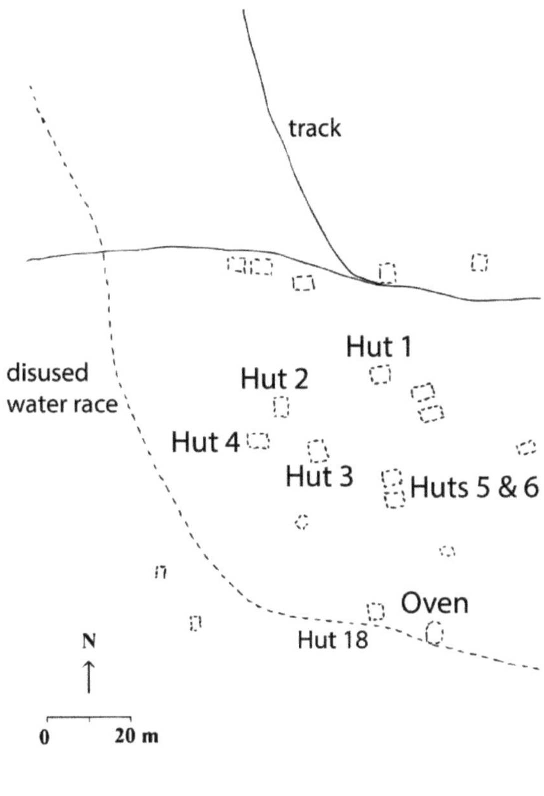

Figure 8.4: Map of ACC (after Smith 2006:544)

Communal Oven and Water Race

A circular stone structure with a diameter of 0.8 m indicated the presence of a Chinese oven. It was located 'about 30 m from the main, more permanent dwellings' (Smith 2006:158). A grid measuring 7 x 7 m was strung in one metre squares and about 4 m² were excavated around the Oven. Part of an adjacent water race which cut through the southern part of the site was also excavated.

No ceramic sherds were recovered from the Oven itself or its immediate surrounds. Three sherds, weighing 22.3 g, were found in the adjacent water race. These had probably washed in to it (Smith 2006:586).

A minimum of three vessels, all British-made white-earthenware includes: a cup with a black transfer-printed grape pattern; and two blue transfer-printed Willow plates with foot diameters of 105-110 mm and 120 mm.

Interpretation of Oven and Water Race

There are no ceramic artefacts from the Oven, although an iron cooking pot was found on the surface near the oven (Smith 2006:585). The lack of artefacts suggests that the site may have been used for festive occasions rather than for daily communal eating, particularly in the latter period of occupation.

The artefacts recovered from the nearby race may have been used at the Oven or have been washed down from a nearby hut site. Smith's map of the site depicts the earth platform of Hut 18 directly to the west of the Oven.

The ceramic sherds are all of British origin. The Willow patterned plates both have a double foot ring, which according to Moir's (1997) diagram were popular in the early nineteenth century. The recovered cup features grapes which were a gothic motif (Speltz 1959:242). Gothic designs were made from 1818-1890 but peaked between 1841 and 1852 (Samford 1997:6).

Hut sites

At least 19 hut sites were identified by Smith (2006). Five were excavated. Table 8.6 details the area surveyed and the internal dimensions of each hut, the excavated area and a brief description.

Hut 1

Hut 1 was located about 80 m to the north of the Oven. Ceramic sherds were found in only one of the excavated squares. Three sherds, weighing 41.3 g, represent a MNV of two.

Both vessels are British transfer-printed white-earthenware: an oval platter with a blue scenic pattern; and an unidentifiable vessel, possibly tea or tableware, with a worn blue pattern.

Hut 2

Hut 2 was located to the west of Hut 1. A pit adjacent to it, which was probably a rubbish dump, was also excavated. No ceramic sherds were found inside the hut but nine, weighing 14.7 g, were excavated from the adjacent pit and two sherds, weighing 1.5 g, were found on the surface. One of the surface sherds, with a green transfer-print, is possibly from the same vessel as an excavated sherd. The MNV associated with Hut 2 is six, of which five are from the pit and the other from the surface.

All the vessels are tea or tableware of non-Chinese origin: one bowl, one cup/bowl, three plate/saucers and one hollowware vessel. Five of the six vessels are transfer-printed white-earthenware: a bowl of unknown size in grey Amoy; a cup or bowl in purple Cable, found on the surface; a small plate or saucer, rim 180 mm, with a brown floral border; a plate or saucer of unknown size with a green foliate pattern; and a plate or saucer in blue Willow. The remaining vessel, identified only as hollowware, has a buff-earthenware body with relief-moulded grape motifs and a clear glaze. It may have been a mug.

Hut 3

Hut 3 was located to the southeast of Hut 2 and northwest of the Oven. One hundred and twenty-eight sherds, weighing 326.2g, represent a MNV of 44. A clay pipe brings the total MNV to 45 (Table 8.7). No fire damaged sherds were recovered.

The containers are both Chinese-made brown-glazed storage jars. No diagnostic sherds are present although there appear to be two sizes of jars from the thickness of the sherds.

Three vessels of Chinese-made tableware were found at Hut 3: two porcelain Winter Green rice bowls, rims 140 mm; and a stoneware hollowware vessel, of which only a small sherd is present. It has a clear-glaze with a green cast, possibly belonging to a blue-underglaze rice bowl, although no decoration is present.

All but two of the non-Chinese vessels are white-earthenware. One of the coloured-body vessels is a blue dyed-body earthenware cup or bowl with a white slip on the interior and a clear-glaze. The other vessel is a buff-earthenware hollow vessel with a Rockingham-type glaze on the exterior and a clear-glaze on the interior. This may have been a mug.

The 34 white-earthenware vessels include: a pudding bowl; an oval dish; two bowls; two cups; nine vessels which are cups or bowls; seven plates; one plate or soup plate; five vessels, either plates or saucers; five saucers and an unidentifiable vessel, possibly tableware, of which the base only is present.

Table 8.6: ACC hut dimensions, area excavated and description (after Smith 2006:160, 546)

Hut No.	Grid size	Int dimen m²	Exc m²	Description
1	6 x 8	8.8	6.0	Built into slope with tamped earth floor and hearth
2	5 x 6	10.0	6.0	Free-standing earth platform with hearth
3	5 x 5	9.0	7.0	Free-standing tented structure with hearth
4	5 x 6	7.5	5.0	Free-standing earth platform with hearth
5	4 x 10	10.5	13.0	Built into slope with hearth

Table 8.7: Number, weight and MNV of ceramic sherds from ACC Hut 3

	No.	% No.	Wt (g)	% Wt	Wt/No.	MNV	No./MNV	% MNV
Food/beverage storage containers								
Chinese	7	5.5	47.6	14.6	6.8	2	3.5	4.4
Non-Chinese	0		0.0			0		0.0
Food/beverage consumption								
Chinese	4	3.1	4.5	1.4	1.1	3	1.3	6.7
Non-Chinese	113	88.3	265.0	81.2	2.3	36	3.1	80.0
Smoking								
Chinese	2	1.6	4.7	1.4	2.4	1	2.0	2.2
Non-Chinese						1		2.2
Other								
Chinese	0		0.0			0		0.0
Non-Chinese	2	1.6	4.4	1.3		2	1.0	4.4
Total	128		326.2			45		

The most common decoration on the white-earthenware is a transfer-print. There are eight nameable transfer-prints on 11 vessels, and a further 20 vessels with unidentifiable patterns, including six which are flown. Vessels having nameable patterns are: a cup or bowl in blue Antique; a cup or bowl in blue Cable, rim 120 mm; a cup in black Antique, rim 85-90 mm; two plates, rims 260 mm, and an oval dish in blue Willow; a dinner plate in green Rhine; a saucer in blue Forest, rim 190 mm; a saucer in green Asiatic Pheasants; a saucer in grey Amoy; and an unidentifiable vessel in blue Fibre.

Vessels with unidentified transfer-prints include: a bowl with an unidentifiable brown transfer-print, foot 80 mm; a bowl with a purple floral pattern, rim 140 mm; five vessels which are either cups or bowls - a green floral pattern, scenic patterns in both brown and grey, a brown vine pattern, and an unidentifiable purple pattern; a cup with a blue scenic pattern and a foliated handle end; four vessels which are either plates or saucers - a black pattern with grapes and leaves, a blue geometric pattern, a purple floral pattern, and a green scrolled pattern similar to Antique; a blue patterned plate with a border of foliated scrolls on a linear background, similar to the Rhine pattern; and a saucer with a green foliate pattern, rim 180 mm.

The six vessels with flown transfer-prints include: a cup or bowl in a black marble pattern; a cup or bowl in a blue foliate pattern; a plate or saucer with a black geometric/foliate pattern; a plate or soup plate in a blue floral/scenic pattern; and two plates with blue patterns, one geometric and the other floral.

The other vessels are: a pudding bowl with a clear-glaze, rim 190 mm; a possible child's alphabet plate with saw-tooth moulding around the rim and further relief-moulding on the marly, possibly script; and a blue sponge-printed saucer with an indeterminate pattern.

The opium pipe bowl sherd is from a circular bowl with ten faceted sides and has a light-orange slip. Two impressed stamps are present, but the sherd is broken, consequently it may have had more (Figure 8.5). One of the stamps has the Chinese characters for water and east, which could relate to the workshop name. The other stamp has two identifiable characters, *nong qing*, relating to playing a musical instrument (translation, Po-yi Chiang, 2 November 2010). Only the mouthpiece is present of the European clay pipe.

Sherds from a pot lid and an unidentifiable vessel were recovered. The pot lid has a rim diameter of 60 mm; no decoration is present. The lid is made of white-earthenware and probably came from Britain or Australia. It may have been a paste lid, either food or toothpaste. The final vessel, also of non-Chinese origin, is made of terracotta. It has a rim diameter of 80-100 mm and may have been a lid. It has relief-moulded decoration, raised dots and possibly a leaf, on the exterior only. The rim is rounded and slightly wider than the body.

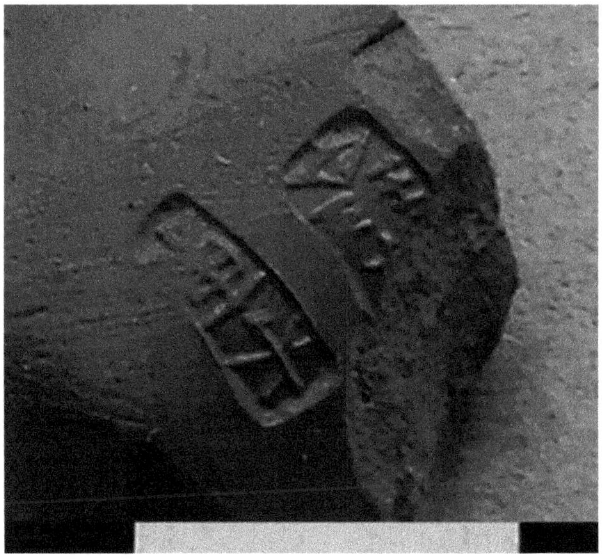

Figure 8.5: Opium pipe bowl sherd with maker's marks from ACC Hut 3 (Cat no. ACC-123)

Hut 4

Hut 4 was located to the west of Hut 3. Three sherds weighing 3.6 g were recovered, all from the same vessel, a blue sponge-printed white-earthenware saucer, foot 85 mm, with blue pooling in the glaze around the foot. The pattern is of an indeterminate nature.

Hut 5

Hut 5 was located to the east of Hut 3 and northwest of the Oven. Three sherds, weighing 12.0 g, were recovered representing a MNV of three. All the vessels are British-made white-earthenware and include: a cup or bowl in blue transfer-printed Fibre pattern; a cup, rim 95 mm, in green transfer-printed Asiatic Pheasants; and a dinner plate with a flown-blue scenic pattern, foot 150 mm.

Interpretation of hut sites

Few ceramic vessels were recovered from Huts 1, 2, 4 or 5, perhaps indicating they were occupied by few people or for a short time. The majority of ceramic vessels were found in Hut 3, which had 'a reasonable spread of artefacts' throughout the hut (Smith 2006:586), suggesting a longer occupation period than those of the other huts. The average weight per sherd, just over two grams, is small, which indicates a normal accumulation over time. Hut 3 is the only hut to have any Chinese-made ceramic vessels. It is also the only hut to have any storage containers, a child's vessel, and smoking pipes. While Huts 1, 2, 4 and 5 may have been occupied by short-term male miners, Hut 3 may have been inhabited by a family.

Summary – ACC

A summary of the number, weight and MNV from each site within ACC can be seen in Table 8.8. The majority of sherds were recovered from Hut 3. The average weight per sherd is higher from Hut 1 as a result of sherds from a thick platter. The number of sherds excavated per area is higher from Hut 3 than other sites. Table 8.9 shows a summary of vessel types and Figure 8.6 a selection of sherds from ACC.

Origin of ceramics

The majority of ceramic artefacts, by number, weight and MNV, recovered from the Adjungbilly Chinese Camp are of non-Chinese origin (Table 8.10). The only Chinese-made vessels are from Hut 3. Most of the other vessels were manufactured in Britain. One vessel, of relief-moulded terracotta, may have been made in Australia.

Dating ceramic sherds

A number of nameable transfer-printed patterns were found including Amoy, Antique, Asiatic Pheasants, Cable, Fibre, Forest, Rhine, and Willow. Although a number of different patterns sighted in reference books are named Amoy, a bowl with the same pattern to the one recovered from Hut 2 was found at the Hibernian Hotel, near Bungendore, NSW, with the maker's mark of Pinder, Bourne and Hope, 1851-1862. The Antique pattern was identified with reference to the New Zealand Historical Ceramics Database (NZHCD 2012). Their plate has the maker's mark of Pinder, Bourne and Co. (1862-1888). The Forest pattern was made by at least three potteries: John Allason, 1838-1841; Joseph Clementson, 1839-1864; and Samuel Alcock and Co, 1826-1859 (Kowalsky & Kowalsky 1999:487). Williams (1978:630) illustrated the Samuel Alcock and Co. pattern, which is similar to that recovered from ACC Hut 3.

Table 8.8: Number, weight and MNV of ceramics from ACC

Site	No.	Wt (g)	Wt/No.	MNV	No./MNV	Exc. Area (m^2)	MNV/Exc Area
Hut 1	3	41.3	13.8	2	1.5	6	0.3
Hut 2	11	16.2	1.5	6	1.8	6	1.0
Hut 3	128	326.2	2.5	44	2.9	7	6.3
Hut 4	3	3.6	1.2	1	3.0	5	0.2
Hut 5	3	12.0	4.0	3	1.0	13	0.2
Race	3	22.3	7.4	3	1.0	unknown	
Total	151	421.6		59			

Note: clay pipe is not in catalogue sherd numbers or weights, therefore not included in MNV numbers for this table

Table 8.9: MNV from Adjungbilly Chinese Camp

Adjungbilly Chinese Camp - MNV									
Function	Origin	Vessel	Race	Hut 1	Hut 2	Hut 3	Hut 4	Hut 5	Total
Food/ beverage containers									
	Chinese	storage jar - small-medium				1			1
	Chinese	storage jar - medium-large				1			1
Food/ beverage consumption									
	Chinese	rice bowl?				1			1
	Chinese	rice bowl				2			2
	Non-Chinese	bowl			1	2			3
	Non-Chinese	cup	1			2		1	4
	Non-Chinese	cup/bowl			1	10		1	12
	Non-Chinese	mixing bowl				1			1
	Non-Chinese	mug?			1	1			2
	Non-Chinese	oval dish				1			1
	Non-Chinese	plate	2			7		1	10
	Non-Chinese	plate/ soup plate				1			1
	Non-Chinese	plate/saucer			3	5			8
	Non-Chinese	platter		1					1
	Non-Chinese	saucer				5	1		6
	Non-Chinese	possible tea/table		1		1			2
Smoking									
	Chinese	opium pipe bowl				1			1
	Non-Chinese	clay pipe				1			1
Other									
	Non-Chinese	pot lid				1			1
	Non-Chinese	unknown				1			1
Total MNV			3	2	6	45	1	3	60

Asiatic Pheasants, Fibre, Rhine, and Willow were all made from the 1830s and into the twentieth century (Coysh & Henrywood, 1982, 1989, and Kowalsky & Kowalsky, 1999). Cable was made from the 1860s onwards (Ward 2006).

The indeterminate nature of the sponge-printed patterns from Huts 3 and 4 suggests a date before 1870 (Kelly et al. 2001:9). The sponge-printed saucer from Hut 4 also has a blue tinge in the pooled glaze.

Profiles of recovered plates, Willow (double foot ring) and flow blue (flat foot) correspond to 1825 and 1845 respectively in Moir's (1997) diagram. The green Rhine plate from Hut 3 and the blue sponge-printed saucer from Hut 4 both have similar profiles to vessels from the Gold Commissioner's quarters at Kiandra (KGQ), dating 1860-1862. The foliated handle found on a cup from Hut 3 is also found on cups from KGQ. The buff-earthenware vessels recovered from Huts 2 and 3 have similar body/glaze/pattern characteristics to a relief-moulded mug with a Rockingham-type glaze found at KGQ. The lower part of that mug has a clear glaze and is similar to one pictured by Brooks (2005:28).

Smith (2006) stated that the glass, nail and miscellaneous artefacts indicated the date of occupation of the site was from the late 1860s possibly into the early 1880s. The ceramics would all fit with this date.

Function of ceramic vessels

The majority of vessels, by number, weight and MNV, are related to food/beverage consumption (Table 8.11).

The only food/beverage containers are two Chinese-made brown-glazed storage jars from Hut 3. They are not identifiable to type, although there is one medium-large and one small-medium jar. The average weight of the sherds from food containers is greater than that of other vessels.

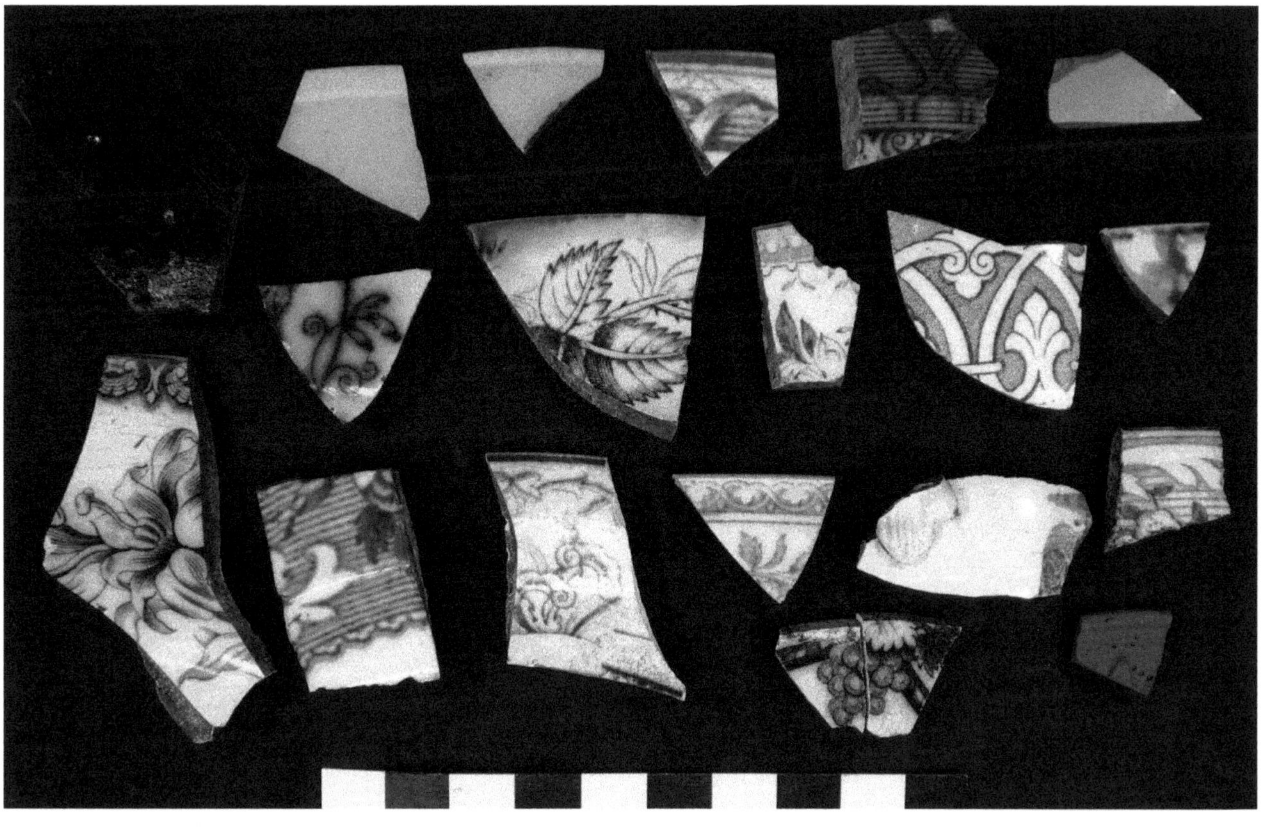

Figure 8.6: Sherds from ACC (top row – Cat nos. 122 brown-glazed storage jar, 114.1, 114.2 celadon rice bowls, 150 Cable, 180 flow blue, 158 dyed-blue body; middle row – Cat nos. 101 flow blue, 126 green transfer, 133 Asiatic Pheasants, 184 Antique, 135 sponge-print; bottom row – Cat nos. 149 flow blue, 129 blue transfer, 175 Amoy, 136 purple transfer, 182 back transfer, 144 foliated handle end – blue transfer, 142 terracotta, 151 Rhine)

Table 8.10: Origin of ceramic vessels from ACC by number, weight and MNV

Origin	No.	% No.	Wt (g)	% Wt	Wt/No.	MNV	No./MNV	% MNV
Chinese	13	8.6	56.8	13.5	4.4	6	2.2	10.2
Non-Chinese	138	91.4	364.8	86.5	2.6	53	2.6	89.8
Total	151		421.6			59		

Note: clay pipe is not in catalogue sherd numbers or weights, therefore not included in MNV numbers for this table

Table 8.11: Function of vessels from ACC by number, weight and MNV

Function	No.	% No.	Wt (g)	% Wt	Wt/No.	MNV	No./MNV	% MNV
Storage	7	4.6	47.6	11.3	6.8	2	3.5	3.4
Consumption	138	91.4	364.3	86.4	2.6	51	2.7	86.4
Smoking	2	1.3	4.7	1.1	2.4	1	2.0	1.7
Other	4	2.6	5.0	1.2	1.3	5	0.8	8.5
Total	151		421.6			59		

Note: clay pipe is not in catalogue sherd numbers or weights, therefore not included in MNV numbers for this table

Decorative types

The only Chinese decoration identified is Winter Green.

Decorative types on white-earthenware from the hut sites include transfer-prints in blue (15 vessels), green (7), brown (4), purple (4), black (3) and grey (3); flown-prints in blue (5) and black (3); and sponge-prints in blue (2). There is also one vessel of blue dyed-body and two of buff-earthenware.

Eight nameable transfer-printed patterns were found at ACC. No bone china or porcelain tea ware vessels were recovered. The only child's vessel, possibly an alphabet plate, is from at Hut 3.

Hut 3 is the only site to have more than one vessel with the same pattern: an oval dish and two dinner plates in blue Willow; a saucer in Asiatic Pheasants and a cup or bowl which was possibly Asiatic Pheasants, both in green; a cup in black Antique and a cup or bowl in blue Antique. There are also two different patterns which feature grape motifs, a cup or bowl in black and a plate or saucer in brown. Grapes were sometimes used as a border pattern in Chinese paintings (Williams 1960:216).

Some vessels have obvious decorative faults, for example the Asiatic Pheasants cup from Hut 5 has a blurred transfer-print and the Antique cup from Hut 3 has part of the transfer missing.

In conclusion, there are no ceramics that suggest the site was occupied beyond the 1880s. The only unusual sherd is from an unidentifiable relief-moulded terracotta vessel. Hut 3 has a substantial assemblage in comparison to the other hut sites, which implies it was occupied longer or by more people, perhaps a family rather than single miners.

DISCUSSION - TUMUT REGION

Both settlements were occupied in the second half of the nineteenth century and the ceramics are predominantly British in origin. Similarities and differences are found in the Chinese ceramic artefacts. Both settlements have celadon and brown-glazed wares but the only blue-underglaze decorated ware is from UACC and the only opium pipe bowl from ACC. Overall, recovered vessels are mainly tea and tableware in similar forms and decorative types. Blue transfer-prints are the most common colour in both camps. While ACC has over double the number of blue to the second most common colour, green, the number of blue transfer-printed vessels at UACC was almost equal to the second most common colour, purple. Although 12 transfer-printed patterns can be named, Rhine and Willow are the only patterns found at both sites. Banded vessels are only found at UACC suggesting that site was occupied to a slightly later date than ACC. Some of the recovered ceramic vessels may have been manufactured in the first half of the nineteenth century, suggesting the method of purchase of these wares by retailers was through auction of second hand goods.

CHAPTER 9

Kiandra Region - Chinese Camp

KIANDRA CHINESE CAMP

The re-examined assemblages from the Kiandra Chinese Camp (KCC) are detailed in this chapter. As outlined in Chapter 6, they were collected by staff and students during a series of ANU field schools and by Lindsay Smith during his research in 1996-1998. Although the results are reported in terms of MNV, there are no whole vessels. The results from isolated miners' huts in the surrounding valley and the Chinese stores in Kiandra Township are in Chapter 10, along with a discussion regarding the Kiandra region.

Kiandra is located in the Kosciuszko National Park, about 90 km southeast of Tumut (Figure 9.1). In 1859-1860 it was the setting for one of the most short-lived gold rushes in Australian history (Moye 1959:ix). More than 10,000 people inhabited the area by March-April 1860 but a year later most left to head north for the rush at Lambing Flat. The Chinese arrived in Kiandra in June 1860 and in some of the following years they outnumbered those of European descent (Moye 1959). The camp became the focus of a large settlement system in the Kiandra Valley and adjacent areas (Smith 2006). Some Chinese moved into the Township in the 1880s and by 1891 only nine Chinese resided at the camp (Smith 2006:168). Most gold was obtained from alluvial deposits, but in the early 1880s hydraulic sluicing was introduced, and by 1900, dredging (Smith 1998:31). Figure 9.2 shows a dredge in front of the Chinese camp at Kiandra.

The main Chinese Camp at Kiandra was located about 1 km east of the township of Kiandra. It was situated to the west of the Eucumbene River at its confluence with Pollock's Creek, in a sub-alpine landscape with relatively barren terrain. A number of features have been identified including over 60 stone foundations and platforms for hut structures, watercourses, tracks, pits, and general evidence of alluvial mining and dredging. Figure 9.3 details the designated hut numbers. Hut 46 is not shown; it is located about 20 m southwest of Hut 37.

Temple Complex

Although two historical references indicate the presence of a Chinese temple in Kiandra neither specify a location. A newspaper article from 1860 covering the arrival of Chinese, reported that they made an offer for a house in order to convert it into a temple (Smith 2006:167).

Two sites, excavated as part of the 2003 field school, are thought to be the Temple and an ancillary building. They are located on the northern end of the main Chinese camp, overlooked by a large granite outcrop. A lower terrace is separated from an upper terrace by an embankment, 70 cm high, which is supported by stones placed along the slope. The lower area contains possible postholes indicating a building almost parallel to the embankment. The upper terrace contains four possible postholes, thought to be the remains of a building.

Temple

The postholes on the lower terrace indicate a building measuring 8.7 x 5.1 m, possibly made of weatherboard, with an earth floor and no hearth. A total of 84 m² were excavated in and around the building; 44 m² to the base of the turf layer and 40 m² to lower levels (Smith 2006:678). A trench excavated in 2001 sliced through part of the Temple building and conjoining ceramic sherds were found from this area, one excavated in 2001 and the other in 2003.

Three hundred and nineteen sherds, weighing 469.3 g, were recovered in 2001 and 2003, representing a MNV of 20. A European clay pipe brings the total MNV to 21 (Table 9.1). These sherds are weathered but not burnt.

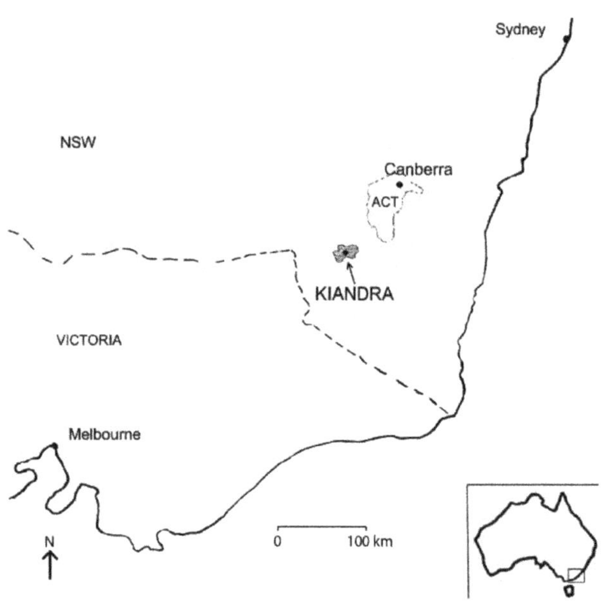

Figure 9.1: Location of Kiandra region (after Smith 2006:4)

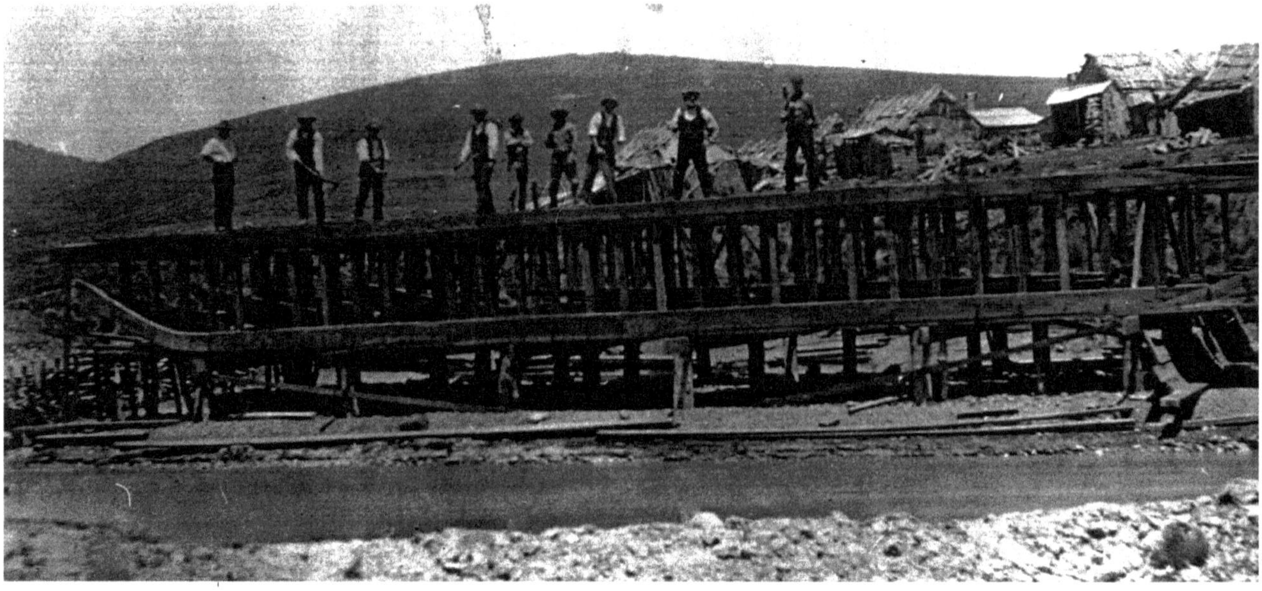

Figure 9.2: Kiandra. Dredge on the Eucumbene River, c.1900 (photograph: National Library of Australia, Pictures Compactus Collection, lent for copying by Mr K. Hueneke. Reproduced with permission of the National Library of Australia)

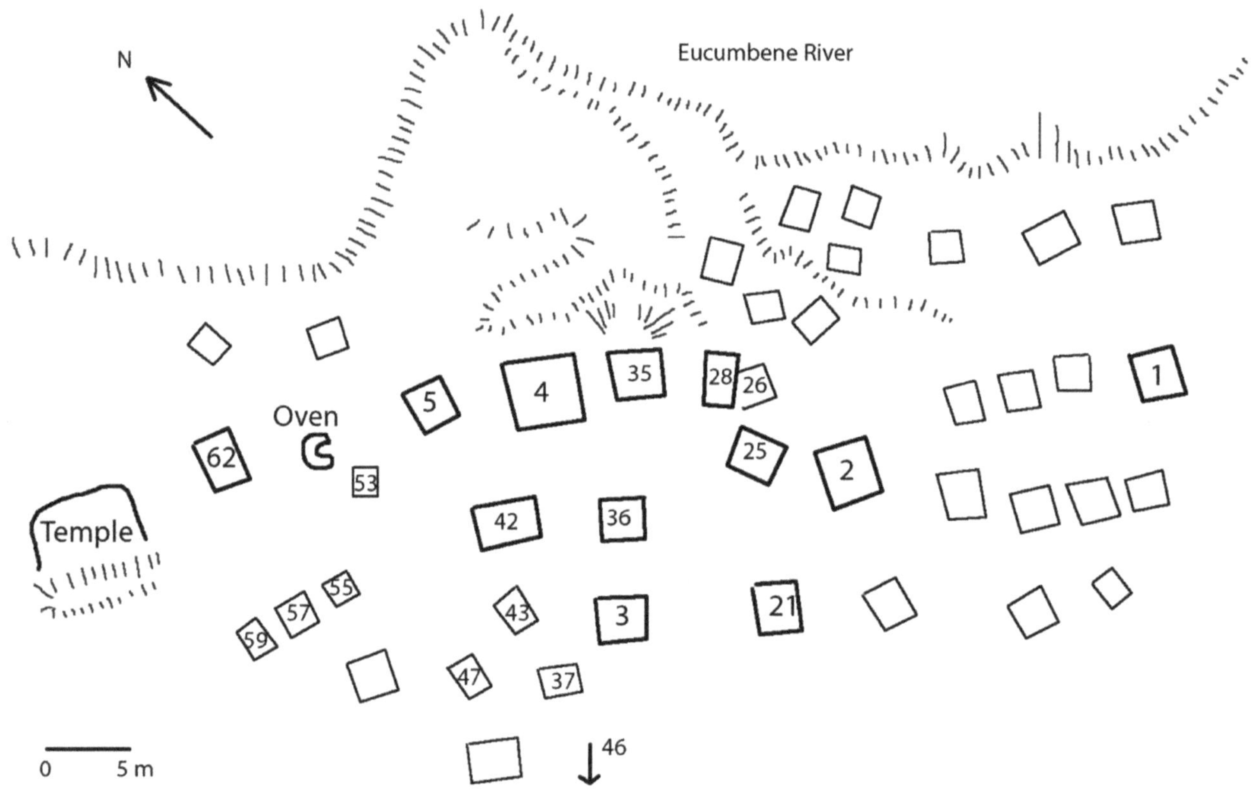

Figure 9.3: Kiandra Chinese Camp with designated hut numbers

All the containers are Chinese: a brown-glazed stoneware wide-mouthed jar, foot 120 mm; a liquor bottle; a medium-large brown-glazed storage jar; and a buff-earthenware wide-mouthed jar lid.

One Chinese-made porcelain vessel, either a cup or rice bowl, was found with Winter Green decoration. The 12 non-Chinese tea/tableware vessels are all white-earthenware, probably all made in Britain: one bowl; one cup; four vessels which are either cups or bowls; a possible mug; two plates; two saucers and one unidentifiable vessel which is possibly tableware. One of the plates is from the surface while all other vessels are from excavations.

Table 9.1: Number, weight and MNV of ceramic sherds from KCC Temple

	No.	% No.	Wt (g)	% Wt	Wt/No.	MNV	No./MNV	% MNV
Food / beverage containers								
Chinese	158	49.5	265.5	56.6	1.7	4	39.5	19.0
Non-Chinese	0		0.0			0		0.0
Food / beverage consumption								
Chinese	3	0.9	0.5	0.1	0.2	1	3.0	4.8
Non-Chinese	128	40.1	117.0	24.9	0.9	12	10.7	57.1
Smoking								
Chinese	10	3.1	5.5	1.2	0.6	2	5.0	9.5
Non-Chinese						1		4.8
Other								
Chinese	0		0.0			0		0.0
Non-Chinese	20	6.3	80.8	17.2	4.0	1	20.0	4.8
Total	319		469.3			21		

Note: clay pipe is not in catalogue sherd numbers or weights

Seven of the non-Chinese vessels have transfer-prints: a cup, rim 120 mm, in a blue pattern with linked circles and intersecting lines; a cup/bowl in blue Fibre; a cup/bowl in brown Crystal; a cup/bowl in a green foliate or scenic pattern; a dinner plate in blue Dresden; and a saucer in blue Cable, rim 180 mm. An unidentifiable vessel, possibly tableware, of which there is only a small sherd, is decorated with a purple geometric pattern. Flown transfer-prints are found on three vessels: a cup/bowl having black swirling dotted lines; a plate with a black floral pattern; and a saucer, rim 180 mm, in a blue foliate pattern. A slightly flown sponge-printed pattern with three purple bands at the rim and a repeated purple leaf motif below is found on a bowl. The possible mug is straight-sided and had a foot diameter of 85 mm, with three indented lines near the base. It may have had a blue decoration, but only a few specks are present.

There are fragments from two Chinese opium pipe bowls in the collection: a light-orange slipped terracotta pipe bowl, with a circular rim and ten faceted sides, from excavations; and one with a brown stoneware body, grey slip, and an octagonal smoking surface and sides, from the surface. A MNV of one European clay pipe was recovered from the surface.

A salt-glazed blacking bottle with a body diameter of 60 mm was excavated.

Temple Ancillary Building

Postholes on the upper terrace indicate that a building, possibly of weatherboard material with a raised floor, measured at least 3.6 x 1.8 m. It is on a levelled area that may have been created by using fill from the lower terrace. The building has no hearth. A total of 55 m² were excavated in and around this structure; 28 m² to the turf layer and 27 m² to lower levels (Smith 2006:678).

Three hundred and sixty-three sherds weighing 666.3 g were found in excavations and surface collections in 2001 and 2003. Together these represent a MNV of 25. A European clay pipe brings the total MNV to 26 (Table 9.2). The recovered sherds are weathered but not burnt.

All the containers are Chinese-made brown-glazed stoneware: a barrel jar, a globular jar, a spouted jar and a liquor bottle.

Table 9.2: Number, weight and MNV of ceramic sherds from KCC Temple Ancillary Building

	No.	% No.	Wt (g)	% Wt	Wt/No.	MNV	No./MNV	% MNV
Food / beverage containers								
Chinese	146	40.2	276.7	41.5	1.9	4	36.5	15.4
Non-Chinese	0		0.0			0		0.0
Food / beverage consumption								
Chinese	4	1.1	10.5	1.6	2.6	2	2.0	7.7
Non-Chinese	201	55.4	375.9	56.4	1.9	17	11.8	65.4
Smoking								
Chinese	12	3.3	3.2	0.5	0.3	2	6.0	7.7
Non-Chinese						1		3.8
Total	363		666.3			26		

Note: clay pipe is not in catalogue sherd numbers or weights

Two vessels of Chinese-made tableware were found: a stoneware rice bowl with a green-blue underglaze Bamboo pattern; and a porcelain bowl of unknown size, having a clear glaze with a green cast but no decoration. There are 17 non-Chinese vessels of white-earthenware, probably all made in Britain: two bowls, six vessels which are cups or bowls, a jug, three plates, one saucer, three vessels which are plates or saucers and one unidentifiable vessel which is possibly tableware. One of the bowls is a surface find and the remainder are from excavation.

Fourteen of the non-Chinese vessels have transfer-prints: a bowl, foot 85 mm, in blue Fibre; a bowl, foot 80 mm, with a purple squiggly-line pattern; three vessels which are cups or bowls having geometric patterns, one in each blue, brown and grey; a jug with a brown geometric pattern featuring lines and dotted bands; plates in blue Albion, blue Eton College and green Rhine; a saucer, rim 180 mm, in grey Rhine; a plate/saucer in blue Magenta; a plate/saucer in blue Willow; and a plate/saucer with a blue geometric pattern. An unidentifiable excavated vessel, represented only by small sherds, features a purple geometric pattern with bands and dots. Other decorative types on the remaining cup/bowl vessels include: a slightly flown-purple sponge-print with three bands at rim with a repeated leaf motif below; a flown-blue floral transfer-print; and a geometric flown-blue transfer-print.

Excavated fragments were found from a minimum number of two opium pipe bowls, one with dark-orange slip and the other light-orange slip. They are not identifiable to type. A MNV of one European clay pipe was excavated.

Interpretation of Temple Complex

It has been suggested that the upper terrace was created by using fill from the lower terrace after that area had been cleared of other dwellings and associated artefacts (Smith 2006:176, 180). This infers that the artefacts recovered from the Temple and Ancillary Building were not associated with those structures but may have been used by occupants of previous huts in that area.

Origin of ceramics

Table 9.3 outlines the origin of vessels by MNV and percentage MNV. The majority of vessels from both sites are non-Chinese, predominantly of British origin. The sponge-printed vessels may have originated from Holland. The blacking bottle may have been made in either Britain or Australia.

Dating ceramic sherds

No dates could be determined from the three unidentifiable makers' marks: a blue transfer-printed partial mark in a garter '...T'; a green transfer-printed Rhine pattern name in a cartouche; and a blue transfer-printed Fibre pattern name in a cartouche. No makers' names are present for the Rhine and Fibre patterns.

Nine nameable transfer-printed patterns were recovered: Temple - Fibre, Crystal, Dresden, and Cable; Ancillary Building - Albion, Eton College, Fibre, Rhine, Magenta, and Willow.

Cable and Fibre are both common patterns at mid to late-nineteenth century sites in Sydney (Casey & Lowe 2009) and in southeast New South Wales. Albion, Rhine and Willow are also commonly found at Australian nineteenth century sites, all produced from at least 1830 and into the twentieth century.

The Crystal pattern, found at Kiandra, has also been recovered from Reservoir Street, Surry Hills NSW, attributed to Hope & Carter, Burslem, 1862-1880 and at St. Patrick's Cathedral, Parramatta, attributed to Pinder, Bourne & Hope, Burslem, 1851-1862 (Casey & Lowe 2009). The pattern has foliated scrolls either side of a trellis-like design consisting of long vertical lines; each line is crossed by a series of short horizontal lines. The interior well features a snowflake. The pattern was also found at a mid-nineteenth century redoubt in Judea, New Zealand (NZHCD 2012).

The Dresden plate at the Temple Ancillary Building was identified with reference to a plate held in the Gundagai Museum, NSW (personal communication, Alan Walden, Gundagai Museum, 19 September 2009). It has the pattern name and mark: 'R (M or H)' in the centre of a garter cartouche; and 'RESDEN/FENTON' in the outer circle. Part of the mark was missing but the pattern name has been interpreted as Dresden. The maker may have been Ralph Malkin who operated in Fenton, Staffordshire, from 1863-1881 (Godden 1991:410). A plate with this pattern and mark has also been found at the Kiandra Hospital, which operated 1860s-1890s.

Eton College is a 'romantic' pattern featuring a man, woman and child in the foreground, a lake in the middle and buildings in the background. It was produced by several makers (Coysh & Henrywood 1982:130). A pattern similar to the ones recovered from Kiandra is illustrated in Neale (2005:49). The maker is unknown but Neale suggested a date of about 1840.

Magenta was identified with reference to a pattern found at Reservoir Street, Surry Hills NSW, attributed to Ralph Malkin, Fenton, 1863-1881 (Casey & Lowe 2009).

Table 9.3: Origin of ceramic sherds by MNV from the KCC Temple Complex

Origin	Temple		Ancillary Building	
	MNV	%	MNV	%
Chinese	7	33.3	8	30.8
Non-Chinese	14	66.7	18	69.2
Total	21		26	

Function of ceramic vessels

The majority of vessels from each site were designed for food/beverage consumption (Table 9.4). Temples at Chinese camps were the focus of traditional festivals and meeting places as well as being places of worship (Wilton 2004:85). Ceramic vessels were used to hold items associated with worship, such as flowers, flags or incense. In addition, food and drink would have been an important part of festivities therefore finds of ceramic storage jars and crockery would be expected. Cup/bowls are the most common vessels along with a few plate/saucers, and a possible serving bowl and a jug from the Temple Ancillary Building. In addition to the tea and tableware, a number of storage containers used for food/beverages are present including a wide-mouthed jar and earthenware lid from the Temple, a spouted jar from the Temple Ancillary Building and liquor bottles from both sites. The blacking bottle (Table 9.4, 'other') found at the Temple may have been used as a vase. No 'unusual' ceramic vessels were found to signify a temple. Although caretakers are associated with temples, it is unlikely that either of these buildings was a residence because neither had a hearth (Smith 2006:709-710).

Table 9.4: Function of vessels from the Temple Complex

Function	Temple		Ancillary Building	
	MNV	%	MNV	%
Storage	4	19.0	4	15.4
Consumption	13	61.9	19	73.1
Smoking	3	14.3	3	11.5
Other	1	4.8		
Total	21		26	

Note: this table includes European clay pipes

Form and decoration

The number of hollowware tea and table vessels is greater than flatware at both sites: Temple - eight and four respectively; and Ancillary Building - 11 and seven. Chinese decorative types include Winter Green at the Temple and Bamboo at the Ancillary Building. There are eleven different patterns on the non-Chinese tea and tableware at the Temple and fifteen at the Ancillary Building, one pattern appearing on two vessels in different colours. The most common colour at both sites is blue.

Does data support the theory that landfill was used from the lower terrace to fill the upper terrace?

If the fill for the upper terrace came from the lower terrace it would be expected to have similar proportions of Chinese and non-Chinese made vessels, and functional categories. Both these expectations are met (Tables 9.3 and 9.4). Although this would appear to support the landfill hypothesis, it may also be expected that many similar decorative types would appear at both sites; but this is generally not the case. Different Chinese tableware decoration is found at each site. Only two of the 25 decorative types of non-Chinese tea and tableware are at both sites.

The Temple Ancillary Building floor may have been raised above the ground. The small size of the sherds with average weights of about two and a half grams or less could have fallen through floor boards. A similar situation is noted for the size of the sherds from the Temple.

Summary

It has previously been suggested that the sherds recovered from the Temple and Ancillary Building are not from vessels used at those sites but are from the former occupation of that area by miners (Smith 2006:176, 180). Further analysis of the ceramics has shown that the sherds came from vessels that had been used in the Temple and Ancillary Building. In addition, it is unlikely that miners residing in that area, prior to the building of the Temple in 1860, would have accumulated the number of vessels excavated.

Communal Oven

A stone structure located within a large stony mound, was thought to be the remains of a Chinese oven. Twelve one-metre squares were excavated within the area encompassing squares O23 to S27 in Grid 1. For this research ceramic artefacts from other squares immediately outside this area, but closer to the Oven than other hut sites, have been included in the total from this site.

The exact size and shape of the oven structure remains unknown as it had collapsed both inwards and outwards; it was possibly 1.5 to 2 m wide (Searle & Smith 2001:60) (Figure 9.4).

Two hundred and two sherds weighing 413.2 g represent a MNV of 34. Two European clay pipes bring the total MNV to 36 (Table 9.5).

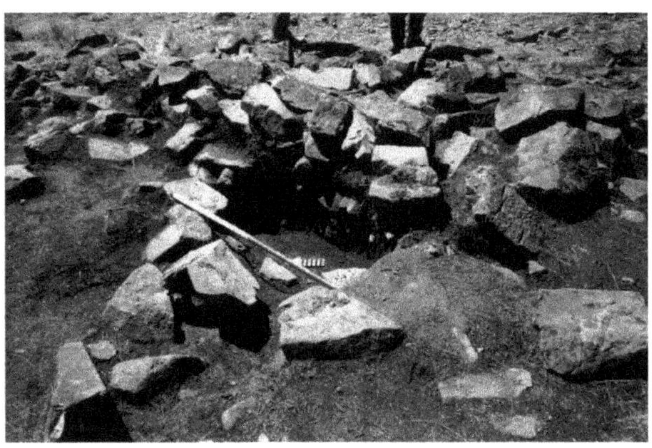

Figure 9.4: KCC Oven after excavation in 2001

Table 9.5: Number, weight and MNV of sherds from KCC Oven

	No.	% No.	Wt (g)	% Wt	Wt/No.	MNV	No./MNV	% MNV
Food / beverage containers								
Chinese	53	26.2	137.2	33.2	2.6	8	6.6	22.2
Non-Chinese	0		0.0			0		0.0
Food / beverage consumption								
Chinese	8	4.0	12.3	3.0	1.5	4	2.0	11.1
Non-Chinese	111	55.0	221.9	53.7	2.0	17	6.5	47.2
Smoking								
Chinese	27	13.4	41.0	9.9	1.5	4	6.8	11.1
Non-Chinese						2		5.6
Other								
Chinese	0		0.0			0		0.0
Non-Chinese	3	1.5	0.8	0.2	0.3	1	3.0	2.8
Total	202		413.2			36		

Note: clay pipes are not in catalogue sherd numbers or weights

All of the food/beverage storage vessels are Chinese in origin. There are six brown-glazed stoneware vessels: a globular jar, a spouted jar, a liquor bottle and three wide-mouthed jars. There is also a buff-earthenware wide-mouthed jar lid. A flanged ginger jar lid, rim 85 mm and flange depth 12 mm, has a clear glaze on the exterior of the lid and unglazed interior. The lids and the spouted jar are from excavation and the other containers are surface finds.

Four tea/tableware vessels are Chinese-made porcelain: a Winter Green bowl, rim 140 mm; another Winter Green vessel which is either a cup or bowl; a Four Seasons decorated liquor cup; and a Four Seasons decorated bowl or serving bowl. The rice bowl and liquor cup are from excavation and the other two are surface finds. The 17 non-Chinese vessels are white-earthenware. They include two bowls, a cup, eight vessels which are cups or bowls, a small plate, three saucers, a vessel which is either a plate or saucer and an unidentifiable vessel, of which the base only is present. A bowl and a cup/bowl are from the surface while the rest are excavated.

Transfer-prints are found on 13 of these vessels: a bowl in brown Crystal, rim 150 mm; a bowl in brown Rhine, foot 80 mm; a cup in blue Rhine; five vessels which were cups or bowls in a blue geometric pattern, a black foliate pattern, a brown floral pattern, a grey floral pattern and purple Crystal; a blue Willow plate, rim 185 mm; a saucer in blue Fibre; a saucer in a purple lattice pattern; and a plate or saucer in blue Wicker, foot 85 mm. There is a base only from an unidentifiable vessel decorated in Asiatic Pheasants. The other decorative types include: a cup/bowl with a flown-blue pattern; a cup/bowl with a flown-purple pattern having a diamond border with white leaves on a purple background; a cup/bowl with a repeated sponge-printed leaf motif in purple; and a blue sponge-printed saucer.

Fragments of four Chinese opium pipe bowls were recovered from the Oven area. These include three light-orange slipped bowls with circular rims and ten faceted sides, one with a rim diameter of 70 mm; and a pipe bowl with a buff-earthenware body, grey slip and an octagonal smoking surface and sides. One of the light-orange slipped pipe bowls is a surface find while the other three are from excavation. Two European clay pipes were excavated but no details of makers are available.

A Holloway's ointment pot lid sherd was retrieved from the surface. It has a black transfer-print with the letters 'OWA' and part of a picture of a seated woman.

Interpretation of Oven site

Only two of the 34 vessels were from outside the original designated Oven area. These are both hollowware vessels of non-Chinese origin, one from square S28 and the other from U24.

Origin of ceramics

There are slightly more non-Chinese made vessels, 55.6 per cent, than Chinese-made, 44.4 per cent, from in and around the Oven. Non-Chinese sherds also dominate by number and weight. They are predominantly of British origin; the possible exceptions are the sponge-printed vessels, which may have been manufactured elsewhere in Europe.

Dating ceramic sherds

A blue transfer-printed Rhine pattern name in a cartouche is the only recovered mark, but this is not datable as the maker's name is absent. The Holloway's pot lid is not datable because the address, usually printed on the lid, is not present on the sherd.

Four of the nameable transfer-printed patterns, Asiatic Pheasants, Fibre, Rhine and Willow, were common throughout the second half of the nineteenth century. The Crystal pattern probably dates to the mid-nineteenth century (Casey & Lowe 2009). The Wicker pattern was identified with reference to the New Zealand historical ceramics database. It has been found at a number of sites in New Zealand with the pattern name but no maker's name (NZHCD 2012). The pattern features a wicker-work border design extending towards a central small floral motif.

Function of ceramic vessels

The majority of vessels from the Oven are tea and tableware. Cup/bowls are the most common vessel/s. A possible Chinese-made serving bowl along with a non-Chinese small plate and saucers, which are the size of Chinese serving dishes, suggest an area of communal eating. The majority of brown-glazed wares are small food/beverage containers, including a ginger jar (lid), spouted jar and wide-mouthed jars. There are also at least six opium/tobacco pipes indicating the Oven site was an area for leisure.

Decorative types

Chinese tea and tableware is decorated in Winter Green and Four Seasons. Fifteen different patterns are found on the non-Chinese tea and tableware, two of which have vessels in different colours: blue and brown Rhine; and brown and purple Crystal. Transfer-prints are the most common decorative type; blue the most common colour.

Summary

The excavation of this site indicates that the Oven was 'indeed an oven, hearth or fireplace' but that 'its original size and shape' remain unknown (Searle & Smith 2001:60). The chronology of the site cannot be determined and the ceramics can only support a general date of mid to late-nineteenth century. The ceramics have established that this site was used communally for eating, drinking and smoking.

Hut sites

Ceramic sherds were collected from 21 hut sites. Figure 9.3 shows the position of huts within the camp. The results of ceramic analysis from each hut are presented in designated numerical order. Tables summarising the number, weight and MNV of sherds are only included for those hut sites with a MNV over five. Sherds from all hut sites are weathered. Table 9.6 summarises the internal dimensions of each hut and the area excavated. Some have a surface collection only.

Hut 1

This hut site was originally referred to as 'Hut 13' in Smith (1998) and 'Hut 2' in Smith (1997), The Chinese of Kiandra, but the hut number was changed to 'Hut 1' after a survey of the whole camp in 2001. One hundred and twenty-nine sherds weighing 2,129.1 g represent a MNV of 17. Two European clay pipes bring the total MNV to 19 (Table 9.7).

All of the storage vessels are Chinese stoneware. They include: a barrel jar and lid, rims 330 mm; a globular jar, foot 115 mm; two brown-glazed wide-mouthed jars, rims 85 mm and 100 mm; and a ginger jar lid, rim 100 mm and flange depth 15 mm, with a worn clear glaze on the exterior top and the rest unglazed. The ginger jar lid and the smaller wide-mouthed jar are surface finds and the others are from excavation.

There are three Chinese-made tea/tableware vessels: a Four Seasons decorated serving bowl, rim 180 mm; a Winter Green liquor cup, rim 45-50 mm; and another celadon-glazed vessel which is either a cup or rice bowl. Only the last vessel is from excavation; the others are from the surface.

The five non-Chinese tea/tableware vessels include a teapot or jug in buff-earthenware with a Rockingham glaze and four vessels of white-earthenware: a bowl with a grey Rhine transfer-print; a cup or bowl with an unidentifiable flown-blue decoration; a saucer with a flown-blue floral pattern; and a saucer decorated with a repeated blue sponge-printed leaf motif. The flow blue saucer and cup/bowl are from excavation; the others are from the surface.

Table 9.6: KCC hut dimensions, area excavated and description (after Smith 2006:183, 619, 624)

Hut No.	Internal dimen m²	Exc m²	Description
1	6.5	21.0	Built into slope; random rubble walls and tamped earth floor
2	6.5	12.0	Free-standing tented structure with stone walls and a tamped floor
3	6.0	15.0	Built into slope with tamped earth floor and hearth
4	10.0	8.0	Free-standing structure with two internal areas each 10 m²; tented structure with tamped earth floor and hearth in one area
5	6.0	6.0	Free-standing structure with tamped earth floor and hearth
21	9.5	0.0	Built into slope
25	9.6	0.0	Built into slope
26	5.5	0.0	Free-standing
28	5.5	0.0	Free-standing
35	11.2	0.0	Built into slope
36	9.6	0.0	Free-standing
37	5.4	0.0	Free-standing
42	7.0	0.0	Built into slope with tamped earth and gravel floor
43	5.0	0.0	Free-standing
46	5.5	0.0	Built into slope
47	5.5	0.0	Free-standing
53	7.6	0.0	Free-standing
55	4.0	0.0	Free-standing
57	9.6	0.0	Free-standing
59	6.2	0.0	Free-standing
62	7.0	9.0	Free-standing with a hearth

Fragments from three excavated Chinese opium pipe bowls with circular rims were recovered: two have smooth sides with a single raised ridge, one with dark-orange slip and the other light-orange slip; and one has ten faceted sides, with a light-orange slip. There are two European clay pipes, one from excavation, but there are no details regarding the makers.

Hut 2

This hut site was originally referred to as 'Hut 8' in Smith (1998) and 'Hut 1' in Smith (1997), The Chinese of Kiandra, but was renumbered to 'Hut 2' in 2001. One hundred and forty-one ceramic sherds weighing 1,134.3 g represent a MNV of 21. A European clay pipe brings the total MNV to 22 (Table 9.8).

Three Chinese-made brown-glazed stoneware storage jars were recovered: a barrel jar, rim 330 mm, a globular jar, foot 120 mm, and another unidentified storage jar, which is a spouted or wide-mouthed jar. The small jar is from excavation while the larger jars are surface finds. A small non-Chinese sherd from a single stoneware Bristol-glazed storage jar or bottle was excavated.

There are six Chinese-made tea/tableware vessels: a porcelain shallow dish, rim 145 mm and foot 85 mm, decorated in blue-underglaze Rocks and Orchid pattern; two Winter Green rice bowls, rim diameters 140 mm; another Winter Green rice bowl, foot 55 mm, with a blue-underglaze maker's mark inside the foot; a spoon in Winter Green; and a porcelain bowl of unknown size featuring a clear glaze with a green cast, but no other decoration. This last sherd is a surface find while all others are from excavation.

Seven non-Chinese tea/tableware vessels were recovered: two bowls, a mug, two vessels which are cups or bowls, a plate, and a saucer. Of these, six are white-earthenware and one is buff-earthenware. The plate is from the surface and the remainder from excavation.

Four vessels have transfer-prints: a bowl with a brown pattern, possibly Rhine; two other vessels which may have been cups or bowls, one of which features a green foliate pattern with a foot diameter of 60 mm and the other has an unidentifiable blue pattern with only blue dots present; and a plate of unknown size with an unidentifiable worn blue transfer-print. The other decorative types include: a bowl, rim 140 mm, with a flown-blue floral pattern; and a saucer, rim 180 mm, with a blue sponge-printed pattern. The buff-earthenware vessel, possibly a mug, has vegetal relief-moulding and a Rockingham-type glaze. The interior and part of the exterior have a clear glaze while part of the exterior has the darker Rockingham-type glaze. A similar mug with floral relief-moulding is illustrated by Brooks (2005:28).

Table 9.7: Number, weight and MNV of sherds from KCC Hut 1

	No.	% No.	Wt (g)	% Wt	Wt/No.	MNV	No./MNV	% MNV
Food / beverage containers								
Chinese	86	66.7	2074.2	97.4	24.1	6	14.3	31.6
Non-Chinese	0		0.0			0		0.0
Food / beverage consumption								
Chinese	4	3.1	11.3	0.5	2.8	3	1.3	15.8
Non-Chinese	13	10.1	24.7	1.2	1.9	5	2.6	26.3
Smoking								
Chinese	26	20.2	18.9	0.9	0.7	3	8.7	15.8
Non-Chinese						2		10.5
Total	129		2129.1			19		

Note: clay pipes are not in catalogue sherd numbers or weights

Table 9.8: Number, weight and MNV of sherds from KCC Hut 2

	No.	% No.	Wt (g)	% Wt	Wt/No.	MNV	No./MNV	% MNV
Food / beverage containers								
Chinese	84	59.6	956.2	84.3	11.4	3	28.0	13.6
Non-Chinese	1	0.7	0.8	0.1	0.8	1	1.0	4.5
Food / beverage consumption								
Chinese	7	5.0	86.5	7.6	12.4	6	1.2	27.3
Non-Chinese	33	23.4	34.8	3.1	1.1	7	4.7	31.8
Smoking								
Chinese	16	11.3	56.0	4.9	3.5	4	4.0	18.2
Non-Chinese						1		4.5
Total	141		1134.3			22		

Note: clay pipe is not in catalogue sherd numbers or weights

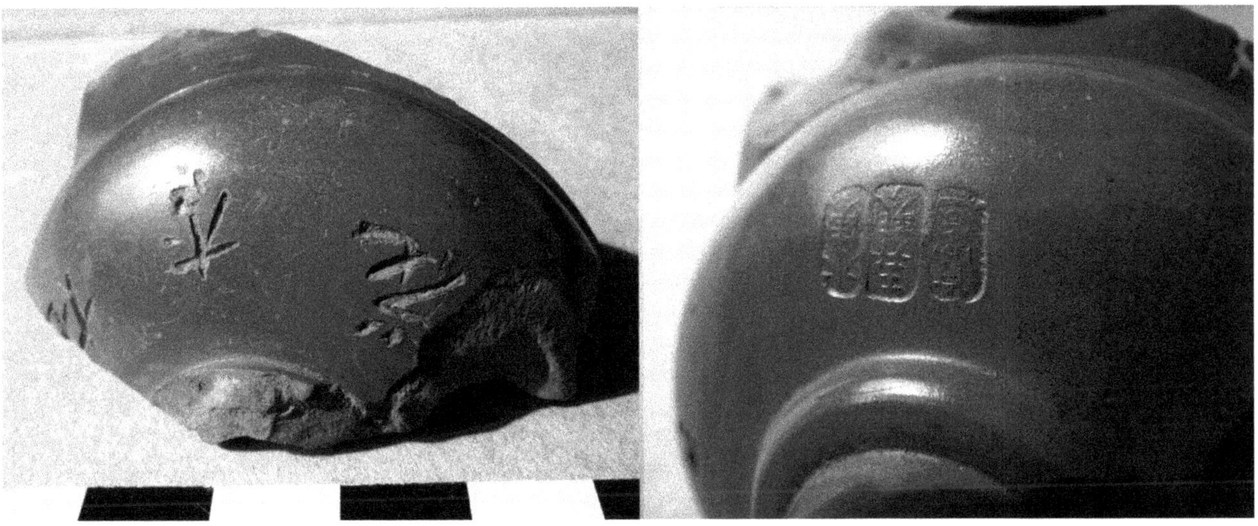

Figure 9.5: Opium pipe bowls from KCC Hut 2 (Cat nos. from left: KCC1998-1096; KCC1998-1097)

Fragments from four opium pipe bowls were recovered from Hut 2. Two have circular rims and smooth sides with a single raised ridge and are dark-orange slipped. Both have marks (Figure 9.5). One of the bowls has three Chinese characters scratched into the lower part - *zai xin hua*, which translates to 'in flower of apricot' (translation, Po-yi Chiang, 2 November 2010). This was done after it was fired, possibly by the owner of the pipe. The other bowl has three impressed rectangular stamps alongside indecipherable characters, which were scratched in the surface during the making of the pipe bowl. The impressed stamps are deciphered as having characters for water and east, possibly the workshop name, *nong qing* (playing musical instrument) and *gao* (high). Another terracotta opium pipe bowl with a light-orange slip could not be identified to type as there are only small sherds. The remaining pipe bowl has a grey stoneware body and a buff-orange slip with a hexagonal smoking surface and sides. The light-orange slipped pipe bowl is from the surface and the others from excavation. A single European clay pipe, manufactured by Davidson, was excavated from Hut 2.

Hut 3

Fifteen squares were excavated: D3-5; E1-4; F1-4; and G1-4. For this research, squares C1 to J7 are considered to be the surface collection associated with Hut 3. Three hundred and twenty-five sherds, weighing 1130.0 g, represent a MNV of 38. A European clay pipe brings the total MNV to 39 (Table 9.9).

Six of the seven storage items are brown-glazed stoneware of Chinese origin: a barrel jar and a barrel jar lid; a globular jar; a wide-mouthed jar, rim 80 mm; a jar, which may have been a spouted or wide-mouthed jar; and a storage jar lid, diameter 90 mm, with a moulded decoration on top and a dark-brown glaze. The interior and flange of the lid are unglazed. The other vessel is a possible ovoid jar with incised horizontal lines around the body. It has a shiny dark-mustard coloured exterior glaze and a matte dark-brown interior glaze. The ovoid jar is a surface find and the others excavated. Although storage jar sherds are generally unburnt, a single sherd from the base of a spouted or wide-mouthed jar was found with an even layer of carbon deposit on the underside. This is a surface find.

Table 9.9: Number, weight and MNV of sherds from KCC Hut 3

	No.	% No.	Wt (g)	% Wt	Wt/No.	MNV	No./MNV	% MNV
Food / beverage containers								
Chinese	78	24.0	589.9	52.2	7.6	7	11.1	17.9
Non-Chinese	0		0.0			0		0.0
Food / beverage consumption								
Chinese	9	2.8	223.9	19.8	24.9	1	9.0	2.6
Non-Chinese	235	72.3	308.9	27.3	1.3	28	8.4	71.8
Smoking								
Chinese	2	0.6	0.5	0.0	0.3	1	2.0	2.6
Non-Chinese						1		2.6
Other								
Chinese	0		0.0			0		0.0
Non-Chinese	1	0.3	6.8	0.6	6.8	1	1.0	2.6
Total	325		1130.0			39		

Note: clay pipe is not in catalogue sherd numbers or weights

A single excavated Chinese-made tableware vessel was recovered: a porcelain Winter Green rice bowl, foot 60 mm, with a blue-underglaze reign-type maker's mark (Figure 9.6).

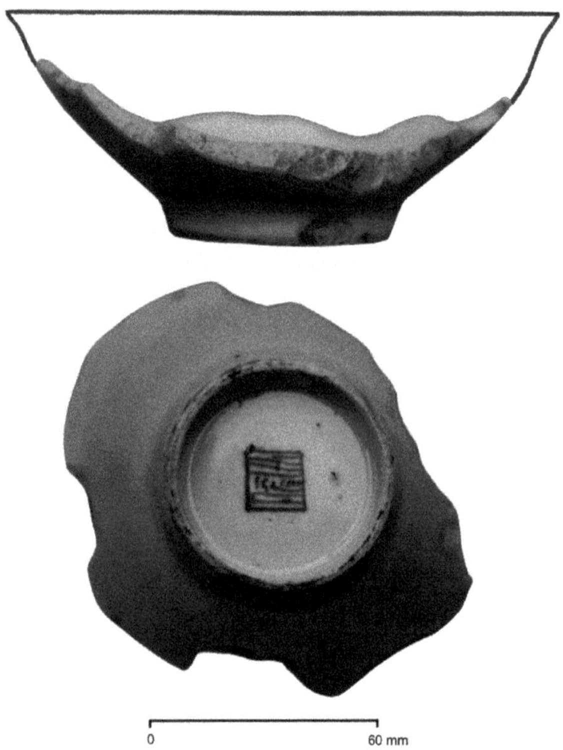

Figure 9.6: Rice bowl with maker's mark from KCC Hut 3 (Cat no. KCC2001-1319)

The 28 tea/tableware vessels of non-Chinese origin are all white-earthenware. They include a bowl, six cups, six cup/bowls, a platter, seven plates, a plate/soup plate, four saucers, a plate/saucer and an unidentifiable vessel, of which the base only is present. Eight vessels are surface finds: a bowl, two cup/bowls, a cup, two plates, a platter and a saucer. The other 20 were excavated.

Transfer-prints are found on the majority of vessels. Nameable transfer-prints are found on 13 vessels: light-blue Albion - a platter and two plates of unknown size; light-blue Asiatic Pheasants - a plate, rim 200 mm, a vessel which is either a plate or soup plate, and a cup, rim 100 mm; Athens - a grey plate and a green cup; Rhine – a blue cup, a green saucer, rim 170 mm, and a grey cup/bowl; blue Willow - two plates of unknown size. Other transfer-printed vessels include: a bowl with a purple seaweed-like pattern; a matching saucer and cup, rim 90 mm, in a green scenic pattern with a geometric border; a cup/bowl with a blue scenic pattern; a cup/bowl with a blue scenic pattern on the exterior and a geometric pattern on the interior; a cup/bowl with a green scenic pattern; a cup/bowl with a grey linear and scrolled pattern; a cup/bowl with an unidentifiable black pattern; a green geometric patterned plate; a green floral patterned saucer, foot 75 mm and cup well 60 mm; a black marble-like patterned plate or saucer; and an unidentifiable vessel, possibly tableware, with a purple floral pattern. An unidentifiable maker's mark featuring an impressed shield with illegible script was found. No decoration, apart from a clear glaze with a blue tinge in the pooling, is present on the sherd. Other decorative types include: a cup with a flown-black floral pattern; a cup with a flown-blue floral pattern; and a purple sponge-printed saucer with a repeated leaf motif.

Small fragments of a Chinese opium pipe bowl with a light-orange slip were recovered from the surface. It is unidentifiable to type. A single European clay pipe was excavated.

A stoneware salt-glazed ink bottle was excavated.

Hut 4

Hut 4 may either have been a single structure or two adjoining occupation areas. Squares associated with surface collection are Q3 to Y13. Excavated squares were: V6, V9; U7-8; T8; and W7-9. Three hundred and twenty-six sherds, weighing 901.3 g, represent a MNV of 27. A European clay pipe brings the total MNV to 28 (Table 9.10).

All of the storage vessels are Chinese in origin. They include seven brown-glazed stoneware vessels: a globular jar, foot 110 mm; a wide-mouthed jar; two liquor bottles; a medium-large storage jar; and two small-medium storage jars. There are also two buff-earthenware unglazed wide-mouthed jar lids.

There are six porcelain Chinese-made tea/tableware vessels: a serving bowl, rim 180 mm, with Four Seasons pattern; two Winter Green rice bowls, rim diameters 140 mm and 150 mm; a rice bowl with Bamboo pattern; and a Winter Green liquor cup, rim 47 mm, foot 17 mm and height of 20 mm, and a mark on the base (Figure 9.7). Two of the celadon-glazed rice bowls are surface finds while the other vessels are from excavation.

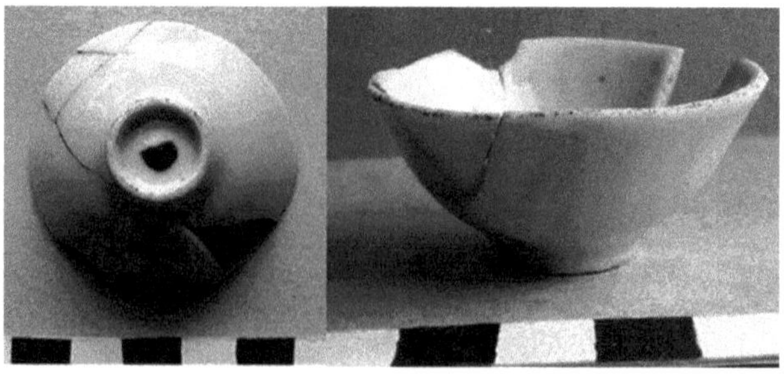

Figure 9.7: Liquor cup from KCC Hut 4 (Cat no. KCC2001-1709)

Table 9.10: Number, weight and MNV of sherds from KCC Hut 4

	No.	% No.	Wt (g)	% Wt	Wt/No.	MNV	No./MNV	% MNV
Food / beverage containers								
Chinese	209	64.1	769.4	85.4	3.7	9	23.2	32.1
Non-Chinese	0	0.0	0.0			0		0.0
Food / beverage consumption								
Chinese	26	8.0	56.3	6.2	2.2	6	4.3	21.4
Non-Chinese	82	25.2	65.4	7.3	0.8	9	9.1	32.1
Smoking								
Chinese	7	2.1	4.5	0.5	0.6	2	3.5	7.1
Non-Chinese						1	0.0	3.6
Other								
Chinese	2	0.6	5.7	0.6	2.9	1	2.0	3.6
Non-Chinese	0		0.0			0		0.0
Total	326		901.3			28		

Note: clay pipe is not in catalogue sherd numbers or weights

The nine non-Chinese tea/tableware vessels include: three bowls; three vessels which are either cups or bowls; two plates; and one saucer. Six vessels are decorated in transfer-prints: a blue Fibre bowl; a bowl with a green geometric pattern; a bowl, rim diameter 180 mm, with a purple geometric pattern; a blue Forest cup/bowl; a purple Cable plate; and a green Rhine plate. Other decorative types include: a flow purple cup/bowl with a floral pattern; a flow blue cup/bowl; and a saucer with a green sponge-printed circular pattern. The saucer and one of the cup/bowls are surface finds; all other vessels are from excavation.

Fragments from a minimum of two Chinese opium pipe bowls were recovered, both with light-orange slip: one pipe bowl from excavation has a circular rim and ten faceted sides; and the other, a surface find, rim 70 mm, is unidentifiable to type. A minimum of one European clay pipe was found, comprising small sherds from the surface.

A small porcelain dish, rim diameter 70 mm, from a Chinese lamp (personal communications, Dr Priscilla Wegars, 31 August 2011) was recovered from the surface. It is partially glazed on the interior; the rim and the exterior are unglazed (Figure 9.8). The dish is thickly potted.

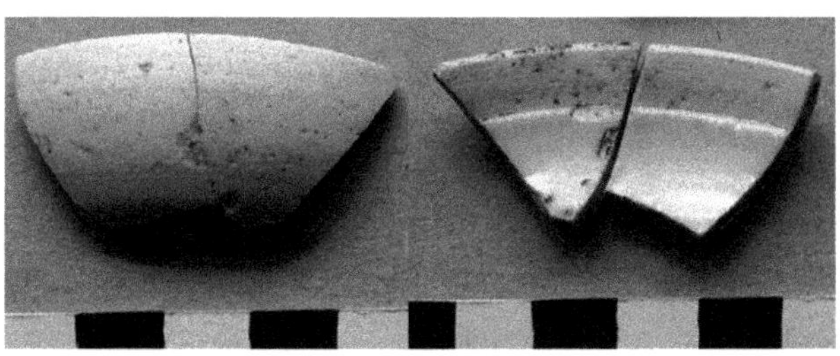

Figure 9.8: Lamp dish sherds from KCC Hut 4 (left: exterior; right: interior; Cat nos. KCC2001-1073, KCC2001-1087)

Hut 5

Hut 5 was located to the northwest of Hut 4. Squares associated with Hut 5 are Q14 to Y21. Excavated squares include: T17; U15-17, U20; and V16. There are 206 sherds, weighing 907.7 g, representing a MNV of 31 (Table 9.11).

The storage containers are all Chinese-made brown-glazed stoneware vessels including: a wide-mouthed jar, a liquor bottle and an unidentified medium-large storage jar. The wide-mouthed jar is a surface find and the other two are from excavation.

Six Chinese-made tea/tableware vessels were recovered: a rice bowl and spoon in Winter Green; and a liquor cup in Four Seasons, rim 45 mm; a small shallow dish with a clear glaze and traces of pink-red enamel on the interior, possibly Four Seasons; a lid with a blue-underglaze decoration, the edge of the lid being unglazed; and a hollowware vessel featuring a clear glaze with green cast on the exterior, and an unglazed interior. There is only a small sherd from this last vessel, which may have been from a teapot, body diameter 90-100 mm. The celadon-glazed rice bowl and Four Seasons liquor cup are from excavation; all the others are surface finds.

All the non-Chinese tea/tableware vessels are white-earthenware and include five bowls; three cups; seven vessels which are either cups or bowls; three plates; a vessel which may have been a plate or soup plate; a saucer; and an unidentifiable vessel that is possibly tableware. Three bowls, three cup/bowls and the unidentifiable vessel are surface finds; the other vessels are from excavation.

Transfer-prints are found on 16 of the 21 vessels. Vessels with nameable transfer-prints include: two bowls in blue in a variation of the Cable pattern, Cable A1 (Figure 9.9), rim diameters 125 mm; a grey Rhine cup with a rim diameter of 100 mm; three vessels which are either cups or bowls, one in blue Antique, one in dark greenish-blue Antique, and the other in green Crystal; a plate in light-blue Asiatic Pheasants, rim 260 mm; a plate in purple Cable; a saucer, rim 175 mm, with a blue Athens pattern. Small sherds only were recovered from an unidentifiable vessel in blue Fibre. Other transfer-printed vessels include: a bowl in a green geometric pattern; a cup in an unidentifiable black pattern; three vessels which are either cups or bowls, one in a blue floral pattern, one in a green floral/foliate pattern, and the other in a purple squiggly-line pattern; and a plate or soup plate in blue which may have been Eton College. Other decorative types include: a bowl with an unidentifiable flown-blue pattern; a cup or bowl in a flown-purple floral pattern; a sponge-printed bowl featuring a purple repeated leaf motif; a blue sponge-printed cup; and a plate with a relief-moulded floral pattern, rim 280 mm, and an impressed base mark of 'PG' (Figure 9.10). This mark may belong to Sampson Bridgwood & Son who used the initials PG to signify 'Parisian Granite' from 1870 onwards (Godden 1991:102). They used a mixture of impressed and printed marks, with printed marks only from 1885 onwards (Godden 1991:102).

Fragments from a single light-orange slipped Chinese opium pipe bowl were recovered but cannot be identified to type.

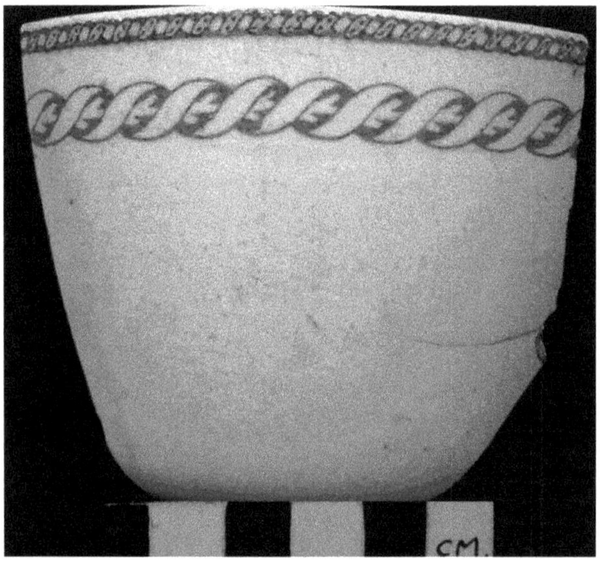

Figure 9.9: Cable A1 (Cat no. KCC2001–1812)

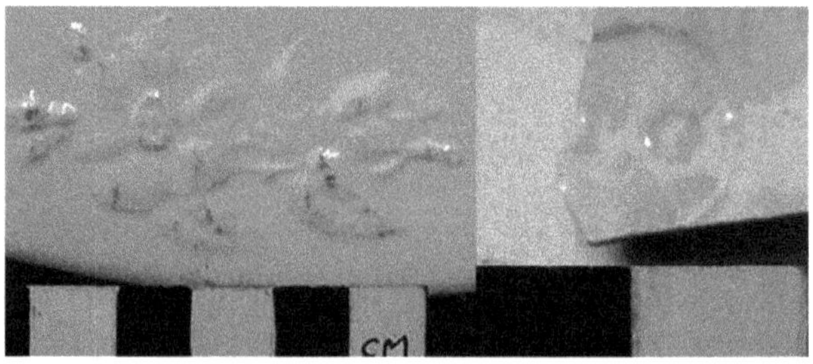

Figure 9.10: Floral pattern and maker's mark on plate from KCC Hut 5 (Cat no. KCC2001-1661)

Hut 21

In 2003, a surface collection was made from an area to the southeast of the 2001 main grid, which covered Hut 21. One hundred and fifty-two sherds, weighing 602.6 g, represent a MNV of 19 (Table 9.12). All sherds are surface finds, collected in 2003 and 2007.

The storage containers are all Chinese-made brown-glazed stoneware: a spouted jar and two other storage jars, one small and the other large, which cannot be identified to type as there are no diagnostic sherds. Two base sherds from a spouted or wide-mouthed jar KCC2003-1298 and KCC2003-1712 have with an even layer of carbon deposit from the underside extending about half way through the sherds.

Two Chinese-made tableware vessels were recovered: a Winter Green rice bowl, rim 140 mm; and a porcelain cylindrical vessel, of which there is only a small sherd. It has a clear glaze with a green cast on the exterior and was unglazed on the interior and may have been part of a teapot, body diameter 90-100 mm.

There are 13 white-earthenware tea and tableware vessels including: one bowl, one cup, five vessels which are cups or bowls, one saucer, three vessels which are plates or saucers and two which are unidentifiable but are also possibly tableware.

Table 9.11: Number, weight and MNV of sherds from KCC Hut 5

	No.	% No.	Wt (g)	% Wt	Wt/No.	MNV	No./MNV	% MNV
Food / beverage containers								
Chinese	70	34.0	277.4	30.6	4.0	3	23.3	9.7
Non-Chinese	0		0.0			0		0.0
Food / beverage consumption								
Chinese	16	7.8	17.8	2.0	1.1	6	2.7	19.4
Non-Chinese	115	55.8	610.9	67.3	5.3	21	5.5	67.7
Smoking								
Chinese	5	2.4	1.6	0.2	0.3	1	5.0	3.2
Non-Chinese						0		0.0
Total	206		907.7			31		

Table 9.12: Number, weight and MNV of sherds from KCC Hut 21

	No.	% No.	Wt (g)	% Wt	Wt/No.	MNV	No./MNV	% MNV
Food / beverage containers								
Chinese	99	65.1	522.8	86.8	5.3	3	33.0	15.8
Non-Chinese	0	0.0	0.0	0.0		0		0.0
Food / beverage consumption								
Chinese	2	1.3	11.0	1.8	5.5	2	1.0	10.5
Non-Chinese	48	31.6	62.8	10.4	1.3	13	3.7	68.4
Smoking								
Chinese	3	2.0	6.0	1.0	2.0	1	3.0	5.3
Non-Chinese						0		0.0
Total	152		602.6			19		

Twelve of the vessels feature transfer-prints: a bowl, foot diameter 80mm with a black scenic pattern; a cup with a brown geometric pattern; a cup/bowl with a blue floral pattern; a cup/bowl with a brown foliate pattern; three cup/bowls with green scenic patterns; a saucer in blue Rhine; a plate/saucer in green Rhine; and a plate/saucer in black Fibre; a plate/saucer in blue floral or scenic pattern; and an unidentifiable vessel with purple Cable. The unidentifiable vessel features a purple sponge-printed decoration, most of which has flaked off.

Fragments from a light-orange slipped pipe bowl, with a circular rim and ten faceted sides, were recovered.

Hut 25

This hut was located within the area examined by Smith (1998) and lies immediately to the west of Hut 2. All sherds are from the surface. Forty-one sherds weighing 327.9 g represent a MNV of seven. A European clay pipe brings the total MNV to eight (Table 9.13).

The three containers are Chinese-made brown-glazed stoneware vessels: a globular jar, rim 120 mm; another large storage jar and a smaller storage jar.

A single sherd from Chinese-made cup or rice bowl was found with Winter Green decoration. There was also one non-Chinese vessel, a white-earthenware bowl, rim 140 mm, which features a green transfer-printed Rhine pattern.

There is a minimum of one opium pipe bowl in light-orange slip with a rim diameter of 75 mm and ten faceted sides. Two impressed stamps with Chinese characters are on the side of the bowl (Figure 9.11); one stamp has the character *hu* (pot) and the other contains the character *shan* (mountain) (translation, Po-yi Chiang, 2 November 2010). A European clay pipe was made by Davidson of Glasgow.

A brown-glazed stoneware sherd from a bottle or jar, possibly an ink bottle, has a shoulder diameter of 110 mm (Figure 9.12). The sherd has straight sides and a shoulder sloping at about 30° towards the neck. The shape resembles an Anglo-style ink bottle. There is patchy glaze on the interior, possibly only drips from the exterior glazing. The sherd is compared to Chinese and British ink bottles (Figures 9.13).

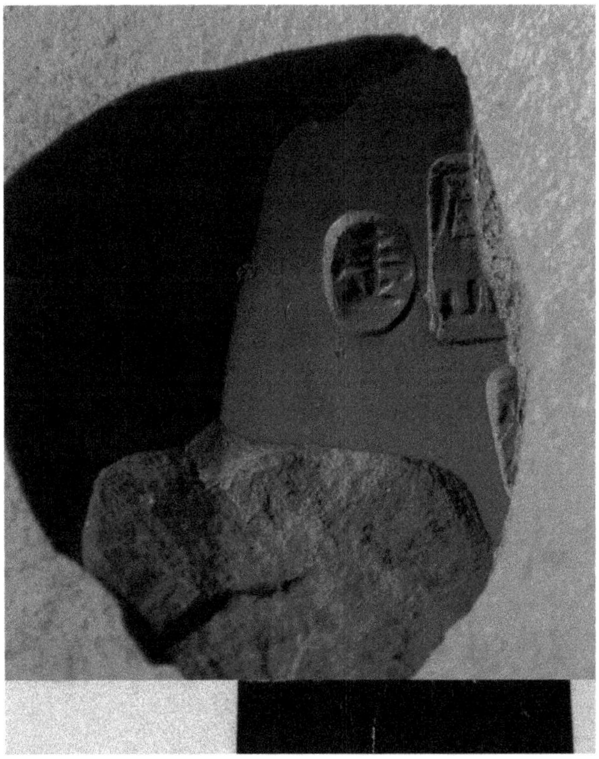

Figure 9.11: Impressed stamps on opium pipe bowl from KCC Hut 25 (Cat no. KCC1998-1076)

Hut 26

Hut 26, to the north of Hut 2, was located within the gridded area examined by Smith (1998). There are 16 sherds, weighing 231.0 g, representing a MNV of five. All finds are from the surface.

A single brown-glazed stoneware globular jar with a rim diameter of less than 120 mm is Chinese-made.

One porcelain rice bowl or cup is decorated with Winter Green. Only small sherds are present of the non-Chinese tea/tableware vessels which are all white-earthenware: a cup/bowl shape has a green transfer-printed geometric scrolled pattern; a saucer features an unidentifiable flown-blue decoration; and an unidentifiable vessel, of which the base only is present, features an unidentifiable blue transfer-print and is possibly tea or tableware.

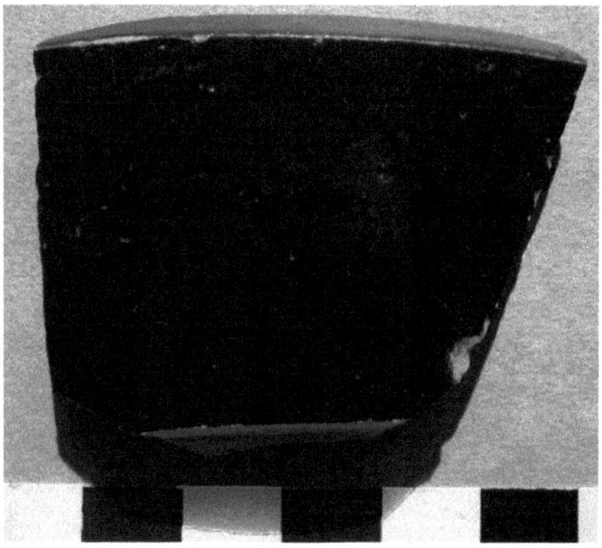

Figure 9.12: Brown-glazed sherd from KCC Hut 25 (Cat no. KCC1998-1055)

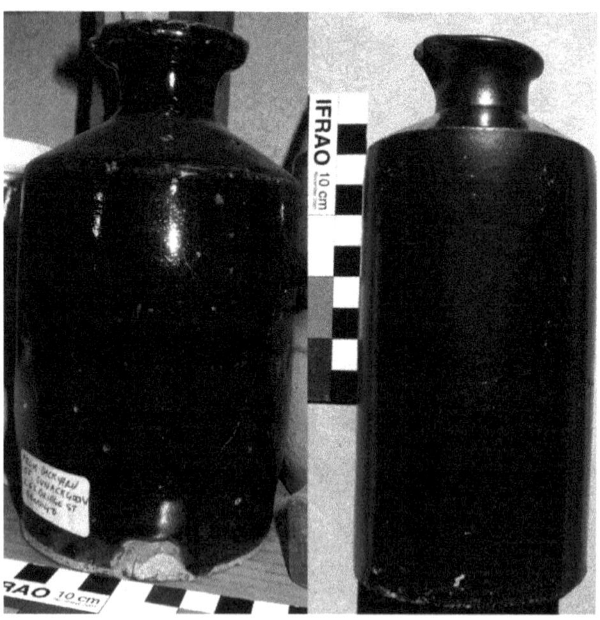

Figure 9.13: Chinese ink bottle (left) and British ink bottle made by Lovatt & Lovatt, Nottingham, post-1895 (private collections)

Hut 28

Hut 28 was located alongside Hut 26, also within the gridded area examined by Smith (1998). Twenty-eight sherds weighing 700.2 g represent a MNV of 12 (Table 9.14), all from the surface.

The storage vessels are all Chinese-made brown-glazed stoneware: a barrel jar, rim 330 mm; a barrel jar lid, rim 350 mm; a liquor bottle; and a small-medium storage jar. A porcelain Winter Green rice bowl, rim 140 mm, and a spoon were recovered. There are two tea/tableware vessels of white earthenware: a cup, rim 90 mm, in light-blue transfer-printed Asiatic Pheasants pattern; and a vessel which is either a small plate or saucer with no decoration present apart from a clear glaze.

A small terracotta sherd is possibly from an Australian-made vessel. It is wheel-thrown, unglazed and has a rim diameter of 45-50 mm. The depth of the sherd to a change in direction of form is 22 mm (Figure 9.14, top left). This sherd may have been from the flange of a lid, possibly from a Water Monkey, a type of water cooler. Visually, the sherd is of a similar size (diameter and thickness), and body, to the lid pictured below the sherd.

Fragments from three Chinese terracotta opium pipe bowls were found: a dark-orange slipped bowl, with a circular rim, smooth sides and a single raised ridge; a light-orange slipped bowl with an octagonal smoking surface and sides, and an indented line parallel to the rim; and a light-orange slipped bowl, of which only the circular smoking surface is present, diameter 75 mm.

Hut 35

Hut 35 was situated immediately to the southeast of Hut 4. Squares associated with Hut 35 are R1 to Y3. There are 126 sherds weighing 287.6 g representing a MNV of 14. A European clay pipe brings the MNV to 15 (Table 9.15). All vessels are surface finds.

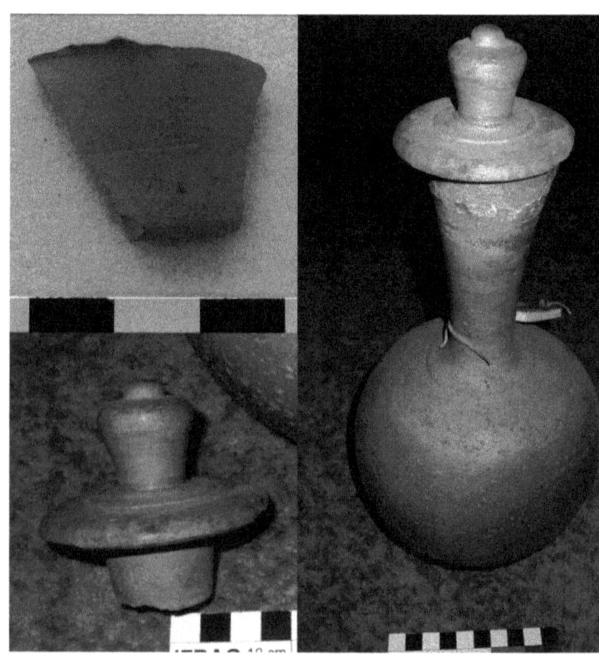

Figure 9.14: Unglazed terracotta sherd from KCC Hut 28 – top left (Cat no. KCC1998-1082); Water Monkey showing the flange of the lid (private collection)

Table 9.13: Number, weight and MNV of sherds from KCC Hut 25

	No.	% No.	Wt (g)	% Wt	Wt/No.	MNV	No./MNV	% MNV
Food / beverage containers								
Chinese	34	82.9	290.2	88.5	8.5	3	11.3	37.5
Non-Chinese	0		0.0			0		0.0
Food / beverage consumption								
Chinese	1	2.4	0.9	0.3	0.9	1	1.0	12.5
Non-Chinese	3	7.3	4.2	1.3	1.4	1	3.0	12.5
Smoking								
Chinese	2	4.9	1.9	0.6	1.0	1	2.0	12.5
Non-Chinese						1	0.0	12.5
Other								
Unknown	1	2.4	30.7	9.4	30.7	1	1.0	12.5
Total	41		327.9			8		

Note: clay pipe is not in catalogue sherd numbers or weights

Table 9.14: Number, weight and MNV of sherds from KCC Hut 28

	No.	% No.	Wt (g)	% Wt	Wt/No.	MNV	No./MNV	% MNV
Food / beverage containers								
Chinese	15	53.6	682.3	97.4	45.5	4	3.8	33.3
Non-Chinese	0		0.0			0		0.0
Food / beverage consumption								
Chinese	3	10.7	5.9	0.8	2.0	2	1.5	16.7
Non-Chinese	4	14.3	2.6	0.4	0.7	3	1.3	25.0
Smoking								
Chinese	6	21.4	9.4	1.3	1.6	3	2.0	25.0
Non-Chinese						0		0.0
Total	28		700.2			12		

Table 9.15: Number, weight and MNV of sherds from KCC Hut 35

	No.	% No.	Wt (g)	% Wt	Wt/No.	MNV	No./MNV	% MNV
Food / beverage containers								
Chinese	44	34.9	186.0	64.7	4.2	3	14.7	20.0
Non-Chinese	0		0.0			0		0.0
Food / beverage consumption								
Chinese	4	3.2	7.3	2.5	1.8	4	1.0	26.7
Non-Chinese	61	48.4	81.4	28.3	1.3	6	10.2	40.0
Smoking								
Chinese	17	13.5	12.9	4.5	0.8	1	17.0	6.7
Non-Chinese						1		6.7
Total	126		287.6			15		

Note: clay pipe is not in catalogue sherd numbers or weights

Two Chinese-made brown-glazed stoneware jars were recovered: a barrel jar, rim 330 mm, and a wide-mouthed jar, rim 85 mm. A possible ovoid storage jar has an exterior shiny mustard glaze and an interior matte brown glaze.

The Chinese-made tableware vessels are: a porcelain liquor cup in Winter Green, rim 45 mm; two porcelain rice bowls or cups in Winter Green; and a stoneware rice bowl with a possible Bamboo pattern. The underglaze is a green-blue colour and the glaze has a green cast.

Six tea and tableware vessels are of non-Chinese origin, five of which have transfer-printed decoration: a bowl, foot diameter 80 mm, with a grey foliate pattern; a cup in grey Rhine; a cup or bowl with a grey scenic pattern; a plate in light blue Asiatic Pheasants; and an unidentifiable vessel, which is possibly tableware, with a green transfer-printed pattern. The last vessel is either a cup or bowl with a blue sponge-printed pattern.

Fragments were found from a single terracotta opium pipe bowl with a light-orange slip, circular rim and ten faceted sides. There was one European clay pipe, of which only small fragments are present.

Hut 36

Hut 36 was located between Huts 3 and 35. Squares associated with it are J1 to P6 and Q1-4. There are 80 sherds, weighing 299.4 g, representing a MNV of 11 (Table 9.16). All vessels are from the surface.

There are two Chinese brown-glazed stoneware storage jars which cannot be identified to type. One is a large jar and the other a spouted or wide-mouthed jar, foot 110 mm.

Sherds from a Chinese porcelain bowl of unknown size have a clear glaze with a green cast, but no decoration. There are seven tea/tableware vessels of non-Chinese origin: one cup; three vessels which are either cups or bowls; one saucer; and two vessels, plates or soup plates. Three vessels have transfer-printed decoration: a cup or bowl in a black floral/geometric pattern; a plate or soup plate in blue Albion; and a saucer in blue Rhine. Vessels with other decorative types include a cup with a flown-blue geometric pattern; a cup/bowl with an unidentifiable flown-blue pattern; a plate/soup plate in a sponge-printed purple pattern with a diamond border with repeated flower motif below; and a cup/bowl with an unidentifiable relief-moulded pattern.

The stem of a light-orange slipped Chinese opium pipe bowl was found. The bowl is not identifiable to type.

Hut 37

Hut 37 was located immediately to the west of Hut 3. Sherds from squares A1 to B10 are deemed to be from Hut 37. There are 12 sherds weighing 23.0 g, accounting for a MNV of four. All sherds are from the surface.

Non-diagnostic sherds are from a small-medium Chinese brown-glazed stoneware storage jar. The Chinese-made tea/tableware includes: a porcelain cylindrical vessel, with a clear glaze and green cast, but no other decoration, which may have been a teapot; and a Winter Green rice bowl or cup. There is one non-Chinese vessel: a plate or saucer with a blue sponge-print.

Hut 42

The hut site, known as the Central Feature, was not excavated as it was inhabited by snakes during the 2001 field season. Squares associated with it are I8 to P18 and K7 to P18. There are 66 sherds weighing 224.9 g representing a MNV of 12. A European clay pipe brings the MNV to 13 (Table 9.17). All sherds are from the surface.

Table 9.16: Number, weight and MNV of sherds from KCC Hut 36

	No.	% No.	Wt (g)	% Wt	Wt/No.	MNV	No./MNV	% MNV
Food / beverage containers								
Chinese	58	72.5	262.1	87.5	4.5	2	29.0	18.2
Non-Chinese	0		0.0			0		0.0
Food / beverage consumption								
Chinese	1	1.3	0.5	0.2	0.5	1	1.0	9.1
Non-Chinese	20	25.0	36.4	12.2	1.8	7	2.9	63.6
Smoking								
Chinese	1	1.3	0.4	0.1	0.4	1	1.0	9.1
Non-Chinese						0		0.0
Total	80		299.4			11		

Table 9.17: Number, weight and MNV of sherds from KCC Hut 42

	No.	% No.	Wt (g)	% Wt	Wt/No.	MNV	No./MNV	% MNV
Food / beverage containers								
Chinese	41	62.1	120.0	53.4	2.9	3	13.7	23.1
Non-Chinese	0		0.0			0		0.0
Food / beverage consumption								
Chinese	5	7.6	7.9	3.5	1.6	1	5.0	7.7
Non-Chinese	17	25.8	22.8	10.1	1.3	6	2.8	46.2
Smoking								
Chinese	2	3.0	4.8	2.1	2.4	1	2.0	7.7
Non-Chinese						1		7.7
Other								
Chinese	0		0.0			0		0.0
Non-Chinese	1	1.5	69.4	30.9	69.4	1	1.0	7.7
Total	66		224.9			13		

Note: clay pipe is not in catalogue sherd numbers or weights

The storage items are all Chinese in origin: a brown-glazed stoneware globular jar and liquor bottle; and a buff-earthenware unglazed wide-mouthed jar lid.

A single porcelain Winter Green rice bowl, rim 150 mm was found. Six tea/tableware vessels of non-Chinese origin are present: a cup in grey Rhine; a cup/bowl in green Fibre; a cup/bowl in a transfer-printed black marble-like pattern; a plate/soup plate in blue Albion; a saucer in purple Cable; and a saucer with a flown-blue pattern with vertical stripes and other decoration.

Flaked sherds from a terracotta opium pipe bowl are not identifiable to type as only small sherds were recovered. There are also small sherds from a single European clay pipe.

A salt-glazed stoneware blacking bottle, body diameter 60-65 mm, was found.

Hut 43

Hut 43 was situated immediately to the northwest of Hut 3. Surface sherds collected from squares C8 to H16 are regarded as coming from Hut 43. Twenty-two sherds, weighing 142.6 g, represent a MNV of four from collections made in 2001 and 2007.

A Chinese brown-glazed stoneware liquor bottle was recovered.

A Chinese-made porcelain Winter Green rice bowl, foot 60 mm, was found, along with an unidentifiable non-Chinese tea/tableware vessel, of which the base only is present, decorated in blue transfer-printed Willow.

The stem of a light-orange slipped Chinese opium pipe bowl was found. The bowl is not identifiable to type.

Hut 46

Hut 46 was located about 20 m southwest of the 2001 grid. Five sherds were collected during 2007, however not all sherds were collected from this hut. The five sherds weigh 334.0 g and represent a MNV of four, all Chinese storage containers. One is a barrel jar lid, rim 330 mm, height 49 mm; and another a brown-glazed globular jar. The third vessel is a blue-underglaze decorated ginger jar with a buff-coloured body and a translucent milky glaze which is greyish-white with white crystals. The final vessel features a dark-honey coloured shiny exterior glaze and a dark-brown matte interior glaze. This sherd may have been from a spouted liquor storage jar.

Hut 47

Hut 47 was located immediately to the west of Hut 43. Surface finds are from grid squares A11 to B16. There are only two sherds weighing 2.7 g, both from a small-medium Chinese-made brown-glazed stoneware storage jar.

Hut 53

This hut was located on the southeast side of the Oven. Sherds found on the surface from grid squares K19 to P21 are considered to be from this hut. Nine sherds weighing 44.4 g represent a MNV of five.

Non-diagnostic sherds are from a small-medium Chinese brown-glazed stoneware storage jar.

Four British white-earthenware tea/tableware vessels include: a bowl in a flown-blue floral pattern, rim 160 mm; a cup/bowl in purple Antique; a plate in green Rhine, foot 110 mm; and a plate/saucer in blue Willow.

Hut 55

Hut 55 was located to the southwest of the Oven. Surface sherds were recovered from squares F20 to H23, in and around the hut. There are five sherds weighing 229.5 g, all from a Chinese-made brown-glazed barrel jar, rim 350 mm.

Hut 57

This hut was immediately to the west of Hut 55. Sherds were found on the surface from squares F24 to I25. Five sherds, weighing 24.9 g, represent for a MNV of two. These are from a medium-large Chinese-made brown-glazed stoneware storage jar and a British cup with an unidentifiable purple transfer-print.

Hut 59

Hut 59 was immediately to the west of Hut 57. Squares C28 to G30 are considered to be part of this hut. There are three sherds, weighing 3.1 g, each from a separate vessel. These include: a small-medium Chinese-made brown-glazed stoneware storage jar and porcelain Winter Green rice bowl or cup; and a non-Chinese made unidentifiable vessel, possibly tea or tableware, with a flown-blue floral pattern.

Hut 62

This hut was located in the eastern corner of KCC2003 Grid 1. Thirty-eight sherds, weighing 62.8 g, represent a MNV of 11, all found in excavation (Table 9.18).

Both of the storage items are Chinese-made: a brown-glazed stoneware spouted jar, rim 45 mm; and a buff-earthenware, unglazed wide-mouthed jar lid, rim 70 mm.

Three Chinese tea/tableware vessels were recovered: two porcelain Winter Green vessels that are either cups or rice bowls; and a porcelain Four Seasons decorated liquor cup, rim 45-50 mm. There are six white-earthenware vessels, probably all of British origin. These include: a bowl of unknown size with pinkish-brown and blue bands; a cup or bowl with a purple transfer-printed foliate pattern; a blue transfer-printed Willow plate, rim 250 mm; a plate or saucer in green Rhine; a plate or saucer with an unidentifiable flown-blue pattern; and an unidentifiable vessel, possibly tea or tableware, featuring a grey transfer-printed pattern that may be Asiatic Pheasants.

Surface collection KCC2003-2

KCC2003 Grid 2 covered an area of 45 x 10 m and was pegged out in 5 x 5 m squares, labelled from A1 to B9. A surface collection was made from the area, which covered at least six hut sites. It was not possible to allocate the sherds to particular huts, therefore they are considered here in total. A large number of sherds are from the northern end of the grid, an area with no hut remains, located at the bottom of a steep cliff. A drainage line developed in this location as a result of dredging. Artefacts may have been moved by water from the top after the camp was vacated. There are 738 sherds weighing 7,551.8 g, which represent a MNV of 78 (Table 9.19).

All the food/beverage containers are Chinese in origin. There are 10 brown-glazed stoneware barrel jars: two with rims of 310 mm, three with rims of 330 mm, two with rims of 350 mm and three of unknown size. There are 10 barrel jar lids but some are different sizes to the jars. Three measure 330 mm in diameter and 45-46 mm in height; two are 350 mm in diameter with a height of 47 mm; three have a diameter of 370 mm and a height of 47-52 mm; one is 390 mm wide and 50 mm high; and one is of unknown diameter but has a height of 51 mm. These lids are all thinly glazed on both sides. Other brown-glazed stoneware includes a globular jar; one wide-mouthed jar, rim 90 mm and two of unknown size; two spouted jars, rims 45 mm; two storage jars which are either spouted or wide-mouthed jars; and two liquor bottles. There are two buff-earthenware wide-mouthed jar lids, one of which has a diameter of 100 mm, the other of unknown size. A ginger jar has a green-glazed exterior and a brown-glazed interior.

There are nine Chinese-made tea/tableware vessels from this area: a Four Seasons porcelain serving bowl, rim 180 mm; a stoneware shallow dish with a green-blue underglaze decoration and an interior unglazed stacking-ring, foot 80 mm; a Bamboo decorated porcelain rice bowl; four porcelain Winter Green rice bowls, all with rim diameters of 140 mm; a Four Seasons liquor cup; and a Winter Green spoon.

There are 31 non-Chinese tea/tableware vessels, two of which are bone china, and 29, white-earthenware. The bone china vessels are: a saucer with a clear glaze only; and a cup, rim 95 mm, with a gold band about 18 mm from the rim. The earthenware vessels include: six bowls, two cups, four cup/bowls, one unidentified hollowware vessel, three plates, one plate or soup plate, four saucers, five plate/saucers, and three unidentified flatware vessels. Of these, 20 vessels are decorated with transfer-prints, seven with flown decoration and two with sponge-prints.

Table 9.18: Number, weight and MNV of sherds from KCC Hut 62

	No.	% No.	Wt (g)	% Wt	Wt/No.	MNV	No./MNV	% MNV
Food / beverage containers								
Chinese	14	36.8	22.1	35.2	1.6	2	7.0	18.2
Non-Chinese	0		0.0			0		0.0
Food / beverage consumption								
Chinese	9	23.7	3.3	5.3	0.4	3	3.0	27.3
Non-Chinese	15	39.5	37.4	59.6	2.5	6	2.5	54.5
Total	38		62.8			11		

Table 9.19: Number, weight and MNV of sherds collected from grid KCC2003-2

	No.	% No.	Wt (g)	% Wt	Wt/No.	MNV	No./MNV	% MNV
Food / beverage containers								
Chinese	577	78.2	7168.4	94.9	12.4	33	17.5	42.3
Non-Chinese	0		0.0			0		0.0
Food / beverage consumption								
Chinese	29	3.9	147.0	1.9	5.1	9	3.2	11.5
Non-Chinese	96	13.0	199.2	2.6	2.1	31	3.1	39.7
Smoking								
Chinese	34	4.6	24.3	0.3	0.7	3	11.3	3.8
Non-Chinese						0		0.0
Other								
Chinese	0		0.0			0		0.0
Non-Chinese	2	0.3	12.9	0.2	6.5	2	1.0	2.6
Total	738		7551.8			78		

The transfer-printed vessels are: a blue Fibre bowl; a bowl with a blue scenic pattern, rim 150 mm; a bowl, rim 130 mm, with a blue scenic pattern on the exterior and a floral pattern on the interior; a bowl with a green scenic pattern; a bowl with a black foliate pattern; a cup with a black marble-like pattern; a cup with a blue linear pattern; a cup/bowl with a brown pattern, possibly geometric; a cup/bowl in black Fibre; a plate or soup plate in blue Eton College; a plate in blue Willow, foot 120 mm; a plate with an unidentifiable blue pattern, that is mostly flaked off; a small plate or saucer having a blue geometric pattern with dots and radiating lines; a plate/saucer with a blue geometric pattern; a plate/saucer with brown scrolled lines with berries on a linear background; a plate/saucer with white shapes on a purple stippled background with geometric border; a saucer with a black marble-like pattern; a saucer in blue Magenta; an unidentifiable flatware vessel with a blurred blue pattern; and an unidentifiable flatware vessel with a purple geometric pattern which has dots inside curving double lines. Six of the flown decorated vessels have transfer-prints and one has a flown sponge-printed pattern. The vessels with flown transfer-prints include: a bowl in a blue floral pattern; a cup/bowl with a blue geometric pattern; an unidentified hollowware vessel with a blue marble-like pattern, rim 65 mm; a plate/saucer in a black floral pattern; a saucer in a purple geometric pattern with overlapping Vs around the rim; and an unidentified flatware vessel with a blurred purple pattern. A saucer has a flown-blue sponge-print. Other sponge-printed vessels include a plate, with a shoulder diameter of 150 mm, having most of the pattern flaked off; and a cup/bowl with a blue sponge-print.

Sherds from three Chinese opium pipe bowls were recovered from this area. Two of the bowls have a circular rim and ten faceted sides; one has a dark-orange slip and the other light-orange slip. The final bowl is an irregular octagonal shape; the rim is straight on four sides and concave on the other four shorter sides. There is an impressed geometric pattern around the rim and the bowl is slipped in light-orange.

A salt-glazed blacking bottle and a paste pot were recovered. The pot has no identifying features and may have been a fish/meat paste pot or a toothpaste pot.

Interpretation of hut sites

Ceramic sherds were collected from 21 individual hut sites within the Kiandra Chinese Camp. In addition, the surface collection, KCC2003-2, covered at least six huts.

Origin of ceramics

Half the hut sites (with a MNV of more than five) have more Chinese than non-Chinese made ceramics (Table 9.20). Figure 9.15 shows the relative positions of these huts and the origin of the majority of their ceramic vessels. The surface collection, KCC2003-2, between the numbered huts and the river, is also predominantly of Chinese origin. A comparison of 'surface finds only' to 'excavated and surface finds' from each of the excavated areas in 2001 Grid 1 confirm similar results in terms of whether Chinese or non-Chinese vessels are dominant. In general, the hut sites with more Chinese-made vessels are further from the Temple Complex than those with more non-Chinese vessels.

Table 9.20: Origin of ceramic vessels from KCC huts by percentage MNV

Hut No.	Origin of ceramics (%)			MNV
	Chinese	Non-Chinese	Unknown	
1	63.2	36.8	0.0	19
2	59.1	40.8	0.0	22
3	23.1	76.9	0.0	39
4	64.2	35.8	0.0	28
5	32.3	67.7	0.0	31
21	31.6	68.4	0.0	19
25	62.5	25.0	12.5	8
28	75.0	25.0	0.0	12
35	53.4	46.6	0.0	15
36	36.4	63.6	0.0	11
42	38.5	61.5	0.0	13
62	45.5	54.5	0.0	11

The majority of storage jars are Chinese in origin. The Bristol-glazed storage jar or bottle may have been manufactured in Britain or Australia. The majority of tea and tableware are British. Some of the non-Chinese made vessels may have been manufactured elsewhere, for example, the sponge-printed vessels in Europe, or the Rockingham-glazed teapot/jug and the terracotta water monkey in Australia. There is one vessel of unknown origin, a possible ink bottle.

Dating ceramic sherds

None of the makers' marks from the huts can be conclusively dated. The Chinese reign-type mark contains no identifiable Chinese characters (personal communication, Po-yi Chiang, 2 November 2010). One of the British marks possibly dates to 1870-1885 (Hut 5).

Albion, Asiatic Pheasants, Cable, Fibre, Rhine and Willow were all relatively common patterns in the second half of the nineteenth century. Crystal, Eton College and Magenta, found at other sites in the camp, date to the mid-nineteenth century or just after.

The Antique pattern was identified with reference to the New Zealand historical ceramics database. Examples of this pattern from New Zealand sites have been manufactured by Pinder, Bourne & Co., 1862-1882 (NZHCD 2012). The pattern features a series of acanthus leaves joined by loops around the marly and classical urns and roses in the well.

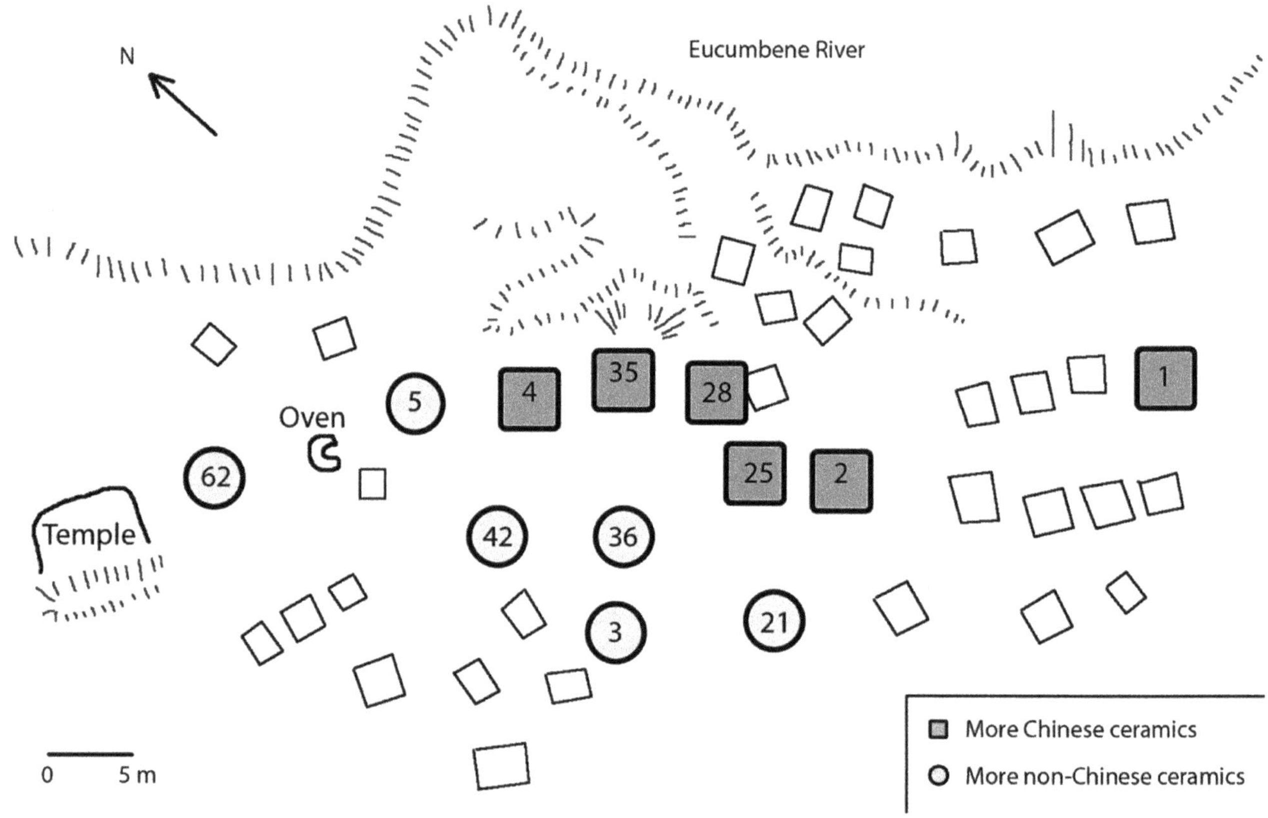

Figure 9.15: Kiandra Chinese Camp showing the hut sites with MNV greater than five and the origin of most of their ceramic vessels

Athens is a pattern name used by several potters for different patterns (Coysh & Henrywood 1982:29). The pattern found at Kiandra is illustrated in Williams (1978:193). The border features oblong vignettes around the marly containing towered buildings framed by scrolls and separated by half-flower designs. The central design consists of swans and a fountain in front of buildings and trees. This pattern was registered by William Adams & Sons in 1849 (Coysh & Henrywood 1982:29).

Three known makers of the Forest pattern are listed by Kowalsky & Kowalsky (1999:487). A similar pattern to those from Kiandra was manufactured by Samuel Alcock & Co., 1826-1859 (Williams 1978:630).

The body and glaze of the sherd from a possible buff-earthenware mug, found at Hut 2, is similar to that of a mug recovered from the Gold Commissioner's quarters at Kiandra. That site dated 1860-1862.

In summary, while many of the ceramics may have been manufactured around mid-nineteenth century, none date later than 1880s. Although no whole plates were recovered, the sherd profiles fit the 1840s to 1860s profiles on Moir's (1997) diagram.

Function of ceramic vessels

Table 9.21 shows the comparative MNV of large and small storage containers and lids. The numbers are too small to make a definitive statement but brown-glazed containers are common across all huts. A single sherd from the base of a spouted or wide-mouthed jar from Hut 3 and two base sherds from Hut 21 have with an even layer of carbon deposit on the underside which extends slightly through the sherds. This could be an indication of the use of a wide-mouthed jar for cooking/heating water or a spouted jar for heating liquor or water.

Table 9.22 details the MNV of Chinese opium pipe bowls and European tobacco pipes found at hut sites which had a total MNV of more than five. The percentage of these pipes as a proportion of total MNV is given. Huts 1, 2, 25 and 28 have a percentage of 20 per cent or higher. Heavy use of opium pipes has been documented in overseas Chinese work camps in USA (Wylie & Fike 1993:291).

The right side of Table 9.22 shows the types of opium pipe bowls. C1 (smooth side with single raised ridge), C3 (ten faceted sides), O1 (octagonal) and H1 (hexagonal) are codes used by Ritchie (1986:373-377). Most of the pipe bowls from KCC hut sites are made of orange terracotta and slipped in either dark-orange or light-orange. A single grey stoneware pipe bowl was recovered. Some pipe bowls cannot be identified to type or decoration, as only flaked sherds are present. Three of the huts sites with a greater proportion of Chinese than non-Chinese made ceramics, Huts 1, 2 and 28, have different pipe bowls to the other huts.

Three opium pipe bowls have makers' marks; all different. In addition to the makers' marks, one opium pipe bowl, from Hut 2, has a mark scratched into the surface, possibly by its owner, 'in flower of apricot'.

Form and decoration

The majority of vessels at KCC were hollowware (Table 9.23). Hut 3, was the only hut not to have more hollowware than flat.

Almost all the huts have at least one vessel of celadon-glazed Winter Green. Three other Chinese patterns identified at the hut sites are: Bamboo, Four Seasons and Rocks and Orchid. Identifiable blue-underglaze patterns are found at the hut sites having more Chinese-made vessels overall. Thirteen British transfer-printed patterns from the hut sites could be named (Table 9.24) and a further 49 patterns are unnameable. Blue is the most common colour found, followed by green and purple (Table 9.25).

Identifying hut occupants

Differences are observed in the assemblages of the hut sites.

The higher percentage of opium/tobacco pipes to food related vessels at Huts 1, 2, 25 and 28 suggest that these were inhabited by one or more single male miners. The assemblages from these huts are similar to those of Chinese work camps in America and New Zealand, where smoking-relating vessels are prevalent.

The huts closer to the Temple Complex have a higher proportion of non-Chinese made ceramic vessels. This could either be an indication of a changing supply source or a gradual change in the makeup of the inhabitants of the camp, namely from single male miners to couples or families. There is only one reference to the presence of a family at KCC, and it includes a European woman and four children (Votes and Proceedings of the NSW Legislative Assembly 1878-1879). Although there is no archaeological evidence that points specifically to females or children, that does not necessarily confirm their absence. Stanin (2004) investigated an overseas Chinese market garden settlement just south of Castlemaine, Victoria, dating 1851-1912, and found no archaeological evidence of women or children, even though it was historically known that some Chinese at the settlement had married European women.

Table 9.21: Number of large/small storage vessels from hut sites at KCC

Hut No.	1	2	3	4	5	21	25	28	35	36	42	62	Total
Large jar/lid	3	2	4	2	1	1	2	2	2	1	1	0	21
Small jar/lid	3	2	3	7	2	2	1	2	1	1	2	2	28
Total	6	4	7	9	3	3	3	4	3	2	3	2	49

Table 9.22: MNV of opium and tobacco pipes from hut sites at KCC

Hut No.	Pipes		Total pipes	% MNV	Dark orange	Light Orange				Grey s/w	Unknown
	Chinese	Non-Chinese			C1	C1	C3	O1	Unknown	H1	
1	3	2	5	26.3	1	1	1				
2	4	1	5	22.7	2				1	1	
3	1	1	2	5.2							1
4	2	1	3	10.8			1		1		
5	1	0	1	3.2							1
21	1	0	1	4.8			1				
25	1	1	2	20.0			1				
28	3	0	3	37.5	1			1	1		
35	1	1	2	13.4			1				
36	1	0	1	9.1					1		
42	1	1	2	15.4							1
62	0	0	0	0.0							

Table 9.23: Number of hollow/flat wares from hut sites at KCC

Hut No.	1	2	3	4	5	21	25	28	35	36	42	62	Total
Hollowware	6	10	14	12	20	9	2	4	8	5	4	5	99
Flatware	2	3	14	3	6	4	0	1	1	3	3	3	43
Total	8	13	28	15	26	13	2	5	9	8	7	8	142

Table 9.24: Nameable patterns from hut sites at KCC

Origin	Pattern	Hut Numbers												
		1	2	3	4	5	21	25	28	35	36	42	62	Other
Chinese	Winter Green	x	x	x	x	x	x	x	x	x		x	x	x
	Four Seasons	x			x	x							x	x
	Bamboo				x					x				x
	Rocks & Orchid		x											
Non-Chinese	Albion			x							x	x		
	Antique					x								x
	Asiatic Pheasants			x		x		x		x				x
	Athens			x		x								
	Cable				x	x	x					x		
	Cable A1					x								
	Crystal					x								
	Eton College					?								x
	Fibre				x	x	x					x		x
	Forest				x									
	Magenta													x
	Rhine	x		x	x	x	x	x		x	x	x		x
	Willow			x									x	x

Key: x = presence; ? = possible presence; Other – huts with MNV < 5

Table 9.25: Colours of transfer-prints from hut sites at KCC

Colour	1	2	3	4	5	21	25	35	36	42	62
Black			x		x	x			x	x	
Blue	x	x	x	x	x	x	x	x	x	x	x
Brown		x				x					
Green			x	x	x	x	x	x			x
Grey	x		x		x			x		x	x
Purple			x	x	x	x			x	x	x

In general, hut assemblages are dominated by hollowware tea and tableware vessels, as would be expected if the inhabitants of the camp were all Chinese. An exception to this pattern is Hut 3. This has the largest proportion of non-Chinese made ceramic vessels and the most flatware, including seven plates. It also has the largest number of cups and saucers. This hut is the only one to have more than one vessel in a particular decorative pattern. In addition, an ink bottle was found, suggesting that at least one occupant was literate.

Two scenarios are suggested: at least one of the occupants of this hut was of European descent; or the occupant/s interacted with people of European descent. Praetzellis and Praetzellis (2001) speculated that overseas Chinese merchants in Sacramento had a large number of non-Chinese tableware vessels because they had to entertain Euro-Americans as part of their business relations. The use of 'familiar symbols of Victorian popular culture' gave the impression of respectability (Praetzellis & Praetzellis 2001:649). The occupant of Hut 3 may have been involved in trade or may have acted as an interpreter for the camp. The Sydney Morning Herald (11 May 1861:7) newspaper correspondent reported that the Chinese in Kiandra were overseen by bosses who had 'some capital' and had 'scraped together sufficient English to enable them to act as interpreters'. Alternatively, the Chinese occupant of Hut 3 may have been living with, or married to, the woman of European descent recorded as living in the camp in 1878. The number of cups and saucers, and the matching crockery suggest a 'respectable' Victorian household.

The other notable hut is Hut 4. It is unusual because it has two rooms. Overall there are more Chinese than non-Chinese made ceramic vessels but separating the assemblage between the two rooms does not clarify any differences in use as similar proportions of vessels are found within each. When compared to the other hut sites, this one has the highest number of storage containers, perhaps suggesting its function as a store; one room for selling and the other a living space or storage room. It is possible that the large number of container sherds found at the northern end of KCC2003-2 washed down along the drainage line, from Hut 4 which is located at the top of the cliff.

Bone Pit and Well

Two other areas were also examined in 2001, the Bone Pit (Grid 3, about 50 m southeast of Grid 1), named because of the amount of animal bone present, and the Well, near the Temple.

Thirty sherds weighing 121.3 g represent a MNV of five from the Bone Pit. There are three Chinese-made stoneware storage jars: two unidentifiable brown-glazed jars, one large and one small; and a jar with a dark-honey coloured external glaze, possibly a spouted liquor storage jar. Two British vessels are: a transfer-printed cup or bowl in blue Fibre; and a plate or saucer with an unidentifiable flown-blue pattern.

Three sherds, weighing 4.6g, came from the Well, representing a MNV of two, both of non-Chinese origin: a plate or saucer with a flown-purple geometric pattern; and a hollowware vessel with a flown-blue floral pattern.

Few ceramic vessels were found in either of these areas which may have been occasional rubbish dumps.

Summary - KCC

The ceramics from Kiandra Chinese Camp have been allotted to individual sites within it in order to reveal differences between those sites. Although it is arguable that surface sherds may have moved from their original place of use, individual vessels were generally found within the confines of a single hut. The major exception to this is the conjoining liquor bottle sherds, Catalogue numbers KCC2001-1489 (Oven) and KCC2001-1093 (Hut 4) found about 15 m apart. That vessel is counted as being associated with the Oven as more pieces of the bottle are found there. The area designated KCC 2001 Grid 1 sloped slightly downhill towards the eastern corner of the grid, that is, from the Oven towards Hut 4.

Chinese-made ceramics

A variety of storage containers are present at KCC (Figure 9.16), including barrel, globular, wide-mouthed, spouted, ginger and ovoid jars, and bottles.

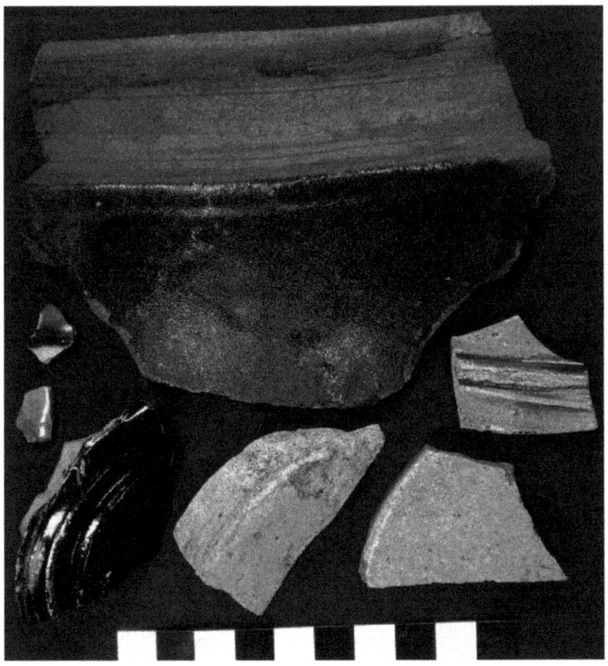

Figure 9.16: Chinese container sherds from KCC (Cat nos. from top: KCC2001-1126 barrel jar, KCC2003-1183 and KCC2003-1282 green-glazed ginger jars, KCC2001-1717 interior bottle seam join, KCC2001-1346.3 storage jar lid, KCC1998-1147 and KCC2001 ginger jar lids)

Differences are seen in the flanged lids associated with blue-underglaze decorated ginger jars. The shapes vary slightly and the lid from Hut 1 is glazed on the top only, while the lid from the Oven is glazed over the whole exterior.

Rice bowls are the most common form of Chinese-made ceramic vessels from the camp. No Chinese tea cups have been identified. Some of the sherds may have been from cups but no rims or feet from tea cups were found. The most common tea and tableware decoration is Winter Green, found at most of the huts and at both communal facilities. It occurs on rice bowls, liquor cups and spoons. Four Seasons pattern is on serving bowls, liquor cups and a small shallow dish. There are no identified rice bowls with this pattern, as the only measurable rims with Four Seasons pattern have diameters of 45 or 180 mm. Figure 9.17 shows a selection of Chinese tableware.

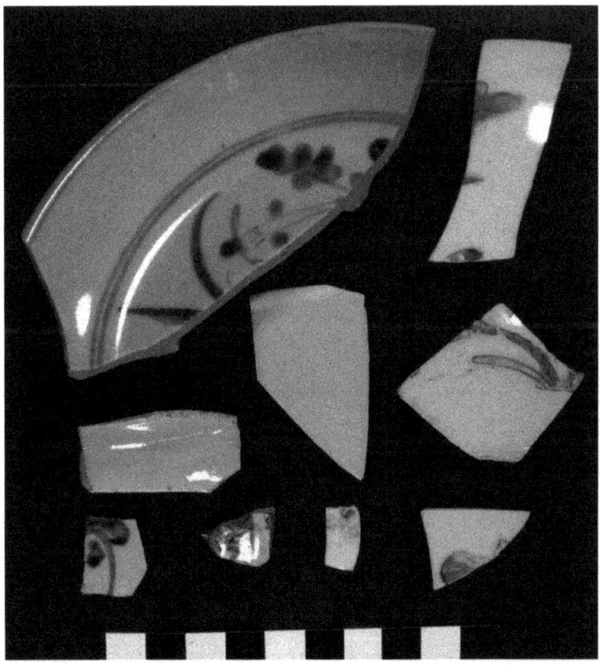

Figure 9.17: Chinese tableware sherds from KCC (Cat nos. from top left: KCC1998-1093 Rocks & Orchid, KCC1998-1123 Four Seasons, KCC1998-1094 Winter Green spoon, KCC2003-1191 Winter Green bowl, KCC2003-1313 Four Seasons, KCC2001-1663 Bamboo bowl, KCC2001-1399 lid, KCC2001-1525 Four Seasons, KCC2003-1308 Four Seasons)

Differences have been noted in the colour of the underglaze for Bamboo patterns on bowls from hut sites to that on a bowl from the Temple Ancillary Building. The bowl from the latter site has a green-blue underglaze while those from the huts are much bluer. An exception is a small sherd that was found between Huts 35 and 36, which may be Bamboo. It has a green-blue underglaze and a green cast to the glaze. A sherd from KCC2003-2 also has a green-blue underglaze but the vessel is a dish with an unidentifiable pattern.

The most common opium pipe bowl type at KCC is light-orange slipped with a circular rim and ten faceted sides. Figure 9.18 shows a selection of sherds. Table 9.26 summarises the Chinese-made MNV at KCC.

Table 9.26: MNV Chinese vessels from KCC

Kiandra Chinese Camp - MNV Chinese vessels

Chinese vessels	Hut 1	Hut 2	Hut 3	Hut 4	Hut 5	Hut 21	Hut 25	Hut 26	Hut 28	Hut 35	Hut 36	Hut 42	Hut 53	Hut 62	Hts <5	BP W	O	T	AB	S	Total
Food/beverage container																					
barrel jar	1	1	1						1	1					1				1	10	17
barrel jar lid	1		1						1						1					10	14
ginger jar															1						1
ginger jar lid	1																1			1	3
globular jar	1	1	1	1			1	1				1			1		1		1	1	11
ovoid jar			1							1											2
wide-mouthed jar	2		1	1	1					1							3	1		3	13
wide-mouthed jar lid				2								1		1			1	1		2	8
spouted jar						1								1			1		1	2	6
storage jar-large				1	1	1	1				1				1	1		1			8
storage jar-small		1	1	2		1	1		1		1		1		3	1				2	15
storage jar lid-small			1																		1
liquor bottle				2	1				1			1			1		1	1	1	2	11
liquor storage jar															1	1					2
Food/beverage consumption																					
bowl/serving bowl		1									1						1		1		4
cup/bowl	1						1	1		2				2	2		1	1			11
lid					1																1
rice bowl		3	1	4	1	1			1	1		1			1		1		1	5	21
serving bowl	1			1																1	3
shallow dish		1			1															1	3
spoon		1			1				1											1	4
teapot?					1	1									1						3
liquor cup	1			1	1					1				1			1			1	7
Smoking																					
opium pipe bowl	3	4	1	2	1	1	1		3	1	1	1			1		4	2	2	3	31
Other																					
lamp				1																	1
Total MNV	12	13	9	18	10	6	5	2	9	8	4	5	1	5	15	3	16	7	8	45	201

Key: Hts <5 – Huts 37, 43, 46, 47, 55, 57, 59, all with less than 5 vessels; BP W – bone pit and well; O – Oven; T – Temple; AB – Ancillary building; S – surface collection KCC2003-2

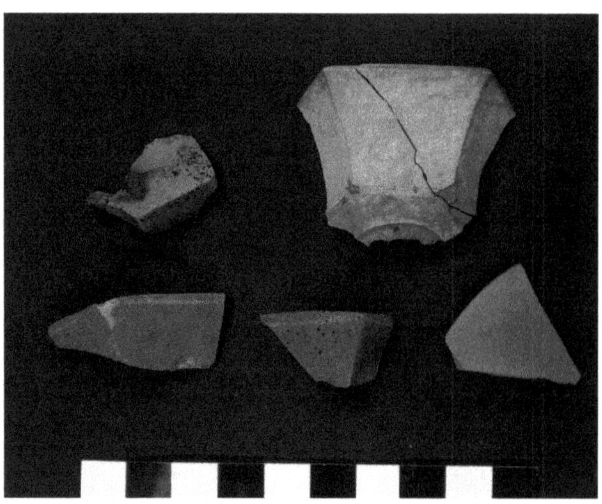

Figure 9.18: Opium pipe bowl sherds from KCC (Cat nos. from top left: KCC2003-1318, KCC1998-1085, KCC2001-1301, KCC2003-1606, KCC2001-1588)

Non-Chinese made ceramics

Transfer-printed patterns are the most common decorative type for non-Chinese tea and tableware. The most common pattern is Rhine, found at ten hut sites, the Temple Ancillary Building and Oven. Other common patterns are Asiatic Pheasants and Fibre. The most common colour is blue. Table 9.27 summarises the non-Chinese MNV at KCC. Figure 9.19 shows a selection of non-Chinese sherds.

Unusual vessels

Three unusual vessels were recovered including two vessel-types previously undocumented at Australia sites. One is a small porcelain dish from a Chinese lamp, from Hut 4. The other is a possible Chinese spouted liquor storage jar; two sherds were found, one from the Bone Pit and the other from Hut 46. In addition, a possible terracotta water monkey, from Hut 28, may have been made in Australia.

Table 9.27: MNV non-Chinese vessels from KCC

Non-Chinese vessels	Hut 1	Hut 2	Hut 3	Hut 4	Hut 5	Hut 21	Hut 25	Hut 26	Hut 28	Hut 35	Hut 36	Hut 42	Hut 53	Hut 62	Hts <5	BP W	O	T	AB	S	Total
Kiandra Chinese Camp - MNV non-Chinese vessels																					
Food/beverage container																					
st.jar/bottle		1																			1
Food/beverage consumption																					
bowl	1	2	1	3	5	1	1			1			1	1			2	1	2	6	28
cup			6		3	1			1	1	1	1			1		1	1		3	20
cup/bowl	1	2	6	3	7	5		1		2	3	2	1	1		1	8	4	6	4	57
flatware																				3	3
hollow ware																1				1	2
jug																			1		1
mug?		1																1			2
pl./soup.pl/dish					1																1
plate		1	7	2	3					1			1	1			1	2	3	3	25
plate/soup plate			1								2	1								1	5
plate/saucer			1			3			1				1	2	1	2	1		3	5	20
platter			1																		1
saucer	2	1	4	1	1	1		1			1	2					3	2	1	5	25
teapot?	1																				1
water monkey lid									1												1
possible tea/table			1		1	2		1		1				1	2		1	1	1		12
Smoking																					
clay pipe	2	1	1	1			1			1		1					2	1	1		12
Other																					
blacking bottle												1						1		1	3
ink bottle			1																		1
ointment pot lid																	1				1
paste pot																				1	1
ink bottle?							1														1
Total MNV	7	9	30	10	21	13	3	3	3	7	7	8	4	6	4	4	20	14	18	33	224

Key: Hts <5 – Huts 37, 43, 46, 47, 55, 57, 59, all with less than 5 vessels; BP W – bone pit and well; O – Oven; T – Temple; AB – Ancillary building; S – surface collection KCC03-2

Water monkeys were produced by potteries in New South Wales, Victoria, South Australia and Queensland (Ford 1995). For example, they appear in an 1883 catalogue, produced by E. D. Middleton, a merchant operating in Melbourne, who advertised water monkeys from Bendigo Pottery, Victoria (Ford 1995:192). They were also advertised in the 1889 Lithgow Pottery catalogue (Ford 1995:82).

Distribution of Chinese / non-Chinese vessels

The huts further from the Temple Complex have greater proportions of Chinese-made ceramic vessels. They have more Chinese tableware patterns and a greater variety of opium pipe bowls. The percentage of their opium and tobacco pipes, compared to the total assemblage, is higher than the other huts. The differences between these two areas of the camp may indicate a change in the supply source over time.

A chi square statistical test determined whether there is a real difference between the assemblages of the huts closer to the Temple Complex in terms of the number of Chinese vessels compared to the number of non-Chinese vessels (Table 9.28). Area 1 includes Huts 1, 2, 4, 25, 28 and 35; Area 2 includes Huts 3, 5, 21, 36, 42 and 62. There is over 99.9 per cent chance that the differences observed between these two areas are real and not a result of vagaries of sampling ($X^2 = 22.71$; $p < .001$) (Drennan 1996).

Although this difference could have resulted from a variety of causes, such as different supply sources to two clan groups, pot hunters scouring one side of the camp only or through earthenware not surviving nearer the river, it is likely that a change in the supply source occurred over time.

Table 9.28: Chi square test comparing the distribution of ceramics in two areas of KCC

Vessel origin	Area 1	Area 2	Total
Chinese	65 (47.19)	39 (56.81)	104
Non-Chinese	38 (55.81)	85 (67.19)	123
Total	103	124	227

In summary, a range of Chinese and non-Chinese made ceramics were found across the camp. Although none are precisely datable, most vessels date to the period 1840-1880. There are no banded patterns or Japanese-made vessels, common in the latter part of the nineteenth century at Australian sites. Most forms from the camp have frequently been found at other overseas Chinese sites. One notable absence is the lack of Chinese tea cups. Indeed, some vessels recorded as cup/bowls may have been tea cups but no rims or feet were found to substantiate this.

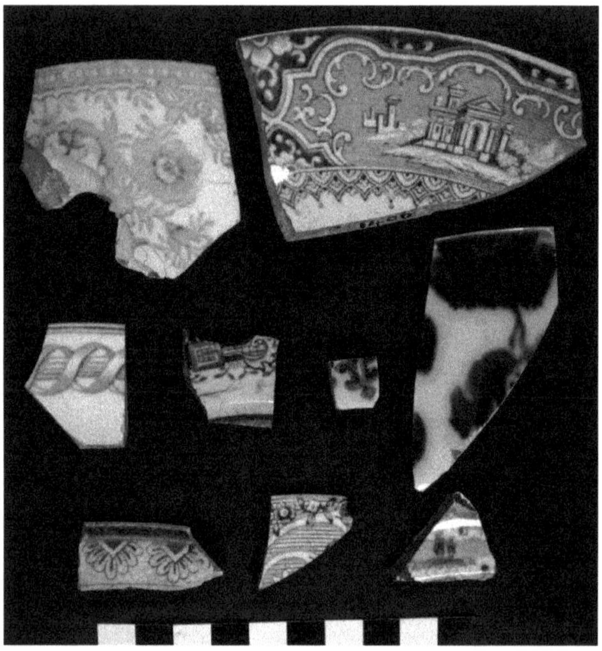

Figure 9.19: Non-Chinese sherds from KCC (Cat nos. from top left: KCC2001-1349.2 Asiatic Pheasants, KCC20001-1406 Athens, KCC2001-Cable, KCC2001 Willow, KCC2003-1772 purple sponge-print, KCC2001-1539 flown-blue, KCC2001-1503 purple transfer-print, KCC2001-1129 Rhine, KCC2001-1426.2 Albion)

CHAPTER 10

Kiandra region - outlying camps and Township stores

The details of ceramics recovered from isolated miners' huts in the Kiandra Valley and the Chinese stores in Kiandra Township are covered in this chapter. Figure 10.1 shows their position relative to the Chinese Camp.

OUTLYING CHINESE CAMPS

Chinese miners' camps in five areas close to Kiandra were examined by Smith (2006) and Gant-Thompson (2008): East Eucumbene River, Eucumbene Crossing, Southwest Eucumbene, Giandarra Gully and New Chum Hill. All these sites have archaeological remains but only Sawyers Hill, in Southwest Eucumbene, and New Chum Hill were historically known. The ceramics from these sites are detailed below. Tables are presented for sites with a minimum number of vessels (MNV) greater than five. Although the results are reported in terms of MNV, there are no whole vessels.

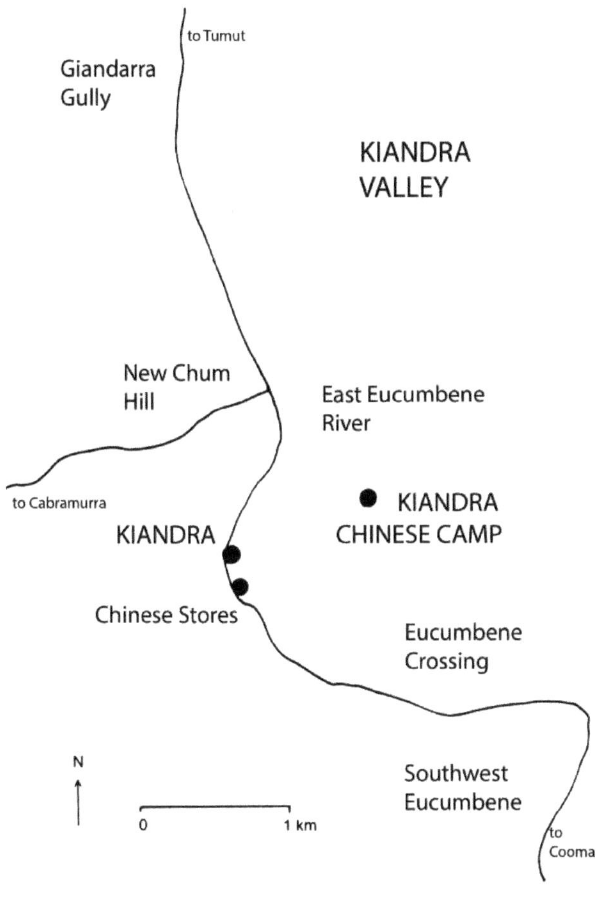

Figure 10.1: Location of Kiandra Valley Chinese settlements (after Gant-Thompson 2008:215)

East Eucumbene River

Smith's 2003 survey of this area identified a small Chinese camp located at Nungar Creek, about 500 m north of the main camp (Smith 2006:674). The camp probably consisted of several tented structures which were occupied from the 1860s to the mid-1870s (Gant-Thompson 2008). The remains of three of these structures were recorded by Smith, designated Huts 1, 2 and 3, while Gant-Thompson (2008) also collected surface artefacts from them as Sites 636, 635 and 637 respectively. In addition, her survey found other Chinese camps nearby, designated 977 and 1058.

Site 635

This was a tented structure built into a slope with random rubble stone footings, tamped earth floor and an internal area of 8.25 m² (Smith 2006:712). Twenty ceramic sherds, weighing 89.4 g, represent a MNV of 11. A European clay pipe brings the total MNV to 12 (Table 10.1).

The only container is a Bristol-glazed stout or ginger beer bottle from Britain or Australia.

Nine white-earthenware tea/tableware vessels of non-Chinese origin include: a bowl in grey Rhine; two cups or bowls with unidentifiable blue transfer-prints; a cup with a flown-blue scenic pattern on the exterior and a floral pattern on the interior; two plates, rim diameters 235 mm, in blue Willow; a plate, rim 255 mm, in brown Rhine; a plate or saucer in brown Crystal; and a saucer with a flown-blue floral/scrolled pattern. An unidentifiable vessel, of which the rim only is present, has a clear glaze and is possibly tea or tableware.

Fragments from two pipes were recovered: a Chinese opium pipe bowl with light-orange slip, unidentifiable to type; and a European clay pipe, made by Davidson of Glasgow.

Site 636

This hut was a tented structure built into a slope with random rubble stone footings, tamped earth floor and an internal area of 10.5 m² (Smith 2006:712). Fifteen ceramic sherds, weighing 88.6 g, represent a MNV of nine. A European clay pipe brings the total MNV to ten (Table 10.2).

There is one container: a Chinese-made brown-glazed stoneware spouted or wide-mouthed jar.

Eight tea/tableware vessels, all white-earthenware, were recovered: a bowl with a blue transfer-printed floral or scenic pattern; a bowl with a green transfer-printed floral pattern; a cup or bowl with a flown- blue floral/scrolled pattern; a cup or bowl with a flown-green floral pattern; a plate in brown Rhine; a plate in blue Willow, rim 235 mm; a saucer in brown Crystal; and a saucer with a blue sponge-print, rim 180 mm.

A single European clay pipe was recovered.

Site 637

This hut was a tented structure built into a slope, with random rubble stone footings, tamped earth floor and an internal area of 8.75 m² (Smith 2006:712). Twelve ceramic sherds, weighing 110.2 g, represent a MNV of seven. A European clay pipe brings the total MNV to eight (Table 10.3).

The container is of Chinese origin: a brown-glazed stoneware spouted jar. Six non-Chinese white-earthenware tea/tableware vessels were recovered: a bowl in purple Sydney; a clear-glazed cup; a plate with a blue transfer-printed floral pattern and raised relief-moulding around the rim; a plate or saucer with a blue transfer-printed geometric pattern; a plate or saucer with red sponge-printed concentric diamonds; and a saucer with a blue sponge-print.

A single vessel was recovered: a European clay pipe, made by C. Crop of London.

Table 10.1: Number, weight and MNV of ceramic sherds from Site 635

	No.	% No.	Wt (g)	% Wt	Wt/No.	MNV	No./MNV	% MNV
Food / beverage containers								
Chinese	0		0.0			0		0.0
Non-Chinese	1	5.0	8.5	9.5	9.5	1	1.0	8.3
Food / beverage consumption								
Chinese	0		0.0			0		0.0
Non-Chinese	18	90.0	80.4	89.9	5.0	9	2.0	75.0
Smoking								
Chinese	1	5.0	0.5	0.6	0.6	1	1.0	8.3
Non-Chinese						1		8.3
Total	20		89.4			12		

Note: clay pipe is not in catalogue sherd numbers or weights

Table 10.2: Number, weight and MNV of ceramic sherds from Site 636

	No.	% No.	Wt (g)	% Wt	Wt/No.	MNV	No./MNV	% MNV
Food / beverage containers								
Chinese	4	26.7	15.8	17.8	4.0	1	4.0	10.0
Non-Chinese	0		0.0			0		0.0
Food / beverage consumption								
Chinese	0		0.0			0		0.0
Non-Chinese	11	73.3	72.8	82.2	6.6	8	1.4	80.0
Smoking								
Chinese	0		0.0			0		0.0
Non-Chinese						1		10.0
Total	15		88.6			10		

Note: clay pipe is not in catalogue sherd numbers or weights

Table 10.3: Number, weight and MNV of ceramic sherds from Site 637

	No.	% No.	Wt (g)	% Wt	Wt/No.	MNV	No./MNV	% MNV
Food / beverage containers								
Chinese	3	25.0	39.0	35.4	13.0	1	3.0	12.5
Non-Chinese	0		0.0			0		0.0
Food / beverage consumption								
Chinese	0		0.0			0		0.0
Non-Chinese	9	75.0	71.2	64.6	7.9	6	1.5	75.0
Smoking								
Chinese	0		0.0			0		0.0
Non-Chinese						1		12.5
Total	12		110.2			8		

Note: clay pipe is not in catalogue sherd numbers or weights

Site 977

Gant-Thompson (2008) located a hut site which she dated 1870-1880s. Nineteen ceramic sherds, weighing 108.7 g, represent a MNV of 4.

Three brown-glazed stoneware containers of Chinese origin include: a wide-mouthed jar, rim 85 mm; a spouted or wide-mouthed jar; and a liquor bottle.

A single non-Chinese cup was found in a flown-blue floral pattern.

Site 1058

This site was occupied after 1870 (Gant-Thompson 2008). Only one sherd weighing 6.4 g was recovered. It comes from a spouted or wide-mouthed jar.

Eucumbene Crossing

Although not historically documented, two small Chinese miners' camps were found by Smith in 2005. They were located 1.3 and 1.4 km southeast of the main camp in a relatively barren sub-alpine landscape (Smith 2006). Five hut sites were recorded by Smith in these two camps. Gant-Thompson (2008) re-examined this area designating the sites 143 (Smith's Huts 1 and 2) and 125 (Smith's Huts 3, 4 and 5). Two other sites were located nearby, designated 127 and 268.

Site 125

These huts were tented structures with random rubble stone footings and tamped earth floors. The internal area of Hut 3 could not be determined; Hut 4 is 11.25 m²; and Hut 5, 10 m². Huts 3 and 5 were free standing and Hut 4 was built into a slope (Smith 2006:729). The site was occupied post-1870 (Gant-Thompson 2008). Eight ceramic sherds, weighing 44.6 g, represent a MNV of six (Table 10.4).

The storage vessels are all Chinese-made and include one large and one small brown-glazed stoneware storage jar and a green-glazed ginger jar.

The tea/tableware includes: a Chinese porcelain Winter Green rice bowl, rim 140-150 mm, and a British small plate or saucer, foot 80 mm. The British-made vessel has a brown transfer-print but only a band of colour is present on the sherd.

A minimum of one opium pipe bowl with light-orange slip cannot be identified to type.

Site 127

This site was located just north of sites 125 and 143 and covers a number of structures. It was occupied from 1860 to the last quarter of the nineteenth century (Gant-Thompson 2008). Sixteen sherds, weighing 64.2 g, represent a MNV of six (Table 10.5).

There is an unidentifiable small-medium Chinese brown-glazed storage jar.

A minimum of five British transfer-printed white-earthenware tea/tableware vessels includes: a cup, rim 100 mm, in blue Aquatic; a cup or bowl in a blue geometric pattern; a plate with a blue floral pattern; and two blue Willow plates, one with rim 230 mm.

Site 143

Huts 1 and 2 were free standing tented structures with random rubble stone footings and tamped earth floors. Hut 1 has an internal area of 11.25 m² but that of Hut 2 could not be determined (Smith 2006:728-729). The site was occupied 1860-1870s (Gant-Thompson 2008). Forty-three ceramic sherds, weighing 697.3 g, represent a MNV of 14 (Table 10.6).

The storage items are all of Chinese origin including five brown-glazed stoneware jars and an unglazed buff-earthenware wide-mouthed jar lid, rim 90 mm. The brown-glazed jars include: two barrel jars, rim 370 mm; a wide-mouthed jar, rim 85 mm; a globular jar; and another jar which was either a spouted or wide-mouthed jar, foot diameter 130 mm.

A single Chinese-made porcelain Winter Green rice bowl has a rim diameter 130-140 mm. The seven non-Chinese tea/tableware vessels are all white-earthenware. Four have transfer-prints: a saucer in black Antique; a saucer with a green floral pattern; a plate/saucer in blue Willow; and a plate/saucer in blue pattern, possibly Dresden. A plate, of which the underside only is present, has a blue tinge in the clear glaze which may be the result of a flown-blue decoration. A bowl and a plate/saucer have the same purple sponge-printed pattern, a band and repeated leaf motif.

Site 268

This site was located in the vicinity of sites 127 and 143, comprising a number of structures. It was occupied 1860-1870s (Gant-Thompson 2008). Twenty-seven sherds, weighing 95.0 g, represent a MNV of 11 (Table 10.7).

One medium-large unidentifiable Chinese brown-glazed storage jar was recovered.

Ten non-Chinese white-earthenware tea/tableware vessels include: one bowl, one cup; four cup/bowls; two plates; one plate/saucer; and an unidentifiable vessel which is possibly tableware. Eight vessels have transfer-printed decoration: a cup in black Antique; three cups or bowls with blue patterns, one floral, one geometric and the other unidentifiable; one cup or bowl with an unidentifiable green pattern; a plate in blue Willow; a plate a green foliate pattern; and an unidentifiable vessel, of which the base only is present, with an unidentifiable green pattern.

Table 10.4: Number, weight and MNV of ceramic sherds from Site 125

	No.	% No.	Wt (g)	% Wt	Wt/No.	MNV	No./MNV	% MNV
Food / beverage containers								
Chinese	4	50.0	8.7	19.5	2.2	3	1.3	50.0
Non-Chinese	0		0.0			0		0.0
Food / beverage consumption								
Chinese	2	25.0	30.7	68.8	15.4	1	2.0	16.7
Non-Chinese	1	12.5	5.2	11.7	5.2	1	1.0	16.7
Smoking								
Chinese	1	12.5	<0.1			1	1.0	16.7
Non-Chinese						0		0.0
Total	8		44.6			6		

Table 10.5: Number, weight and MNV of ceramic sherds from Site 127

	No.	% No.	Wt (g)	% Wt	Wt/No.	MNV	No./MNV	% MNV
Food / beverage containers								
Chinese	1	6.3	11.6	18.1	11.6	1	1.0	16.7
Non-Chinese	0		0.0			0		0.0
Food / beverage consumption								
Chinese	0		0.0			0		0.0
Non-Chinese	15	93.8	52.6	81.9	3.5	5	3.0	83.3
Total	16		64.2			6		

Table 10.6: Number, weight and MNV of ceramic sherds from Site 143

	No.	% No.	Wt (g)	% Wt	Wt/No.	MNV	No./MNV	% MNV
Food / beverage containers								
Chinese	30	69.8	648.5	93.0	21.6	6	5.0	42.9
Non-Chinese	0		0.0			0		0.0
Food / beverage consumption								
Chinese	1	2.3	6.8	1.0	6.8	1	1.0	7.1
Non-Chinese	12	27.9	42.0	6.0	3.5	7	1.7	50.0
Total	43		697.3			14		

Table 10.7: Number, weight and MNV of ceramic sherds from Site 268

	No.	% No.	Wt (g)	% Wt	Wt/No.	MNV	No./MNV	% MNV
Food / beverage containers								
Chinese	1	3.7	13.9	14.6	13.9	1	1.0	9.1
Non-Chinese	0		0.0			0		0.0
Food / beverage consumption								
Chinese	0		0.0			0		0.0
Non-Chinese	26	96.3	81.1	85.4	3.1	10	2.6	90.9
Total	27		95.0			11		

Vessels with other decorative types are: a bowl with a black sponge-printed pattern; and a plate with a flown-blue pattern.

Southwest Eucumbene

The sites at Southwest Eucumbene River were located about 2.5 km southeast of KCC (Smith 2006; Gant-Thompson 2008). Sherds were collected from hut sites and a water race.

Site 698

This site contained five huts and was occupied post-1870 (Gant-Thompson 2008). A total of 12 sherds, weighing 198.4 g, were recovered. A MNV of five includes: an unidentifiable small-medium Chinese brown-glazed storage jar; a Bristol-glazed bottle; a cup or bowl with a green transfer-printed foliate pattern; a plate with a relief-moulded pattern; and a dark-orange slipped opium pipe bowl, unidentifiable to type.

Site 740

This camp was located at the base of Sawyers Hill and was occupied by about 20 people in the early 1860s (Smith 2006:738). Fifty-three sherds, weighing 298.4 g, represent a MNV of seven (Table 10.8).

Sherds from three food/beverage storage vessels were found: two Chinese brown-glazed stoneware wide-mouthed jars; and a Bristol-glazed bottle with a cork seal finish and a body diameter of 80 mm.

Two white-earthenware saucers of non-Chinese origin were recovered: one, rim 180 mm, with a green transfer-printed ivy pattern; and the other, rim 160 mm, with a repeated blue sponge-printed floral motif with red bands either side.

Fragments from two Chinese opium pipe bowls were found: a light-orange slipped bowl with a circular rim and ten faceted sides; and a dark-orange slipped bowl, rim diameter 80 mm, having a smooth side with a single raised ridge, and an impressed square stamp featuring Chinese characters (Figure 10.2). Most of the stamp is illegible but its right-hand side has characters for water and east, possibly denoting the workshop name (translation, Po-yi Chiang, 2 November 2010).

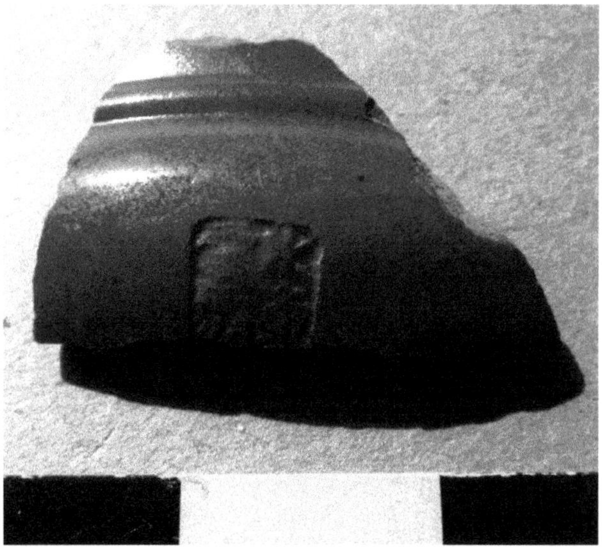

Figure 10.2: Opium pipe bowl sherd from Site 740 (Cat no. KV-1015)

Site 809

Sherds were recovered from along a water race, located about 5 m east of a cluster of Chinese huts. This is the only place in Kiandra, to date, where the Chinese Double Happiness pattern has been found. Nine sherds, weighing 29.6 g, represent a MNV of three: a stoneware rice bowl in Double Happiness; a saucer, rim 180 mm, with a blue transfer-printed floral pattern; and a saucer, foot and cup well diameters 60 mm, with a flown-blue floral pattern. A foot ring, possibly from the transfer-printed saucer, has a blue tinge in the pooled glaze.

Site 1038

This site was a small hut platform, which was occupied 1860-1870s. Eleven sherds, weighing 26.9 g, represent three vessels, all white-earthenware: a cup or bowl with an unidentifiable flown-blue pattern; a cup or bowl with a blue transfer-printed pattern; and a saucer in grey Rhine.

Giandarra Gully

These sites were located just north of New Chum Hill. They include several small Chinese miners' huts (Gant-Thompson 2008).

Site 1079

Site 1079 was occupied from the 1870s (Gant-Thompson 2008). Two sherds, weighing 1.6 g, are from a Chinese Winter Green rice bowl.

Table 10.8: Number, weight and MNV of ceramic sherds from Site 740

	No.	% No.	Wt (g)	% Wt	Wt/No.	MNV	No./MNV	% MNV
Food / beverage containers								
Chinese	22	41.5	60.7	20.3	2.8	2	11.0	28.6
Non-Chinese	2	3.8	151.1	50.6	75.6	1	2.0	14.3
Food / beverage consumption								
Chinese	0		0.0			0		0.0
Non-Chinese	23	43.4	77.8	26.1	3.4	2	11.5	28.6
Smoking								
Chinese	6	11.3	8.8	2.9	1.5	2	3.0	28.6
Non-Chinese						0		0.0
Total	53		298.4			7		

Site 1082

This site was occupied from the 1880s. Four sherds, weighing 360.3 g, are from two vessels: a Chinese brown-glazed barrel jar, rim 350 mm; and a porcelain relief-moulded cup of unknown origin.

New Chum Hill

These sites are located about 2 km northwest of the main camp. A newspaper report from 1886 details a disturbance involving Chinese and European miners working for the Kiandra Gold-mining Company on New Chum Hill, which resulted in the Chinese leaving the claim (The Sydney Morning Herald, 14 September 1886:7).

Site 485

This site covered two huts which were occupied 1875-1900 (Gant-Thompson 2008). Twelve sherds, weighing 27.8 g, represent a MNV of seven (Table 10.9).

All vessels are tea or tableware, one being Chinese-made, a Winter Green rice bowl with a partial blue-underglaze maker's mark (Figure 10.3). The non-Chinese vessels include: a cup or bowl with a blue floral transfer-print; a cup/bowl with a green geometric transfer-print; a cup/bowl with a red geometric transfer-print; a medium plate with two thin blue bands; a possible teapot with a buff-earthenware body and Rockingham glaze; and a bone china clear-glazed unidentifiable vessel that is possibly tea/tableware.

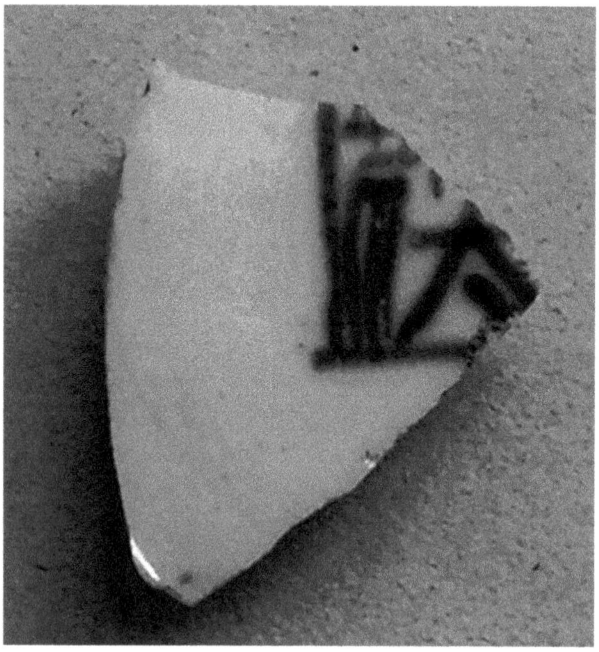

Figure 10.3: Maker's mark on celadon rice bowl from Site 485(Cat no. KV-1307)

Site 742

This was a hut on New Chum Hill with an unknown occupation date (Gant-Thompson 2008). Five sherds, weighing 51.2 g, represent a MNV of five. These include a Chinese brown-glazed wide-mouthed jar; a bone china cup or bowl; a plate or saucer with a green transfer-printed floral pattern; and two unidentifiable vessels which were possibly tableware, one is clear-glazed porcelain and the other clear-glazed white-earthenware.

Site 1069

Site 1069 comprised 11 Chinese huts and was occupied from the 1880s (Gant-Thompson 2008). There was also later European occupation adjacent to these huts. One hundred and two sherds, weighing 554.5 g, represent a MNV of 31 (Table 10.10).

A single container was recovered: a Chinese brown-glazed spouted or wide-mouthed jar.

Twenty-eight tea/tableware vessels include 25 non-Chinese (15 of white-earthenware, one red-earthenware, seven bone china and two porcelain) and three porcelain vessels of unknown origin: a clear-glazed cup and saucer, rim 160 mm, possibly Japanese; and an item, possibly a lid handle from an unidentifiable vessel, which had a hand-painted blue line.

The white-earthenware vessels include: a clear-glazed bowl, foot 120 mm; three vessels that are either cups or bowls, one with an unidentifiable relief-moulded pattern, one with a blue floral/foliate transfer-printed pattern and the other with a blue band-and-line pattern; two clear-glazed cups; a cup with foliated relief-moulding; a cup with a blue floral transfer-print; a cup with a grey floral transfer-print; two plates or saucers with relief-moulded scrolled leaves; a plate/saucer with a blue band-and-line pattern; a plate with relief-moulded Lily of the Valley, rim 220 mm; and two plates with green transfer-printed floral/foliate patterns, one with rim 255 mm. A possible teapot in red-earthenware with a black glaze has a foot diameter of 100 mm.

The bone china includes: a clear-glazed cup; a cup with an enamelled floral pattern; a clear-glazed plate or saucer; a plate with a floral enamelled pattern; a saucer with a blue band at rim and thin gilt band below; and two clear-glazed saucers, one with rim 150 mm. The porcelain vessels are: a cup with a blue band and three gilt lines at the rim; and a hollowware vessel with a blue transfer-print and brown enamelled leaves.

Two stoneware vessels, a salt-glaze ink or blacking bottle and a Bristol-glazed ink bottle were recovered.

Table 10.9: Number, weight and MNV of ceramic sherds from Site 485

	No.	% No.	Wt (g)	% Wt	Wt/No.	MNV	No./MNV	% MNV
Food / beverage consumption								
Chinese	1	8.3	3.1	11.2	3.1	1	1.0	14.3
Non-Chinese	11	91.7	24.7	88.8	2.2	6	1.8	85.7
Total	12		27.8			7		

Table 10.10: Number, weight and MNV of ceramic sherds from Site 1069

	No.	% No.	Wt (g)	% Wt	Wt/No.	MNV	No./MNV	% MNV
Food / beverage containers								
Chinese	1	1.0	5.0	0.9	5.0	1	1.0	3.2
Non-Chinese	0		0.0			0	0.0	0.0
Food / beverage consumption								
Chinese	0		0.0			0		0.0
Non-Chinese	95	93.1	440.2	79.4	4.6	25	3.8	80.6
Unknown	4	3.9	7.8	1.4	2.0	3	1.3	9.7
Other								
Chinese	0		0.0				0.0	0.0
Non-Chinese	2	2.0	101.5	18.3	50.8	2	1.0	6.5
Total	102		554.5			31		

Summary – outlying camps

Origin of ceramics

Table 10.11 details the origin of ceramic vessels from the peripheral sites at Kiandra. Some sites within each area have a higher proportion of Chinese-made vessels while others have a higher proportion of non-Chinese vessels. For example, Site 740, at Sawyers Hill, known from documents to have been occupied in the early 1860s, has predominantly Chinese vessels. New Chum Hill, which was known from documents to be occupied from the late 1870s-1880s, has predominantly British vessels. This scenario mirrors KCC itself, where the later hut sites have predominantly British vessels. The various sites within each area may indicate that miners stayed for short periods at different times, firstly during the main rush in the 1860s and secondly in later years.

Dating ceramic sherds

Seven maker's marks were found on vessels from outlying camps (Table 10.12). Only two marks could be dated, both from Site 1069, New Chum Hill, and both probably date after 1890. This particular Meakin mark does not appear in references but 'England' suggests a post 1890 date (Birks 2002). These marked vessels may have belonged to the later European occupation rather than the Chinese huts.

Function of ceramic vessels

Table 10.13 features the function of vessels in terms of storage containers, food/beverage consumption and smoking, by MNV. Tea and tableware dominate the majority of sites. The assemblage with the highest proportion of pipes in comparison to other vessels is from Site 740, known to have been occupied in the 1860s. No pipes were recovered from New Chum Hill, the sites occupied from 1880s onwards.

Form and decoration

A range of large and small Chinese brown-glazed storage jars were recovered from the outlying camps, but no large non-Chinese storage jars were found, although there are some ceramic bottles.

There are only four Winter Green rice bowls recovered from hut sites and a Double Happiness rice bowl from Southwest Eucumbene River water race.

Non-Chinese tea and tableware decorative types include: transfer-prints, sponge-prints, flown patterns, relief-moulded, banded and gilt. Two transfer-prints, Aquatic from Site 127, and Sydney, from Site 637, are not found at the main camp. They have been recovered from other sites in Kiandra: Aquatic and Sydney from a Township site excavated in 1998 dating from 1860 into the twentieth century; and Sydney from Kiandra Hospital dating 1860-1890. Figure 10.4 shows a selection of sherds, and Table 10.14 the MNV, from the Kiandra Valley sites.

A range of ceramic vessels were recovered indicating some sites were occupied at the start of the rush in the 1860s and other sites at a later date, possibly from the late 1870s onwards.

Table 10.11: Origin of ceramic vessels by MNV

Origin	East Eucumbene					Eucumbene Crossing				SW Eucumbene				Giandarra Gully		New Chum Hill			Total
	site 635	site 636	site 637	site 977	site 1058	site 125	site 127	site 143	site 268	site 698	site 740	site 809	site 1038	site 1079	site 1082	site 485	site 742	site 1069	
Chinese	1	1	1	3	1	5	1	7	1	3	4	1	0	1	1	1	1	1	34
Non-Chinese	11	9	7	1	0	1	5	7	10	2	3	2	3	0	0	6	3	27	97
Unknown	0	0	0	0	0	0	0	0	0	0	0	0	0	0	1	0	1	3	5
Total	12	10	8	4	1	6	6	14	11	5	7	3	3	1	2	7	5	31	136

Note: clay pipes are included in table

Table 10.12: Makers' marks from Kiandra Valley sites

Site No.	Cat No.	Origin	Vessel	Details
485	1307	China	celadon-glazed bowl	Blue-underglaze reign-type maker's mark
740	1015	China	terracotta opium pipe bowl, with circular flared rim, smooth sides and single raised ridge, dark-orange slip	Impressed square stamp with Chinese characters for water and east on right-hand side, rest of stamp illegible
635	1115	Britain	white-earthenware unidentifiable vessel	Partial grey transfer-printed cartouche on underside of Rhine patterned vessel; partial unidentifiable impressed mark over cartouche
742	1032	Britain	white-earthenware unidentifiable vessel	Black transfer-printed maker's mark 'NSTONE C' (ironstone china) in circular band with scenic picture in the centre, no decoration present on reverse; unidentifiable
1069	1242	Britain	white-earthenware plate/saucer	Black transfer-printed partial maker's mark 'ONE' (ironstone?), unidentifiable
1069	1352	Britain	white-earthenware plate/saucer	Blue/green transfer makers mark - "......OOD....." and Crown. Wood & Son, 1891-1907 (Godden 1991:689)
1069	1363	Britain	white-earthenware saucer	Green transfer-printed maker's mark 'J & G ME..... HANL...... ENGLAND', J & G Meakin - post 1890 (Birks 2002)

Table 10.13: Function of ceramic vessels by MNV

Function	East Eucumbene					Eucumbene Crossing				SW Eucumbene				Giandarra Gully		New Chum Hill			Total
	site 635	site 636	site 637	site 977	site 1058	site 125	site 127	site 143	site 268	site 698	site 740	site 809	site 1038	site 1079	site 1082	site 485	site 742	site 1069	
Storage	1	1	1	3	1	3	1	6	1	2	3	0	0	0	1	0	1	1	26
Consumption	9	8	6	1	0	2	5	8	10	3	2	3	3	1	1	6	4	28	100
Smoking	2	1	1	0	0	1	0	0	0	1	2	0	0	0	0	0	0	0	8
Other	0	0	0	0	0	0	0	0	0	0	0	0	0	0	0	0	0	2	2
Total	12	10	8	4	1	6	6	14	11	6	7	3	3	1	2	6	5	31	136

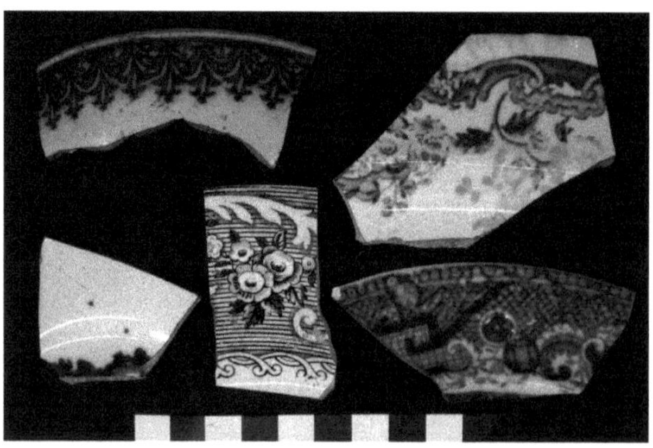

Figure 10.4: Sherds from Kiandra Valley sites (Site and Cat nos. from top left: Site 143 KV-1441 purple sponge-print, Site 637 KV-1137 blue transfer-print, and KV-1129 blue sponge-print, Site 635 KV-1118 Rhine and KV-1119 Willow)

Table 10.14: MNV from the camps in the Kiandra Valley

Vessel	East Eucumbene					Eucumbene Crossing				SW Eucumbene				Giandarra Gully		New Chum Hill			Total
	site 635	site 636	site 637	site 977	site 1058	site 125	site 127	site 143	site 268	site 698	site 740	site 809	site 1038	site 1079	site 1082	site 485	site 742	site 1069	
Food/beverage containers																			
Chinese																			
barrel jar								2							1				3
ginger jar						1													1
globular jar								1											1
wide-mouthed jar				1				1			2						1		5
wide-mouthed jar lid								1											1
spouted jar			1																1
storage jar - small					1	1	1			1								1	5
storage jar - large		1		1		1		1	1										5
liquor bottle				1															1
Non-Chinese																			
storage jar - large																			0
bottle	1									1	1								3
Food/beverage consumption																			
Chinese																			
rice bowl						1		1				1		1		1			5
Non-Chinese																			
bowl-large									1										1
bowl	1	2	1					1										1	6
cup	1		1	1			1		1									8	13
cup/bowl	2	2					1		4	1			2			3	1	3	19
hollow																		1	1
plate	3	2	1				3	1	2	1						1		4	18
plate/saucer	1		2			1		3	1		1							4	13
saucer	1	2	1					2			1	2	1				1	3	14
teapot?																1		1	2
teapot lid																			0
jelly mould																			0
possible tea/table									1							1	1		3
cup															1			1	2
cup/bowl																	1		1
hollow																		1	1
saucer																		1	1
Smoking																			
Chinese																			
opium pipe bowl	1					1				1	2								5
Non-Chinese																			
clay pipe	1	1	1																3
Other																			
Non-Chinese																			
ink bottle																		1	1
blacking/ink bottle																		1	1
Total MNV	12	10	8	4	1	6	6	14	11	5	7	3	3	1	2	7	5	31	136

CHINESE IN KIANDRA TOWNSHIP

Two Chinese stores existed in Kiandra Township from the late-nineteenth into the early-twentieth century, before being destroyed by fire in 1916 (Heffernan 1996). One belonged to Tom Ah Yan and the other, firstly to George Ah Chee and then to Henry Ping Kee. They were located next to each other in Town Section 14 on the main street. Tom Ah Yan bought his land in 1882, the first recorded Chinese person to purchase town land. He had married Catherine Wortz, a German woman, in 1868 and by 1882 they had three children (Smith 1996:21). It is not known where they were living prior to purchasing the store. George Ah Chee, a miner, purchased his land in 1895 and sold it to Ping Kee in 1902 (Heffernan 1996). The latter's store was also a boarding house, 'the town residence of the Chinese from Nine Mile' (Hueneke 1987:53). This site was surveyed by the ANU field school in 1996. Although the results are reported in terms of MNV, there are no whole vessels.

The allotment of Ah Chee/ Ping Kee

This assemblage is from grid squares A12 to J18. It has been divided into two: the front of the allotment; and the rubbish dump at the back of the allotment. Most of the sherds within this area are from in and around the rubbish dump.

Allotment front

Twenty-eight sherds, weighing 36.4 g, represent a MNV of seven. A European clay pipe brings the total MNV to eight (Table 10.15).

The storage vessels are both Chinese brown-glazed stoneware and include: a spouted or wide-mouthed jar and a liquor bottle.

Two porcelain Winter Green vessels were found towards the front of the property: a liquor cup, rim 50 mm; and a spoon. Three non-Chinese white-earthenware tea/ tableware vessels were recovered: a dinner plate with blue transfer-printed Albion decoration; a cup or bowl with a green transfer-printed foliate pattern; and a cup or bowl with an indeterminate blue sponge-printed pattern.

Small fragments of a European clay pipe were also recovered.

Allotment back – rubbish dump area

Two hundred and seventy sherds, weighing 3,089.9 g, represent a MNV of 47. A European clay pipe brings the total MNV to 48 (Table 10.16).

A minimum of ten Chinese storage items were recovered including seven brown-glazed stoneware vessels: three barrel jars; one barrel jar lid; a spouted or wide-mouthed jar; a liquor bottle and a straight-sided jar lid, rim 260 mm; two buff-earthenware wide-mouthed jar lids and a blue-underglaze decorated ginger jar. The barrel jar lid measures 370 mm in diameter. Two of the barrel jars have rims of 330 mm and feet of 370 mm; the other jar is slightly larger, foot 400 mm. A single non-Chinese Bristol-glazed ginger beer or stout bottle with a body diameter of 90 mm was found. Sherds from a large storage jar of unknown origin and dimensions were recovered. The jar has an exterior light-brown glaze over the upper part of the body and an interior clear lead-glaze.

A minimum of seven Chinese tea/tableware vessels were recovered, all porcelain: a serving bowl with Four Seasons decoration, rim 180 mm; three Winter Green rice bowls, rims 140 mm; a Winter Green tea cup, rim 75 mm; a Winter Green liquor cup; and a Winter Green spoon. Twenty-three tableware vessels are of non-Chinese origin, mostly from Britain. These include: one cup; six vessels which are either cups or bowls; four plates; seven vessels which are either plates or saucers; four saucers; and a possible teapot. Two of these vessels are bone china, one red-earthenware and 20 are white-earthenware. The bone china vessels are a floral relief-moulded cup and a saucer without any decoration present, both with a clear-glaze only. The red-earthenware vessel is a possible teapot with a blue slip on the exterior.

White-earthenware decoration includes sponge and transfer-prints, linear banding, relief-moulding and a decal. There are 12 vessels decorated with transfer-prints, four of which can be named: a dinner plate in blue Asiatic Pheasants; a dinner plate in grey Rhine; and plates or saucers in blue Gem and Willow. Other transfer-printed vessels include: four cups or bowls in blue floral, two green floral and green floral/scenic patterns; two plates or saucers in green geometric/floral and brown fern-like patterns; and two saucers in blue scrolls/foliate and green geometric patterns. Sponge-prints are on two cups or bowls: one with a repeated green star motif on the exterior and red bands on the exterior and interior rim; and the other with an unidentifiable blue pattern. Two linear banded patterns are found on three vessels: a plate/saucer with two thin blue bands at the rim; and a plate and a saucer with one thick and one thin red band at the rim. Relief-moulded patterns are on two vessels: a plate with a floral or foliate pattern; and a plate/saucer with key frets and hearts. There is a polychrome floral decal on a plate/saucer. Two porcelain cups of unknown origin, rim diameters 85 and 90 mm, are from this area. These may be Japanese in origin. A porcelain vessel, which may be a serving plate or dish, with cut away sections and embossed dots and gold band at rim, is possibly of European or Japanese origin.

Fragments from two Chinese opium pipe bowls were recovered: a light-orange slipped bowl, rim diameter 70 mm, with ten faceted sides; and a bowl, rim 60 mm, not identifiable to type. This last pipe bowl was under-fired; the clay is chalky and slip is a buff-orange colour. Small fragments from a European clay pipe were found.

The allotment of Ah Yan

Seven hundred sherds weighing 2,799.9 g were from grid squares A1 to J11. There were no sherds found in grid row 11 quadrants I and II, creating an arbitrary division of one metre between the assemblages of Ah Yan and Ah Chee/Ping Kee. Ah Yan's assemblage has been divided into four areas: in and around a pair of chimneys (Area 1), a deflated area (Area 2), and three other isolated assemblages designated Areas 3, 4 and 5.

Area 1 – Chimney

One hundred and nineteen sherds, weighing 551.7 g, represent a MNV of 14. A European clay pipe brings the MNV to 15 (Table 10.17).

Three Chinese-made stoneware vessels include: a barrel jar lid, rim 350 mm and height 39 mm; a brown-glazed liquor bottle; and a green-glazed ginger jar with an interior brown glaze. There are two large non-Chinese storage jars: one is salt-glazed on the exterior with an interior under-fired clear glaze; and the other a Bristol-glazed demijohn. A large storage jar, of which the base only is present, has an interior shiny clear-glaze and is of unknown origin. This last jar may be similar to that found in Ah Chee's allotment.

A single Chinese-made porcelain rice bowl with Winter Green decoration was found from this area. Four white-earthenware tableware vessels of non-Chinese origin were recovered: a cup or bowl with a gilt band on the edge of the rim; a cup with a green floral transfer-print; a plate with a flown-blue floral transfer-print; and a plate with a floral relief-moulded pattern. A porcelain plate or saucer was found decorated with a worn red or pink band at the rim is possibly of European or Japanese origin.

A single European clay pipe was recovered from this area.

Table 10.15: Number, weight and MNV of sherds from the front of Ah Chee/Ping Kee allotment

	No.	% No.	Wt(g)	% Wt	Wt/No.	MNV	No./MNV	% MNV
Food / beverage containers								
Chinese	7	25.0	21.8	59.9	3.1	2	3.5	25.0
Non-Chinese	0		0.0			0		0.0
Food / beverage consumption								
Chinese	5	17.9	3.0	8.2	0.6	2	2.5	25.0
Non-Chinese	16	57.1	11.6	31.9	0.7	3	5.3	37.5
Smoking								
Chinese	0		0.0			0		0.0
Non-Chinese						1		12.5
Total	28		36.4			8		

Note: clay pipe is not in catalogue sherd numbers or weights

Table 10.16: Number, weight and MNV of sherds from the back of Ah Chee/Ping Kee allotment

	No.	% No.	Wt(g)	% Wt	Wt/No.	MNV	No./MNV	% MNV
Food / beverage containers								
Chinese	123	45.6	2589.0	83.8	21.0	10	12.3	20.8
Non-Chinese	2	0.7	64.9	2.1	32.5	1	2.0	2.1
Unknown	3	1.1	153.6	5.0	51.2	1	3.0	2.1
Food / beverage consumption								
Chinese	19	7.0	97.7	3.2	5.1	7	2.7	14.6
Non-Chinese	106	39.3	120.1	3.9	1.1	23	4.6	47.9
Unknown	6	2.2	46.8	1.5	7.8	3	2.0	6.3
Smoking								
Chinese	11	4.1	17.8	0.6	1.6	2	5.5	4.2
Non-Chinese						1		2.1
Total	270		3089.9			48		

Note: clay pipe is not in catalogue sherd numbers or weights

Table 10.17: Number, weight and MNV of sherds from Area 1 - Ah Yan assemblage

	No.	% No.	Wt(g)	% Wt	Wt/No.	MNV	No./MNV	% MNV
Food / beverage containers								
Chinese	6	5.0	79.3	14.4	13.2	3	2.0	20.0
Non-Chinese	3	2.5	106.3	19.3		2	1.5	13.3
Unknown	2	1.7	251.2	45.5	125.6	1	2.0	6.7
Food / beverage consumption								
Chinese	6	5.0	26.1	4.7	4.4	1	6.0	6.7
Non-Chinese	98	82.4	78.1	14.2	0.8	4	24.5	26.7
Unknown	4	3.4	10.7	1.9	2.7	3	1.3	20.0
Smoking								
Chinese	0		0.0			0		0.0
Non-Chinese						1		6.7
Total	119		551.7			15		

Note: clay pipe is not in catalogue sherd numbers or weights

Area 2 - deflated area

A total of 537 sherds, weighing 2,169.2 g, accounts for a MNV of 47. A single European clay pipe brings the total MNV to 48 (Table 10.18).

A minimum number of ten Chinese-made stoneware container vessels are from the deflated area including: four barrel jar lids, three with diameters 370 mm and one 390 mm; two liquor bottles; a wide-mouthed jar, rim 110 mm; a large brown-glazed storage jar of which no diagnostic sherds are present; a green-glazed ginger jar; and a blue-underglaze decorated ginger jar. A large salt-glazed storage jar of either British or Australian origin was recovered.

Seven Chinese-made tea/tableware vessels include: a serving bowl, rim 180 mm, decorated in Four Seasons pattern; a tea cup, rim 70 mm, a spoon and three vessels which are cups or rice bowls, all porcelain Winter Green; and a stoneware rice bowl with a blue-green underglaze block-printed decoration.

There are 24 non-Chinese tea/tableware vessels: 22 are white-earthenware; one, a clear-glazed cup or bowl, is bone china; and the other a buff-stoneware possible jug with a Langley-type rust-brown glaze. The white-earthenware includes: one bowl; two cups; three vessels which are cups or bowls; a mixing bowl; four plates; six vessels which were plates or saucers; three saucers; and two possible tableware vessels.

Transfer-prints are found on 12 vessels. Three plates or saucers have nameable patterns: Athens in blue; and Rhine in brown and grey. Other transfer-printed vessels include: a breakfast cup in purple with a scenic exterior and floral interior; a bowl with a brown geometric/foliate transfer-printed pattern, rim diameter 180 mm; a cup or bowl in a brown geometric pattern; plates or saucers in a blue pattern with a meandering line and bands, a grey floral pattern and a purple geometric pattern; a saucer, rim 180 mm, with a brown geometric/floral pattern; and two unidentifiable vessels, possibly tableware, one with a green pattern and the other in black with orange enamel on the thick rim.

Other decorative types include flown-blue, linear-banding, relief-moulding, glaze, and decal. A saucer with a flown-blue transfer-printed Melbourne pattern was made by W. H. Grindley. Two vessels have banded decoration: a plate of unknown size has two thin blue bands at the rim; and a cup has a single red band one the exterior and interior. Three vessels have relief-moulded patterns: a plate, rim 230 mm, with dots and leaves; a plate of unknown size with a foliate pattern; and a saucer with foliated scrolls. A plate, rim 245 mm, has a clear-glaze only, as did the mixing/pudding bowl, rim 170 mm. A cup or bowl has a pale-yellow glaze and another, a blue and green decal.

A porcelain cup, rim 85 mm, and saucer, both with a clear-glaze only were found in this area. They are possibly of European or Japanese origin.

Fragments from a minimum of two smoking-related vessels were found: a Chinese opium pipe bowl with a light-orange slip, a circular rim and ten faceted sides; and a European clay pipe.

Two stoneware salt-glazed ink bottles of British or Australian origin were recovered.

Areas 3, 4 and 5 – isolated assemblages

These areas are related to separate storage buildings, known historically from oral recollections and photographs. They are located at the back of Ah Yan's store (Figure 10.5).

Table 10.18: Number, weight and MNV of sherds from Area 2 – Ah Yan assemblage

	No.	% No.	Wt(g)	% Wt	Wt/No.	MNV	No./MNV	% MNV
Food / beverage containers								
Chinese	293	54.6	1873	86.3	6.4	10	29.3	20.8
Non-Chinese	3	0.6	6.9	0.3	2.3	1	3.0	2.1
Food / beverage consumption								
Chinese	17	3.2	24.5	1.1	1.4	7	2.4	14.6
Non-Chinese	213	39.7	222.4	10.3	1.0	24	8.9	50.0
Unknown	6	1.1	32.8	1.5	5.5	2	3.0	4.2
Smoking								
Chinese	3	0.6	1.7	0.1	0.6	1	3.0	2.1
Non-Chinese						1		2.1
Other								
Chinese	0		0.0			0		0.0
Non-Chinese	2	0.4	7.9	0.4	4.0	2	1.0	4.2
Total	537		2169.2			48		

Note: clay pipe is not in catalogue sherd numbers or weights

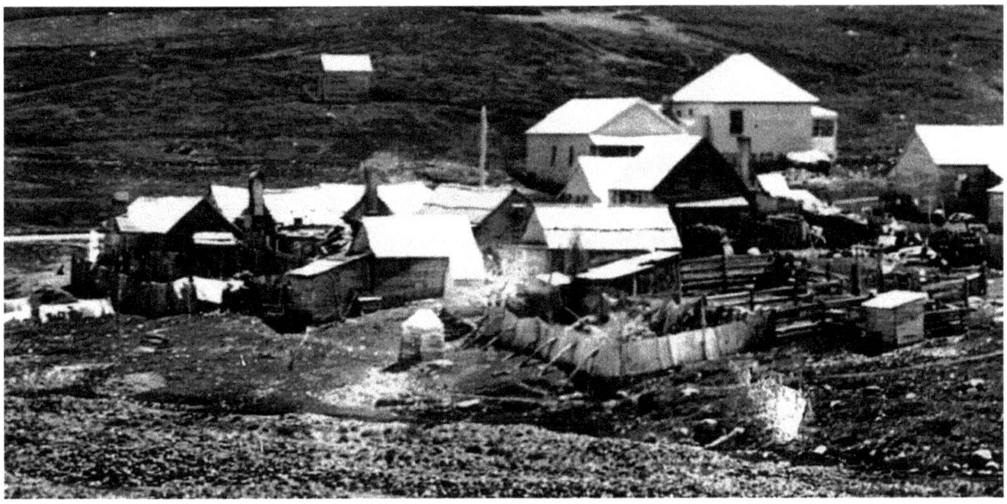

Figure 10.5: The back of the Chinese stores, Kiandra Town Section 14, c. 1903 (detail from photograph: Kiandra. T. Plumb album, National Library of Australia, Pictures Compactus Collection. Reproduced with permission of the National Library of Australia)

Fifteen sherds, weighing 34.1 g, representing a MNV of five, are from Area 3, squares I9 to J10. Area 4, squares H5 and I5, has a total of 19 sherds, weighing 21.6 g with a MNV of four. Area 5, squares I3 and J3, has a total of 10 sherds, weighing 23.3 g and a MNV of three. A comparison of these can be seen in Table 10.19.

Table 10.19: MNV from Areas 3, 4 and 5 – Ah Yan assemblage

	Area 3	Area 4	Area5
Food / beverage containers			
Chinese	2	2	2
Non-Chinese	0	0	0
Food / beverage consumption			
Chinese	2	0	0
Non-Chinese	1	2	1
Total	5	4	3

Four Chinese-made vessels and one of non-Chinese origin are from Area 3: two brown-glazed stoneware storage jars of different sizes; a porcelain Winter Green dish, foot 110 mm; a porcelain Winter Green rice bowl, rim 140 mm; and a blue-banded plate with two thin bands at the rim and shoulder.

Area 4 has a MNV of four: two Chinese-made brown-glazed stoneware storage jars of different size; an English bone china plate or saucer with a fluted body and clear-glaze; and a white-earthenware plate with relief-moulded decoration possibly from Britain or Japan.

A minimum number of three vessels are from Area 5. There are two Chinese brown-glazed stoneware storage jars of different sizes and a clear-glazed plate or saucer of non-Chinese origin, with blue transfer-printed decoration.

Summary – Kiandra stores

Origin of ceramics

The origin of the recovered ceramic vessels is shown as a percentage MNV (Table 10.20). All sites have a greater or equal amount of non-Chinese ceramic vessels.

Table 10.20: Origin of ceramic vessels from the Kiandra Chinese stores by percentage MNV

Origin	Chee/Kee Front	Chee/Kee Rubbish dump	Yan Area 1	Yan Area 2
Chinese	50.0	39.6	26.7	37.5
Non-Chinese	50.0	52.1	46.7	58.3
Unknown	0.0	8.3	26.7	4.2

Dating of ceramic sherds

No makers' marks were recovered from either store in 1996. In 1997 the ANU field school excavated an area across the Snowy Mountains Highway from the stores. A sherd from that surface collection conjoins to a sherd from Ah Yan's assemblage. It is from a saucer, the style of which was registered by W.H. Grindley in 1902 and used for the Melbourne pattern (Van Buskirk 2002:98, 110).

Seven nameable transfer-printed patterns were recovered from the stores: Albion, Asiatic Pheasants, Athens, Gem, Melbourne, Rhine and Willow. All but the Gem and Melbourne patterns are also found at KCC. Gem is a common nineteenth century pattern with at least ten known makers (Kowalsky & Kowalsky 1999:488). The pattern at Kiandra is illustrated in Williams (1978:69) but the maker is unknown.

The assemblage contains decorative types found in late-nineteenth/early-twentieth century sites including decal and banded vessels, and clear-glazed porcelain cups and saucers possibly of Japanese origin.

Langley ware was first made in Bendigo by Jack Gare in 1913 (O'Hoy 1989). These wares have a smooth upper body separated from a textured lower body by a beaded band. A similar vessel to the one from Ah Yan's Store was found at the Kiandra Township hotel/residence surveyed by the ANU field school in 1995.

Function of ceramic vessels

Table 10.21 summarises the function by percentage MNV of the store sites. Food/beverage consumption vessels comprise the highest percentage across all areas, followed by storage containers and smoking-related vessels.

Table 10.21: Function of ceramic vessels from Kiandra Chinese stores by percentage MNV

Function	Chee/Kee Front	Chee/Kee Rubbish dump	Yan Area 1	Yan Area 2
Containers	25.0	25.0	40.0	22.9
Consumption	62.5	68.8	53.3	68.8
Smoking	12.5	6.3	6.7	4.2
Other	0.0	0.0	0.0	4.2

Form and decoration

All the common storage jar forms found at KCC are also at the stores. A type of storage jar not found at KCC has an exterior light-brown glaze which does not extend to the base and an interior clear-glaze. No dimensions of the jar are available from the sherds.

The Chinese-made tea and tableware includes forms not found at KCC: a celadon-glazed dish and tea cups. Similar forms of non-Chinese made tea and tableware were recovered although with different decorative types. Both Gem and Melbourne have been recovered from other sites in Kiandra Township; Gem from a residence surveyed by ANU field school in 1998 and Melbourne from a hotel/residence surveyed in 1995.

Only one type of opium pipe bowl has been identified: a flared circular rim with ten faceted sides, all decorated with light-orange slip although one pipe bowl is under-fired and is a buff-orange colour.

Figure 10.6 shows a selection of sherds, and Table 10.22, the MNV from the Kiandra stores.

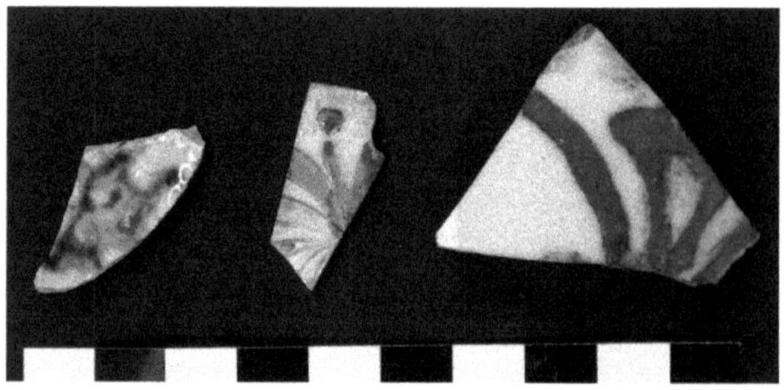

Figure 10.6: Chinese tableware sherds from the Kiandra Township stores (Cat nos. from left: KAP96-62.2 block-print, KAP96-161 and KAP96-184 Four Seasons)

Discussion – Kiandra Region

The earlier date of the Chinese Camp and some of the outlying camps in comparison to the stores can be seen through the distribution and quantity of non-Chinese ceramics. Although some similar vessels were recovered, the store sites have decorative types associated with late-nineteenth/early-twentieth century that have not been found at KCC or most of the outlying camps.

Although no Chinese-made ceramic vessels can be dated, there are differences between the assemblages. The huts at KCC, which have a greater percentage of Chinese-made vessels, have all but one of the blue-underglaze rice bowls and a greater variety of opium pipe bowls. The exception is a Bamboo rice bowl recovered from the Temple Ancillary Building. This bowl differs from the others as its underglaze is a green-blue rather than a clear blue. The bowl from Ah Yan's store has a similar green-blue underglaze colour but is block-printed rather than hand-painted. In addition, some Winter Green forms found at the stores are not present at the camps. While there are many similar storage jars from all sites there are exceptions. Ovoid jars and the possible spouted liquor storage jar are only from KCC while the jars with a light-brown exterior glaze and clear-glazed interior are only found at the stores. An impressed stamp on an opium pipe bowl, containing the Chinese characters for water and east, was found at both KCC Hut 2 and a hut at Sawyers Hill.

The location of the previous residences of George Ah Chee and Tom Ah Yan is not known. It is known that Ah Yan married Catherine Wortz in 1868 (Smith 1997:150). At the main camp, one hut displays a number of matching transfer-printed vessels and a greater percentage of flatware than the other huts. It is speculated that this hut may have been occupied by a European woman or a Chinese man who entertained Europeans. In 1878, a European woman with one son and three daughters was recorded as living at the Chinese Camp (Votes and Proceedings of the NSW Legislative Assembly 1878-1879). A reasonable conclusion based on this evidence is that Hut 3 may have been the residence of Tom Ah Yan and Catherine Wortz.

Table 10.22: MNV from Kiandra Township stores

Origin	Vessel	Chee/Kee Front	Chee/Kee Rubbish dump	Yan Area 1	Yan Area 2	Yan Area 3	Yan Area 4	Yan Area 5	Total
Kiandra Township Chinese Stores - MNV									
Food / beverage container									
Chinese	barrel jar		3						3
Chinese	barrel jar lid		1	1	4				6
Chinese	ginger jar		1	1	2				4
Chinese	wide-mouthed jar				1				1
Chinese	wide-mouthed jar lid		2						2
Chinese	straight-sided jar lid		1						1
Chinese	storage jar - large				1	1	1	1	4
Chinese	storage jar - small	1	1			1	1	1	5
Chinese	liquor bottle	1	1	1	2				5
Non-Chinese	beer/stout bottle		1						1
Non-Chinese	storage jar-large			2	1				3
Unknown	storage jar-large		1	1					2
Food / beverage consumption									
Chinese	cup/bowl				3				3
Chinese	rice bowl		3	1	1	1			6
Chinese	serving bowl		1		1				2
Chinese	shallow dish					1			1
Chinese	spoon	1	1		1				3
Chinese	tea cup		1		1				2
Chinese	wine/liquor cup	1	1						2
Non-Chinese	bowl				1				1
Non-Chinese	cup		1	1	2				4
Non-Chinese	cup/bowl	2	6	1	4				13
Non-Chinese	jug?				1				1
Non-Chinese	mixing bowl				1				1
Non-Chinese	plate	1	4	2	4	1	1		13
Non-Chinese	plate/saucer		7		6		1	1	15
Non-Chinese	saucer		4		3				7
Non-Chinese	teapot?		1						1
Non-Chinese	possible tea/table				2				2
Unknown	cup		2	2	1				5
Unknown	saucer			1	1				2
Unknown	serving plate/dish		1						1
Smoking									
Chinese	opium pipe bowl		2		1				3
Non-Chinese	clay pipe	1	1	1	1				4
Other									
Non-Chinese	ink bottle				2				2
Total MNV		8	48	15	48	5	4	3	131

CHAPTER 11

Technical analyses

Some of the sherds recovered from sites in southeast NSW were further analysed using macroscopic and microscopic methods. The selection of samples and the methodology are outlined in Chapter 6. The results of the analyses are presented in this chapter, ordered by Chinese samples, non-Chinese samples and samples of unknown origin. A statistical analysis and discussion follows the results.

CHINESE FOOD/BEVERAGE CONTAINERS

Globular jars

Table 11.1 lists the globular jar sherds examined and their find site.

Table 11.1: Globular jars samples

Sample	Settlement	Site	Cat no.
1	Flanagan's Point	Hut 4	122.2
2	Jembaicumbene	Hut 2	181.1
3	Jembaicumbene	Hut 2	700.18
4	Jembaicumbene	Surface	1259.1
5	Jembaicumbene	Store	1220.2
6	Jembaicumbene	Hut 5	1252
7	Kiandra	Hut 3	1348.3
8	Kiandra	Hut 42	1308
9	Kiandra	Hut 2	1026
10	Kiandra	KCC2003-2	1328
11	Eucumbene Crossing	Site 143	1065

Jar characteristics

All the jars in the study are fragmentary, but sherds indicate that they were thrown on potters' wheels. As detailed in Chapter 4, most of the rims are chipped in parts, and unchipped areas have been wiped of glaze. All have a dark-brown, shiny or matte, glaze which is lighter where thinner. Pin-holing is present on most sherds. Samples 8 and 11 have buff-coloured clay after firing, while the other samples have varying grey bodies produced by differing amounts of oxygen in the kiln atmosphere. All samples contain visible inclusions, particularly quartz. The apparent porosity of the globular jar sherds ranges from 5.71 per cent to 8.74 per cent with an average of 7.26 per cent, and median, 7.35 per cent.

Compositional analysis

Electron microprobe analysis (EMPA) was used throughout this study to determine the composition of the ceramic bodies and glazes. The results for the averaged body and glaze composition of 11 globular jars are shown in Tables 11.2 and 11.3.

The glaze on all the globular jars contains phosphorous and manganese oxides, which are not present in the body, in addition to increased amounts of magnesium, calcium and iron oxides. Sample 6 has a small percentage of lead oxide which is not present in any of the other globular jar samples. The main flux, an oxide which helps melt other ingredients and promotes ceramic fusion, in all these glazes is calcium oxide.

The ceramic body of all globular jar samples exhibit areas of vitrification, open and closed pores and many inclusions, 5-300 μm in size. The most common inclusions are quartz, iron oxide, rutile, ilmenite and zircon. Samples 6, 8, 10 and 11, contain monazite, composed of cerium (Ce), lanthanum (La), and neodymium (Nd), while Samples 4, 5 and 9 contain xenotime (YPO_4). Glaze layers are distinct from the body, as seen in Sample 3 (Figure 11.1). Gas bubbles that formed in the body during the vitrification process and travelled towards the surface were trapped in the glaze layer when the firing stopped.

The photomicrographs of the fired ceramic bodies throughout this chapter show a mix of melted and unmelted particles and pores. When compounds are mixed to form a clay body, the mixture is not homogenous. Firing causes some areas within the clay to melt before others resulting in the gradual increase over time in the formation of eutectics, that is, a proportion of two or more substances that melt at the lowest temperature possible. As the temperature increases the pores in the body change from being jagged to round. When the ceramic body cools, the clay mixture pulls away from the unmelted particles forming circumferential cracks as it shrinks (personal communication, Dr Anthony Flynn, 7 January 2011).

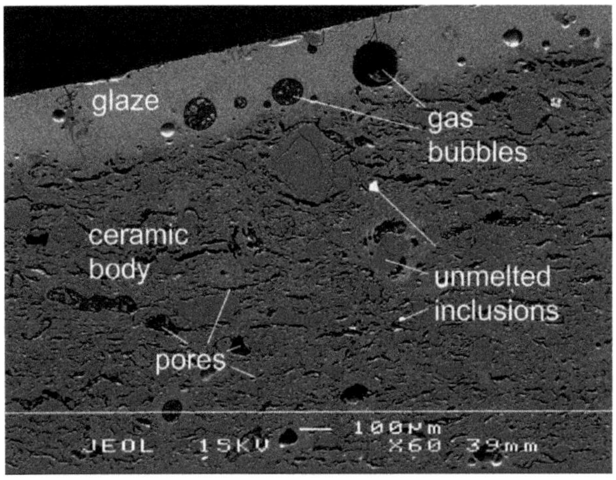

Figure 11.1: SEM photomicrograph of Sample 3 at x60 magnification

Table 11.2: Averaged body composition of globular jars

Sample	Oxide concentrations (wt%)										
	Na_2O	MgO	Al_2O_3	SiO_2	P_2O_5	SO_3	K_2O	CaO	TiO_2	MnO	FeO
1	0.23	0.46	23.04	69.78	0.04	0.07	2.84	0.20	0.80	0.00	2.54
2	0.22	0.55	20.19	73.62	0.02	0.04	2.23	0.17	0.90	0.00	2.06
3	0.21	0.39	22.42	71.32	0.00	0.02	2.86	0.08	0.88	0.00	1.82
4	0.20	0.47	25.18	67.58	0.00	0.02	3.00	0.07	1.12	0.00	2.36
5	0.15	0.53	20.17	73.73	0.02	0.09	2.05	0.10	0.92	0.00	2.24
6	0.20	0.47	21.94	71.44	0.02	0.00	2.86	0.25	0.75	0.00	2.07
7	0.20	0.43	23.28	70.10	0.00	0.06	2.75	0.22	0.71	0.00	2.25
8	0.21	0.42	23.39	70.16	0.02	0.05	2.79	0.13	0.70	0.00	2.13
9	0.18	0.47	24.91	67.81	0.05	0.06	2.90	0.22	0.86	0.00	2.54
10	0.20	0.45	22.51	70.51	0.02	0.00	3.06	0.18	0.79	0.00	2.28
11	0.18	0.42	23.76	69.30	0.00	0.07	3.10	0.15	0.73	0.00	2.29

Table 11.3: Averaged glaze composition of globular jars

Sample	Oxide concentrations (wt%)											
	Na_2O	MgO	Al_2O_3	SiO_2	P_2O_5	SO_3	K_2O	CaO	TiO_2	MnO	FeO	PbO
1	0.37	2.09	16.18	60.20	1.10	0.14	2.60	9.61	1.32	0.19	6.20	0.00
2	0.43	2.27	18.79	57.47	1.06	0.08	2.71	8.31	1.42	0.27	7.19	0.00
3	0.39	2.64	17.07	57.20	1.34	0.00	2.73	9.36	1.51	0.43	7.33	0.00
4	0.36	2.61	17.84	57.54	1.46	0.00	2.82	8.76	1.30	0.46	6.85	0.00
5	0.36	2.60	17.45	58.63	1.49	0.02	2.59	8.36	1.19	0.28	7.03	0.00
6	0.37	2.61	17.12	56.01	1.21	0.00	2.63	9.78	1.79	0.44	7.41	0.63
7	0.45	2.59	18.36	55.97	1.39	0.04	2.60	9.40	1.50	0.42	7.28	0.00
8	0.41	2.47	18.15	57.54	1.36	0.04	2.72	8.87	1.29	0.44	6.71	0.00
9	0.41	2.43	18.97	55.67	1.32	0.02	2.56	9.64	1.43	0.37	7.18	0.00
10	0.38	2.72	18.22	56.82	1.45	0.00	2.62	8.96	1.37	0.47	6.99	0.00
11	0.35	2.66	16.63	57.69	1.28	0.06	2.79	10.08	1.30	0.41	6.75	0.00

Glaze firing temperatures

After expressing each glaze composition as a unity formula, and using the firing temperatures suggested by Hamer and Hamer (1997:379), five of the samples (3, 6, 7, 9 and 11) have glazes appropriate to a firing temperature in the range 1100°C - 1200°C. The other six samples (1, 2, 4, 5, 8 and 10) have glazes appropriate to be fired between 1200°C and 1300°C.

Wide-mouthed jars

Table 11.4 lists the wide-mouthed jar sherds examined and their find site.

Table 11.4: Wide-mouthed jar samples

Sample	Settlement	Site	Cat no.
12	Jembaicumbene	Hut 4	787.1
13	Jembaicumbene	Store	1238.1
14	Southwest Eucumbene	Site 740	1029
15	Jembaicumbene	Hut 2	756.8
16	Jembaicumbene	Hut 2	721.4
17	Kiandra	Hut 1	112.2
18	Kiandra	KCC2003-2	1126B
19	Kiandra	Hut 3	1346.4
20	Kiandra Township	Yan's Store	92
21	East Eucumbene River	Site 977	1059
22	Eucumbene Crossing	Huts 1-2	1098
23	Jembaicumbene	Hut 2	721.2

Jar characteristics

Again, all the wide-mouthed jars in this study are fragmentary, but sherds indicate they were all wheel-thrown. All the examined jar samples are from a jar rim or shoulder and while the overall form of the rims is similar, there are slight variations in the mouth-width to neck-depth ratio and in the profile of the shoulders and rolled rims. Two samples, 12 and 14, are from the shoulder of jars. Both have a shoulder angle of about 40°, which corresponds to illustrations by Ritchie (1986:239). Sample 12 has an angular shoulder resulting in a lighter brown glaze at that point, while Sample 14 has a smooth curved shoulder with no change in the glaze colour. The rims on six jars (Samples 17, 18, 20, 21, 22 and 23) have a rolled lip that does not touch the neck (Figure 11.2). Samples 15, 16 and 19 have rims that are made in the same way but roll over to touch the neck. The rim of Sample 13 is thicker than the body and rounded but appears similar to 15, 16 and 19 unless observing the cross-section.

Glaze colour ranges from dark to light brown and appearance from shiny to matte, the colour being lighter where the glaze is thinner. Pin-holing is seen on the exterior of all the sherds and the interior glaze is patchy. Sample 21 has a buff-coloured oxidised body and all the other samples have varying shades of grey resulting from the amount of reduction during firing, and the iron content of their body. The apparent porosity of the wide-mouthed jar sherds ranges from 0.70 per cent to 9.29 per cent. The average is 5.69 per cent and the median 5.80 per cent.

Compositional analysis

The results for the averaged body and glaze composition of 12 wide-mouthed jars are shown in Tables 11.5 and 11.6. Phosphorous and manganese oxides are present in the glaze, but not the body, along with increased amounts of magnesium, calcium and iron oxides. Calcium oxide is the main flux.

The glaze layer of wide-mouthed jars is distinct from the body. The glaze of Sample 18 contains trapped gas bubbles and unmelted quartz (Figure 11.3). The ceramic body is made up of the partially melted clay minerals around unmelted inclusions which are predominantly quartz. Numerous pores of various shapes and sizes are present. Samples 15, 19 and 21 have noticeably larger quartz particles, ranging from 150-200 µm, than are seen in the other wide-mouthed jars. Apart from quartz, the most common inclusions in each of these jars are zircon, rutile, ilmenite and iron oxide. Samples 12 and 20 have inclusions of monazite (Ce, La and Nd composition) and Samples 12 and 22, xenotime.

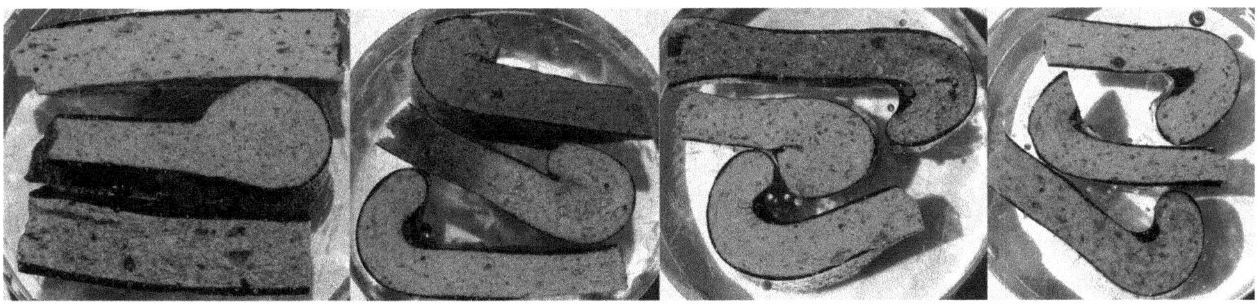

Figure 11.2: Wide-mouthed jar samples (from top of left disk 1: 12, 13, 14; disk 2: 15, 16 17; disk 3: 18, 19 20; disk 4: 21, 22 23)

Table 11.5: Averaged body composition of wide-mouthed jars

Sample	Oxide concentrations (wt%)										
	Na_2O	MgO	Al_2O_3	SiO_2	P_2O_5	SO_3	K_2O	CaO	TiO_2	MnO	FeO
12	0.24	0.62	26.52	66.65	0.06	0.09	2.57	0.16	0.94	0.00	2.15
13	0.23	0.58	20.97	73.08	0.02	0.14	2.41	0.05	0.81	0.00	1.71
14	0.25	0.70	25.11	66.93	0.07	0.02	3.09	0.14	0.99	0.00	2.70
15	0.20	0.49	18.63	75.47	0.03	0.06	1.97	0.10	1.00	0.00	2.05
16	0.20	0.54	21.05	72.44	0.12	0.13	2.13	0.07	1.15	0.00	2.17
17	0.21	0.55	23.58	69.85	0.02	0.09	2.74	0.09	1.03	0.00	1.84
18	0.21	0.48	20.86	72.92	0.02	0.10	2.16	0.15	0.79	0.01	2.30
19	0.21	0.53	20.13	73.79	0.09	0.02	2.09	0.17	1.06	0.00	1.91
20	0.24	0.58	26.05	67.53	0.05	0.02	2.22	0.00	1.20	0.00	2.11
21	0.18	0.53	21.46	72.39	0.03	0.04	2.17	0.10	1.15	0.01	1.94
22	0.22	0.52	24.34	69.50	0.04	0.08	2.47	0.01	0.97	0.00	1.85
23	0.23	0.58	22.43	70.66	0.00	0.04	2.86	0.07	0.99	0.00	2.14

Table 11.6: Averaged glaze composition of wide-mouthed jars

Sample	Oxide concentrations (wt%)										
	Na_2O	MgO	Al_2O_3	SiO_2	P_2O_5	SO_3	K_2O	CaO	TiO_2	MnO	FeO
12	0.42	2.80	17.86	57.05	2.21	0.02	2.85	9.26	1.17	0.48	5.88
13	0.37	2.62	15.36	62.57	1.64	0.08	3.00	7.77	1.07	0.55	4.97
14	0.32	3.10	15.71	59.89	2.20	0.02	2.77	8.07	1.14	0.71	6.07
15	0.43	2.52	15.34	62.66	1.47	0.00	2.36	7.83	1.27	0.54	5.58
16	0.41	2.28	17.77	61.23	1.37	0.02	2.53	7.53	1.29	0.50	5.07
17	0.37	3.14	16.53	58.29	2.07	0.02	2.78	8.48	1.28	0.72	6.32
18	0.41	3.23	17.31	57.80	2.30	0.00	2.73	8.10	1.18	0.92	6.02
19	0.38	2.58	18.04	61.20	1.90	0.00	2.67	6.40	1.17	0.52	5.14
20	0.44	3.13	18.71	56.00	2.37	0.00	2.48	8.72	1.19	0.90	6.06
21	0.39	3.93	17.09	52.39	3.88	0.04	2.54	10.95	1.31	1.18	6.30
22	0.43	3.25	17.76	56.80	2.02	0.04	2.80	8.71	1.28	0.74	6.17
23	0.30	2.47	16.67	62.46	1.58	0.02	3.02	6.24	1.22	0.59	5.43

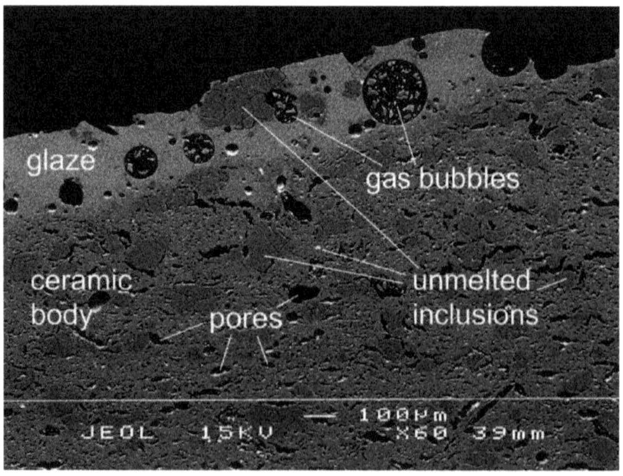

Figure 11.3: SEM photomicrograph of Sample 18 at x60 magnification

Glaze firing temperatures

Using the firing temperatures suggested by Hamer and Hamer (1997:379), Samples 20, 21 and 22 have glazes appropriate to a firing temperature range 1100°C - 1200°C. All other samples have glazes appropriate to be fired between 1200°C and 1300°C.

Wide-mouthed jar lids

Table 11.7 lists the wide-mouthed jar lid sherds examined and their find site.

Lid characteristics

The wide-mouthed jar lids in this study were all thrown using a potter's wheel, probably off a mound of clay, as outlined in Chapter 4. The lids are all a saucer-like shape and have cutting wire marks on the base consistent with being removed from a mound of clay. The samples chosen for examination all have rims present, confirming their identification as wide-mouthed jar lids. Two samples, 32 (KCC2003-1198) and 33 (KAP96-341), have a similar rim profile with a groove on the underside of the lid. Sample 26 (JCC-186) and a sherd from KCC Hut 4 (KCC2001-1411 not examined in SEM) have similar cutting wire marks on their bases. All the wide-mouthed jar lids are unglazed. All the lids appear to have the same amount of weathering.

Table 11.7: Wide-mouthed jar lid samples

Sample	Settlement	Site	Cat no.
24	Jembaicumbene	Store	777.3
25	Jembaicumbene	Hut 2	721.6
26	Jembaicumbene	Hut 2	186
27	Jembaicumbene	Grid 10	1243
28	Jembaicumbene	Grid12	1254.2
29	Kiandra	Hut 42	1331
30	Kiandra	Oven	1376
31	Kiandra	Hut 62	1277
32	Kiandra	KCC2003-2	1198
33	Kiandra Township	Chee's Store	341
34	Eucumbene Crossing	Site 143	1067

The colour of the bodies includes buff, orange, red and grey. Once again these colours are produced by the extent of oxidation or reduction during firing and the amount of iron oxide in the bodies. All the bodies have visible inclusions, particularly quartz. The apparent porosity of the shouldered-food jar lid sherds ranges from 6.03 per cent to 16.17 per cent. The average is 10.74 per cent and the median 10.72 per cent.

Compositional analysis

The results for the averaged body composition of 12 wide-mouthed jar lids are shown in Table 11.8. Sample 26 has high sodium and calcium oxide readings in comparison with the other samples. Sample 32 exhibits large inclusions, particularly of quartz, some measuring 300-400 µm (Figure 11.4). Many pores are also present. All the wide-mouthed jar lid samples contain zircon inclusions; rutile, ilmenite and iron oxide are also common inclusions. Less common inclusions are monazite (Ce, La and Nd composition) in Samples 24, 29 and 31, and xenotime in Sample 29.

Table 11.8: Averaged body composition of wide-mouthed jar lids

Sample	Oxide concentrations (wt%)										
	Na_2O	MgO	Al_2O_3	SiO_2	P_2O_5	SO_3	K_2O	CaO	TiO_2	MnO	FeO
24	0.13	0.43	20.35	74.47	0.06	0.06	1.72	0.09	0.85	0.00	1.84
25	0.17	0.55	20.40	72.95	0.05	0.09	2.02	0.16	0.94	0.00	2.67
26	1.35	0.40	16.10	77.15	0.02	0.19	1.79	0.61	0.72	0.00	1.67
27	0.16	0.48	21.66	72.45	0.02	0.04	1.64	0.10	0.95	0.00	2.50
28	0.28	0.48	22.00	71.55	0.03	0.04	2.50	0.13	0.96	0.00	2.03
29	0.15	0.44	18.91	75.20	0.05	0.12	1.65	0.13	1.11	0.00	2.24
30	0.17	0.43	19.57	74.31	0.03	0.02	1.81	0.13	1.19	0.00	2.34
31	0.17	0.42	16.24	79.01	0.00	0.02	1.29	0.06	0.78	0.00	2.01
32	0.19	0.45	16.61	78.23	0.06	0.02	1.70	0.12	0.94	0.00	1.68
33	0.20	0.47	16.83	76.62	0.08	0.09	1.22	0.14	2.18	0.00	2.17
34	0.23	0.61	19.32	74.42	0.02	0.06	1.92	0.11	1.13	0.00	2.18

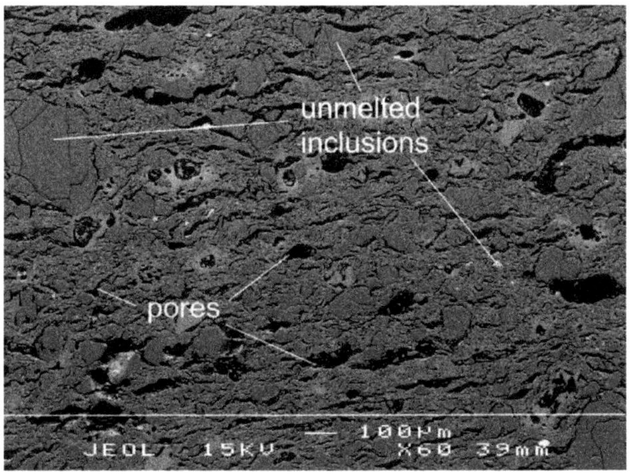

Figure 11.4: SEM photomicrograph of Sample 32 at x60 magnification

Liquor bottles

Table 11.9 lists the bottle sherds examined and their find site.

Table 11.9: Liquor bottle samples

Sample	Settlement	Site	Cat no.
36	Jembaicumbene	Hut 2	700.4
37	Jembaicumbene	Hut 2	721.7
38	Jembaicumbene	Hut 2	735.6
39	Jembaicumbene	Hut 2	759.3
40	Jembaicumbene	Hut 5	1250.1
41	Kiandra	Oven	1489
42	Kiandra	Hut 35	1602.2
43	Kiandra	Temple Anc.Blg.	1851
44	Kiandra	KCC2003-2	1582.3
45	Kiandra	KCC2003-2	1565
46	Kiandra Township	Yan's Store	56
47	East Eucumbene River	Site 977	1058
48	Narrandera	Chinatown	10005
49	Narrandera	Chinatown	10031

Bottle characteristics

No whole bottles were found but sherds indicate that they were thrown with the aid of a mould and of the standard Chinese bottle shape. Many join seams from the middle of bottles were found, however only one join seam from the neck of a bottle was seen (JCC Hut 2). The samples chosen to be examined in detail are either from the rim (4 sherds) or foot (10 sherds) of a bottle. In general the rims of the bottles are 4-5 mm thick. Sample 43, a rim sherd from the Temple Ancillary Building at Kiandra is 7 mm thick. All of the bases in this study have a diameter of 85-90 mm. Generally the thickness of the foot ring is 5-6 mm, but Sample 47, from East Eucumbene, Site 977, is 8 mm thick.

Most bottle sherds have a shiny or lustrous exterior glaze; only a few have a duller matte glaze. There are various shades of brown glaze; all lighter where the glaze is thinner. The glaze on a small number of samples is almost black. Pin-holing in the glaze is seen on most samples. All but one of the bottles, are glazed to the base, i.e. to the bottom of the foot, and under the base, with the foot ring left unglazed. Sample 47 is only glazed to the top of the foot and appears to be unglazed under the base, although the whole base is not present. A number of the bottles glazed to the base have chipped edges around the foot ring, suggesting that they have been forcibly removed from the kiln shelf after the glaze has run and stuck. Sample 49 has sand stuck in the glaze on underside of the base, illustrating the method of placing vessels on sand when firing. The interior glaze on all bottles varies in thickness and coverage due to the glazing method, which appears to be the same as the shouldered food jars.

All but one of the sherds, have varying shades of buff to grey body depending on the extent of oxidation or reduction. Sample 42 has an orange-pink body. The texture of the body from Samples 43 and 47 is different to those of the other samples, which are represented by Sample 48 (Figure 11.5). The apparent porosity of the bottle sherds ranges from 4.34 per cent to 9.73 per cent. The average is 7.32 per cent and the median 7.70 per cent.

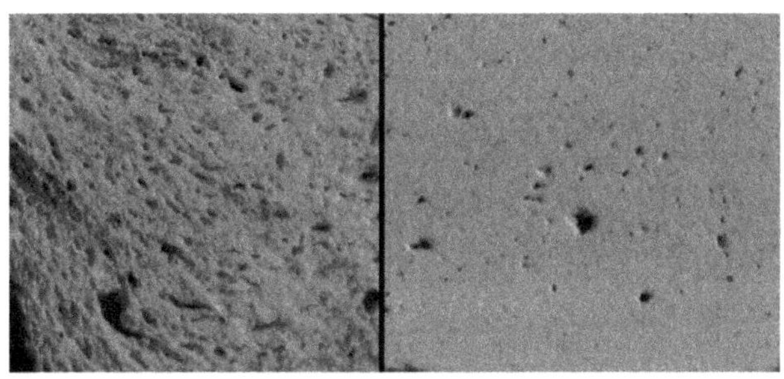

Figure 11.5: Samples 47 (left) and 48 showing the difference in body texture

Compositional analysis

The results for the averaged body and glaze composition of 14 liquor bottles are shown in Tables 11.10 and 11.11.

Generally, the glaze has greater amounts of magnesium and calcium oxides than the body. Two samples, 43 and 47, have phosphorous and manganese oxides in their glazes along with a higher percentage of iron oxide.

Most of the bottles have similar bodies to Sample 44 (Figure 11.6). The exceptions are Samples 43 (Figure 11.7) and 47 which have larger pieces of unmelted quartz and larger pores. Although the pores are smaller in Sample 44 and most other samples, they are more numerous.

Common inclusions apart from quartz are rutile, ilmenite, iron oxide and zircon. Sample 40 has monazite (Ce, La and Nd composition) and Samples 37, 39, 47 and 48, xenotime.

Table 11.10: Averaged body composition of liquor bottles

Sample	Oxide concentrations (wt%)										
	Na_2O	MgO	Al_2O_3	SiO_2	P_2O_5	SO_3	K_2O	CaO	TiO_2	MnO	FeO
36	0.29	0.37	25.19	69.35	0.00	0.00	1.89	0.47	1.11	0.00	1.33
37	0.28	0.35	28.55	65.93	0.00	0.06	1.92	0.20	1.38	0.00	1.33
38	0.34	0.37	26.92	67.55	0.00	0.06	1.82	0.36	1.32	0.00	1.26
39	0.35	0.35	26.92	67.91	0.03	0.00	1.65	0.60	1.23	0.00	0.96
40	0.36	0.45	25.02	69.14	0.03	0.04	1.98	0.35	1.35	0.00	1.28
41	0.33	0.45	26.19	68.22	0.02	0.02	1.94	0.34	1.25	0.00	1.24
42	0.36	0.36	27.99	66.41	0.02	0.02	1.56	0.40	1.39	0.00	1.49
43	0.20	0.54	23.18	70.17	0.00	0.02	2.87	0.14	0.90	0.00	1.98
44	0.37	0.36	30.19	64.12	0.00	0.02	1.81	0.24	1.39	0.00	1.50
45	0.25	0.34	25.63	69.10	0.00	0.04	1.75	0.49	1.18	0.00	1.22
46	0.28	0.40	27.85	66.83	0.02	0.06	1.92	0.21	1.33	0.00	1.10
47	0.20	0.59	24.59	69.03	0.00	0.05	2.73	0.14	0.97	0.00	1.70
48	0.33	0.42	27.88	66.57	0.04	0.02	2.15	0.32	1.18	0.00	1.09
49	0.36	0.46	28.27	65.42	0.04	0.02	2.05	0.68	1.26	0.00	1.44

Table 11.11: Averaged glaze composition of liquor bottles

Sample	Oxide concentrations (wt%)										
	Na_2O	MgO	Al_2O_3	SiO_2	P_2O_5	SO_3	K_2O	CaO	TiO_2	MnO	FeO
36	0.55	1.79	20.05	64.92	0.05	0.02	2.61	5.74	0.81	0.00	3.46
37	0.63	2.00	18.67	65.15	0.09	0.03	2.92	5.68	1.05	0.00	3.78
38	0.63	1.48	22.40	65.05	0.13	0.00	2.69	3.26	1.01	0.02	3.33
39	0.91	2.42	15.88	65.86	0.12	0.02	3.00	6.00	0.88	0.02	4.89
40	0.71	2.36	15.63	66.75	0.15	0.03	2.78	6.20	0.88	0.00	4.51
41	0.68	2.12	17.27	66.75	0.02	0.05	2.83	5.23	0.92	0.00	4.13
42	0.77	1.77	19.05	65.34	0.05	0.00	2.48	5.82	1.02	0.00	3.70
43	0.36	3.00	16.03	57.20	2.32	0.00	2.67	9.86	1.31	0.59	6.66
44	1.04	2.82	14.58	64.99	0.10	0.04	3.01	7.28	0.80	0.00	5.34
45	0.49	1.89	19.54	65.79	0.20	0.08	2.64	4.73	0.88	0.00	3.76
46	0.62	1.75	20.28	66.10	0.09	0.02	3.32	3.45	0.96	0.05	3.36
47	0.50	2.73	16.16	58.74	2.02	0.10	2.89	8.60	1.32	0.51	6.43
48	0.78	2.61	15.91	65.63	0.16	0.02	2.98	6.87	0.72	0.04	4.28
49	0.75	2.82	15.51	65.63	0.13	0.03	2.87	6.36	0.94	0.00	4.96

Glaze firing temperatures

Using the firing temperature range for unity formulae, suggested by Hamer and Hamer (1997:379), most of the glazes are suitable to be fired over 1200°C. Sample 43 has a glaze suitable for less than 1200°C.

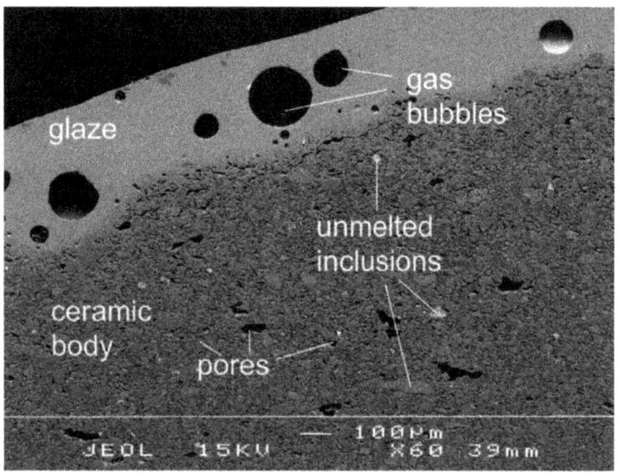

Figure 11.6: SEM photomicrograph of Sample 44 at x60 magnification

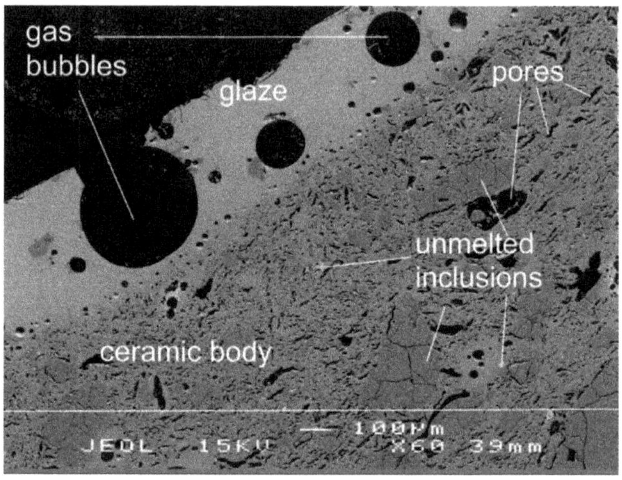

Figure 11.7: SEM photomicrograph of Sample 43 at x60 magnification

Ginger jars

The apparent porosity of thirteen sherds from ginger jars was tested, four from blue-underglaze decorated jars and nine from green-glazed jars. The glaze composition of two blue-decorated sherds and three green-glazed sherds was determined using EMPA. Table 11.12 lists the ginger jar sherds examined, their find site, apparent porosity and if their glaze composition was determined.

Jar characteristics

The majority of ginger jar sherds were small but all were from round, rather than straight-sided, jars. The blue-underglaze jars have a clear glaze over all their body while the green-glazed ones have a brown-glazed interior. It was noted that in all cases when the lower section of a ginger jar glaze was present on sherds, there was evidence of running. Lids have the same glaze as their jar types. The blue-underglaze jar lids are flat with a flanged side and the green-glazed lids are domed. The apparent porosity of the blue-underglaze samples ranges from 4.33 to 11.16 per cent with an average of 7.60 per cent. The green-glazed samples range from 5.17 to 11.54 per cent with an average of 7.90 per cent. Crystals were observed in the glaze on many of the jar sherds.

Compositional analysis

The averaged glaze composition of green-glazed and blue-underglaze decorated ginger jars is shown in Table 11.13, by oxide weight percentage, along with the standard deviation (St. Dev.). All the tested glazes are different. The only common characteristic is the low percentage of alumina in comparison to silica, indicating a fluid glaze. This is the reason why these jars are only glazed on the upper section; the glaze is likely to run when fired. The colorant in the green glazes is copper (Cu) and two of them contain lead (Pb). The green-glazed ginger jar from KCC is probably older than the other two. This jar has more lead and no zinc (Zn) in comparison to the jars from the Township. The blue-underglaze decorated jars have differing amounts of flux; the JCC sample has considerably more calcium, while the Kiandra Store sample has more sodium.

Glaze firing temperatures

Four of the ginger jar glazes are suitable to be fired to around 1000°C. The glaze of Ah Yan's blue jar does not fit within the guidelines proposed by Hamer and Hamer (1997:379).

Table 11.12: Ginger jar samples

Decoration	Settlement	Site	Cat no.	App.porosity	Glaze comp.
Blue-underglaze	Kiandra Township	Residence	10373	4.99	no
Blue-underglaze	Jembaicumbene	Hut 4	788.4	11.16	no
Blue-underglaze	Jembaicumbene	Surface	1260.2	4.33	yes
Blue-underglaze	Kiandra Township	Chee's Store	384	9.94	no
Blue-underglaze	Kiandra Township	Yan's Store	74	n/a	yes
Green glaze	Jembaicumbene	Hut 2	725.1	5.17	no
Green glaze	Jembaicumbene	Store	1206.2	8.84	no
Green glaze	Kiandra	KCC2003-2	1183, 1282	6.90	yes
Green glaze	Kiandra Township	Yan's Store	306	7.19	yes
Green glaze	Kiandra Township	Residence	1015	9.89	yes
Green glaze	Kiandra Township	Residence	1291	11.54	no
Green glaze	Kiandra Valley	Site 125	1073	6.03	no
Green glaze	Narrandera	Chinatown	1826	7.42	no
Green glaze	Narrandera	Chinatown	10046	8.10	no

Table 11.13: Averaged glaze composition of green-glazed and blue-underglaze decorated ginger jars

Site			Oxide concentrations (wt%)												
			Na_2O	MgO	Al_2O_3	SiO_2	P_2O_5	SO_3	K_2O	CaO	TiO_2	FeO	CuO	ZnO	PbO
JCC Surface	Blue	Average	1.03	3.31	5.55	70.25	0.82	0.00	2.02	13.98	0.41	2.39			
		St.Dev.	0.10	0.01	0.39	0.47	0.11	0.00	0.05	0.60	0.15	0.01			
Ah Yan Store	Blue	Average	4.02	1.54	2.32	84.11	0.00	0.68	1.71	4.45	0.00	1.02			
		St.Dev.	0.21	0.12	0.16	2.43	0.00	0.18	0.13	2.24	0.00	0.09			
Ah Yan Store	Green	Average	4.20	1.84	2.09	61.84	0.16	0.22	1.67	15.71	0.00	0.83	9.14	2.06	0.00
		St.Dev.	0.27	0.10	0.06	0.23	0.28	0.03	0.10	0.22	0.00	0.03	0.25	0.15	0.00
Kiandra Town residence	Green	Average	2.42	3.09	3.68	63.50	0.62	0.00	2.39	13.24	0.00	1.50	4.75	1.68	2.70
		St.Dev.	0.15	0.11	0.35	0.51	0.11	0.00	0.11	0.41	0.00	0.07	0.39	0.14	0.21
KCC Surface	Green	Average	1.57	3.56	3.97	54.36	1.15	0.00	1.82	16.40	0.36	1.76	5.94	0.00	8.99
		St.Dev.	0.08	0.06	0.16	0.42	0.04	0.00	0.06	0.37	0.02	0.22	0.11	0.00	0.07

Celadon-glazed Winter Green

Table 11.14 lists the celadon-glazed sherds examined, their find site and their form. Samples 70 and 72 are from excavations at sites across the road from the Chinese stores in Kiandra Township.

Table 11.14: Celadon-glazed samples

Sample	Settlement	Site	Cat no.	Form
50	ACC	Hut 3	114	Rice bowl
51	JCC	Hut 1	10006	Rice bowl
52	JCC	Hut 2	193	Rice bowl
53	JCC	Hut 5	1248	Cup/Bowl
54	JCC	Store	1205.2	Rice bowl
55	JCC	Store	1208.2	Rice bowl
56	JCC	grid 14	1247	Rice bowl
57	KCC	Hut 2	1092	Rice bowl
58	KCC	Hut 2	1095	Rice bowl
59	KCC	Hut 4	1613	Rice bowl
60	KCC	Hut 4	1791	Rice bowl
61	KCC	Hut 4	1818	Rice bowl
62	EC	Site 143	1070	Rice bowl
63	KCC	Hut 21	1343	Rice bowl
64	KCC	Hut 42	1364	Rice bowl
65	KCC	Oven	1578	Rice bowl
66	KCC	KCC2003-2	1089	Rice bowl
67	KCC	KCC2003-2	1191	Rice bowl
68	KCC	KCC2003-2	1685	Rice bowl
69	KT	Chee's Store	144	Tea cup
70	KT	Residence	1133	Rice bowl
71	KT	Chee's Store	383	Rice bowl
72	KT	Residence	1726	Rice bowl
73	EC	Site 125	1079	Rice bowl
74	KCC	Hut 43	10007	Rice bowl
75	NC	Chinatown	596	Rice bowl
76	NC	Chinatown	10060	Rice bowl
155	KT	Yan's Store	190	Dish

Key: ACC-Adjungbilly Chinese Camp; JCC-Jembaicumbene Chinese Camp; KCC-Kiandra Chinese Camp; EC-Eucumbene Crossing; KT-Kiandra Township; NC-Narrandera Chinatown

Vessel characteristics

The sherds examined include 24 rims (23 rice bowls and one cup), three foot ring sherds (two rice bowls and a dish) and one fragment from either a rice bowl or cup. The celadon glazed vessels were all wheel-thrown.

There are slightly different profiles in the group of 23 rice bowl rim sherds. Samples 51, 55, 56, 59, 60 and 75 have a tapered rim with a flat edge. The other samples have rounded rims, the ends of which are slightly thicker producing the effect of a light band of glaze on the exterior rim where the glaze is thinner. This thickened clay rim section varies in length. Samples 58, 63, 65 and 73 have a thickened band of about 2 mm to 3 mm. The other rims have a thickened band of 4 mm to 5 mm. The colour of this lighter band is similar to the interior colour of the bowl. The majority of the rim edges are a rusty-orange colour. Wood (1978:12) stated that this is a characteristic of celadon-glazed wares from Zhejiang province, occurring when their body is re-oxidised.

The glaze on the bowls is about twice as thick on the exterior compared to the interior, suggesting that the vessels were first dipped in the glaze to get a total coverage and then re-dipped on the exterior only. The rim and foot would then have been wiped to remove the glaze so that the bowls would not stick together when stacked in the kiln for firing. Bubbles are seen in the glaze. In some cases the gas bubbles have broken the surface and not healed, causing pin-holing. Iron spotting is another fault observed in the glaze on the majority of samples. The celadon-glazed porcelain in this study is translucent to about 3.5-4.0 mm thickness when shining a light through it. The thicker areas, for example near the foot rings of larger vessels, are not translucent. The body colour of all sherds is a grey-white. The apparent porosity of the celadon-glazed sherds ranges from 0.09 per cent to 0.67 per cent. The average is 0.20 per cent and the median 0.17 per cent.

Compositional analysis

The results for the averaged body and glaze composition of 28 celadon-glazed vessels are shown in Tables 11.15 and 11.16. The body is composed of a siliceous material which is high in silica, low in alumina and has negligible titanium oxide. The glaze contains a higher percentage of calcium oxide, which is the main flux.

The numerous bubbles trapped in the glaze, along with unmelted quartz (silica) inclusions, are illustrated in Sample 60 (Figure 11.8). All the porcelain bodies exhibit extensive vitrification. The pores are rounded. There are large amounts of unmelted silica, generally less than 100 µm, in each of the Chinese porcelain body samples examined. Very small zircons and unmelted titanium oxide are present in about half the samples. A few samples contain inclusions of monazite, xenotime or iron oxide.

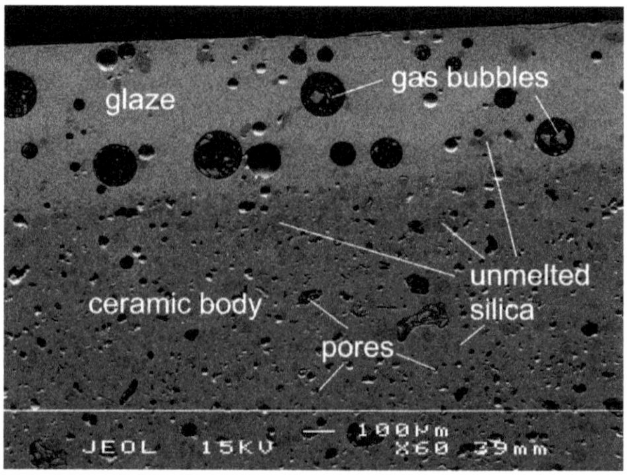

Figure 11.8: SEM photomicrograph of Sample 60 at x60 magnification

Table 11.15: Averaged body composition of celadon-glazed porcelain

Sample	Oxide concentrations (wt%)										
	Na_2O	MgO	Al_2O_3	SiO_2	P_2O_5	SO_3	K_2O	CaO	TiO_2	MnO	FeO
50	1.30	0.28	20.75	71.73	0.00	0.05	3.46	1.22	0.02	0.02	1.17
51	1.44	0.28	21.03	71.30	0.00	0.02	3.59	0.80	0.09	0.00	1.45
52	1.39	0.27	20.50	72.02	0.00	0.00	3.57	0.82	0.02	0.00	1.41
53	1.15	0.23	20.24	72.48	0.00	0.02	3.86	0.98	0.02	0.00	1.02
54	1.00	0.26	21.55	70.93	0.00	0.07	3.82	1.19	0.00	0.00	1.18
55	1.22	0.24	19.88	73.18	0.00	0.04	3.44	0.85	0.02	0.00	1.13
56	1.08	0.31	18.91	73.96	0.00	0.04	3.24	1.18	0.03	0.00	1.25
57	0.72	0.31	19.25	74.47	0.00	0.09	3.31	0.74	0.02	0.00	1.09
58	1.06	0.34	17.91	75.15	0.00	0.02	3.24	1.31	0.00	0.00	0.97
59	1.18	0.24	19.10	73.78	0.00	0.04	3.91	0.71	0.00	0.00	1.04
60	1.22	0.31	18.72	74.43	0.00	0.06	3.02	1.05	0.04	0.00	1.15
61	1.26	0.27	19.41	73.71	0.00	0.04	3.03	1.09	0.00	0.02	1.17
62	1.07	0.32	20.43	72.67	0.00	0.06	3.38	0.84	0.00	0.00	1.23
63	0.97	0.34	16.72	76.69	0.02	0.07	3.10	1.15	0.02	0.00	0.92
64	0.73	0.21	19.70	74.02	0.00	0.08	3.29	0.93	0.00	0.00	1.04
65	0.97	0.31	20.70	72.13	0.00	0.07	3.03	1.65	0.02	0.00	1.12
66	0.89	0.27	20.19	73.07	0.00	0.02	3.68	0.69	0.00	0.02	1.17
67	0.87	0.37	20.62	72.65	0.00	0.02	2.96	1.36	0.04	0.00	1.11
68	1.09	0.27	22.27	70.14	0.05	0.06	3.47	1.29	0.00	0.02	1.34
69	1.41	0.21	19.48	74.15	0.00	0.07	2.71	0.87	0.04	0.00	1.06
70	1.31	0.25	19.20	72.89	0.00	0.04	4.07	1.10	0.00	0.02	1.12
71	0.86	0.36	18.92	74.59	0.00	0.10	2.74	1.19	0.07	0.00	1.17
72	0.84	0.32	18.40	74.87	0.00	0.06	3.22	1.19	0.00	0.00	1.10
73	0.86	0.30	22.98	69.94	0.00	0.10	3.42	0.84	0.06	0.00	1.50
74	1.14	0.23	20.55	72.44	0.00	0.06	3.59	0.89	0.02	0.02	1.06
75	1.54	0.27	21.06	70.62	0.00	0.06	3.47	1.59	0.04	0.00	1.35
76	1.10	0.24	20.45	73.17	0.00	0.07	3.00	0.93	0.00	0.00	1.04
155	0.89	0.29	19.26	73.52	0.00	0.04	4.02	0.79	0.02	0.00	1.18

Table 11.16: Averaged glaze composition of celadon-glazed porcelain

Sample	Oxide concentrations (wt%)										
	Na_2O	MgO	Al_2O_3	SiO_2	P_2O_5	SO_3	K_2O	CaO	TiO_2	MnO	FeO
50	1.32	0.46	13.87	68.05	0.00	0.06	2.58	12.88	0.00	0.00	0.78
51	2.19	0.37	16.10	69.60	0.00	0.00	4.15	6.65	0.00	0.00	0.94
52	1.77	0.29	14.42	70.94	0.00	0.02	3.74	8.03	0.02	0.00	0.77
53	1.31	0.38	13.64	69.64	0.00	0.00	3.49	10.84	0.02	0.00	0.68
54	1.14	0.40	14.77	69.26	0.00	0.03	3.34	10.29	0.02	0.00	0.75
55	1.58	0.56	13.72	66.60	0.00	0.02	3.03	13.63	0.00	0.00	0.86
56	1.50	0.37	15.89	69.24	0.00	0.01	3.38	8.59	0.00	0.02	1.00
57	0.96	0.49	15.51	68.85	0.01	0.03	3.44	9.72	0.04	0.00	0.95
58	1.32	0.55	15.18	65.57	0.00	0.03	2.83	13.68	0.00	0.00	0.84
59	1.36	0.40	14.19	68.94	0.00	0.00	3.46	10.52	0.25	0.00	0.88
60	1.50	0.51	14.61	68.88	0.00	0.02	2.74	11.01	0.02	0.00	0.71
61	1.54	0.54	14.69	68.86	0.00	0.03	2.72	10.78	0.00	0.00	0.84
62	1.20	0.40	13.84	69.53	0.00	0.04	2.97	11.18	0.00	0.00	0.84
63	1.37	0.54	15.20	65.60	0.00	0.02	2.75	13.65	0.04	0.00	0.83
64	0.90	0.43	14.43	68.24	0.00	0.00	2.72	12.32	0.04	0.00	0.92
65	1.26	0.55	15.26	69.50	0.00	0.02	3.20	9.26	0.00	0.02	0.93
66	1.09	0.39	15.56	69.65	0.00	0.08	3.62	8.66	0.00	0.00	0.95
67	1.25	0.46	14.90	69.16	0.00	0.02	3.34	10.08	0.00	0.00	0.79
68	1.19	0.44	14.40	66.94	0.04	0.00	2.87	13.30	0.00	0.00	0.82
69	2.25	0.29	15.98	69.48	0.00	0.00	3.10	8.09	0.03	0.00	0.78
70	1.14	0.49	12.95	67.63	0.00	0.00	2.56	14.40	0.02	0.00	0.81
71	1.43	0.50	15.25	65.65	0.03	0.05	2.88	13.22	0.00	0.00	0.99
72	0.89	0.58	13.08	67.71	0.00	0.02	2.35	14.40	0.00	0.03	0.94
73	1.04	0.37	14.02	69.94	0.00	0.05	3.21	10.46	0.00	0.00	0.91
74	1.27	0.41	14.59	67.81	0.02	0.00	2.86	12.22	0.02	0.00	0.80
75	1.89	0.41	15.36	66.62	0.00	0.02	3.06	11.84	0.00	0.00	0.80
76	1.57	0.52	14.88	68.02	0.00	0.02	3.10	10.82	0.00	0.03	1.04
155	1.05	0.63	14.98	67.93	0.00	0.00	3.39	11.04	0.03	0.00	0.94

Glaze firing temperatures

The unity formulae for the celadon-glazed samples suggest that all are suitable to be fired to 1300°C.

Four Seasons Wares

The apparent porosity of eight sherds from Four Seasons decorated vessels was tested and the glaze composition of two sherds was determined using EMPA.

Table 11.17 lists the Four Seasons sherds examined, their find site, apparent porosity and if their glaze composition was determined.

Vessel characteristics

There were no whole vessels. The glaze has a green tint and the overglaze colours conform to previous researchers descriptions. The average apparent porosity overall is 0.30 per cent but the average of the bowls alone is 0.19 per cent, illustrating the extent of vitrification.

Compositional Analysis

The glaze from two Four Seasons decorated bowls was examined using EMPA (Table 11.18). Both contain lead oxide.

Glaze firing temperatures

Both analysed glazes were suitable to be fired in excess of 1200°C (Hamer & Hamer 1997:379).

Blue-underglaze wares

The apparent porosity of 14 sherds from blue-underglaze decorated vessels was tested and the glaze composition of four was determined using EMPA.

Table 11.19 lists the blue-underglaze sherds examined, their find site, apparent porosity and if their glaze composition was determined.

Vessel characteristics

Again, no whole vessels were present. The Bamboo and Double Happiness patterns appear on porcelain and stoneware bodies but the Rocks and Orchid dishes are both porcelain. The glazes have either a blue or green tint and the blue-underglaze colour varies from bright blue to olive-green. Two samples analysed from Kiandra, from the Temple ancillary building and Hut 35, have a green-blue underglaze and green tint to their glazes. All the other samples have blue underglaze with blue tints to their glaze.

Compositional analysis

The glazes of four blue-underglaze decorated vessels were examined using EMPA (Table 11.20). The Rocks and Orchid glaze, the only one to contain lead and zinc oxides, is on a dish; the other vessels are bowls.

Glaze firing temperatures

The analysed glazes of the three rice bowls are suitable to be fired between 1200°C and 1300°C. The glaze on the Rocks and Orchid dish is suitable for about 1200°C.

Table 11.17: Four Seasons decorated samples

Vessel	Decoration	Settlement	Site	Cat no.	App.porosity	Glaze comp.
Serving bowl	Four Seasons	Kiandra	Hut 1	1123	0.08	no
Serving bowl	Four Seasons	Kiandra	Hut 4	1627	0.15	no
Serving bowl	Four Seasons	Kiandra	KCC2003-2	1672	0.27	no
Serving bowl	Four Seasons	Kiandra Township	Yan's Store	37	0.15	no
Serving bowl	Four Seasons	Kiandra Township	Residence	10313	0.14	no
Serving bowl	Four Seasons	Narrandera	Chinatown	610	0.42	no
Serving bowl	Four Seasons	Kiandra Township	Chee's Store	356	n/a	yes
Liquor cup	Four Seasons	Kiandra	Hut 62	1128	0.11	yes
Spoon	Four Seasons	Jembaicumbene	Surface	1258	1.08	no

Table 11.18: Averaged glaze composition of Four Seasons decorated bowls

Site	Decoration		Oxide concentrations (wt%)									
			Na_2O	MgO	Al_2O_3	SiO_2	K_2O	CaO	TiO_2	MnO	FeO	PbO
Ah Chee Store	Four Seasons	Average	0.76	0.43	11.57	69.26	2.71	5.14	0.00	0.00	1.30	8.83
		St.Dev.	0.09	0.02	0.12	0.27	0.02	0.05	0.00	0.00	0.06	0.26
KCC Hut 62	Four Seasons	Average	1.84	0.44	13.63	67.53	4.77	4.52	0.00	0.00	1.78	5.33
		St.Dev.	0.13	0.04	0.14	0.62	0.20	0.25	0.00	0.00	0.11	0.66

Table 11.19: Blue-underglaze decorated samples

Vessel	Decoration	Settlement	Site	Cat no.	App.porosity	Glaze comp.
Rice bowl	Double Happiness	Jembaicumbene	Hut 2	169	6.48	no
Rice bowl	Double Happiness	Kiandra Valley	Site 809	1435	5.99	no
Rice bowl	Double Happiness	Narrandera	Chinatown	614	0.14	no
Rice bowl	Double Happiness	Narrandera	Chinatown	10044	2.76	no
Rice bowl	Bamboo	Jembaicumbene	Hut 2	705	1.02	no
Rice bowl	Bamboo	Jembaicumbene	Temple	786	0.18	no
Rice bowl	Bamboo	Kiandra	Hut 4	1652, 1663	0.36	yes
Rice bowl	Bamboo	Kiandra	Temple Ancillary Blg	1760	0.63	no
Rice bowl	Bamboo	Kiandra	KCC2003-2	1319	0.20	no
Rice bowl	possible Bamboo	Jembaicumbene	Store	1205.3	2.64	yes
Dish	Rocks and Orchid	Jembaicumbene	Hut 2	199, 747	0.17	yes
Dish	Rocks and Orchid	Kiandra	Hut 2	1093	0.09	no
Teapot lid	Blue underglaze	Jembaicumbene	Hut 2	189.3	0.24	no
Rice bowl	Blue underglaze	Kiandra Township	Yan's Store	62.2	6.89	no
Rice bowl	Blue underglaze	Kiandra	Hut 35	1587	n/a	yes

Table 11.20: Averaged composition of blue-underglaze decorated glazes

Site	Decoration		Oxide concentrations (wt%)										
			Na_2O	MgO	Al_2O_3	SiO_2	K_2O	CaO	TiO_2	MnO	FeO	ZnO	PbO
KCC Hut 35	Blue-u/g	Average	0.00	0.37	10.59	76.79	2.98	8.21	0.00	0.00	0.74	0.00	0.00
		St.Dev.	0.00	0.04	0.23	0.27	0.09	0.12	0.00	0.00	0.02	0.00	0.00
KCC Hut 4	Bamboo	Average	0.82	0.43	12.51	71.20	3.80	10.17	0.00	0.00	0.98	0.00	0.00
		St.Dev.	0.10	0.04	0.25	0.10	0.08	0.13	0.00	0.00	0.15	0.00	0.00
JCC Store	possible Bamboo	Average	0.00	0.33	12.50	73.45	4.80	8.08	0.00	0.00	0.74	0.00	0.00
		St.Dev.	0.00	0.02	0.37	0.67	0.12	0.39	0.00	0.00	0.05	0.00	0.00
JCC Hut 2	Rocks & Orchid	Average	0.77	1.91	11.71	63.80	3.68	12.93	0.00	0.00	2.48	1.06	1.51
		St.Dev.	0.19	0.45	0.39	2.06	0.58	1.40	0.00	0.00	0.49	0.42	0.42

NON-CHINESE SAMPLES

Fourteen samples, described below, were included in the study as a comparison to the Chinese samples for the determination of provenance. Five sherds may have originated in Britain or Australia, seven are definitely from Britain and two are from Japan. Table 11.21 lists the sherds, the vessels from which they came and their apparent porosity. The apparent porosity shows a range of bodies. The average for the porcelain, bone china and stoneware is 0.19, 0.49 and 0.17 respectively. The earthenware has 10.24 and 4.60 per cent apparent porosity. The averaged body composition for these samples is seen in Table 11.22. The glaze composition for the earthenware and porcelain samples is detailed in Table 11.23. No readings were taken for the stoneware salt glaze.

British/Australian stoneware

Sample 120 (JCC-162) is from a large stoneware storage jar from JCC Hut 2. The two-tone Bristol glaze features the whitish zinc glaze over all the exterior and interior, and the honey-coloured manganese glaze on the upper section of the exterior. Samples 125 (JCC-185.2), 126 (KCC2003-1622) and 127 (KAP96-207.3) are all from stoneware salt-glazed blacking bottles, found respectively at JCC Hut 2; KCC Temple; and Ah Yan's Kiandra Store. All of these vessels are wheel-thrown and all have inclusions of zircon, rutile and iron oxide in their ceramic bodies. Sample 126 also contains monazite (Ce, La and Nd composition). Bristol and salt glazes were commonly used for utilitarian wares in the nineteenth century in Britain and Australia.

Table 11.21: Apparent porosity of non-Chinese samples

Sample	Vessel	Body type and decoration	App.Por.
120	Large st jar	Stoneware - Bristol glaze	0.09
125	Blacking bottle	Stoneware - salt glaze	0.08
126	Blacking bottle	Stoneware - salt glaze	0.32
127	Blacking bottle	Stoneware - salt glaze	0.19
114	Teapot?	E/ware - Rockingham glaze	10.24
147	Plate	E/ware - clear glaze	4.60
128	Unidentifiable	Bone china - clear glaze	0.48
129	Saucer	Bone china - clear glaze	0.50
130	Saucer	Bone china - clear glaze	0.63
132	Saucer	Bone china - clear glaze	0.10
133	Saucer	Bone china - clear glaze	0.77
134	Hollowware	Porcelain - clear glaze	0.10
135	Hollowware	Porcelain - clear glaze	0.28

Table 11.22: Averaged body composition for non-Chinese samples

Sample	Oxide concentrations (wt%)										
	Na_2O	MgO	Al_2O_3	SiO_2	P_2O_5	SO_3	K_2O	CaO	TiO_2	MnO	FeO
120	0.48	0.90	25.85	66.34	0.02	0.04	3.17	0.33	1.17	0.00	1.70
125	0.40	0.92	27.31	64.66	0.00	0.03	3.39	0.26	1.22	0.00	1.81
126	0.54	0.85	25.70	66.52	0.00	0.03	3.18	0.22	1.13	0.00	1.83
127	0.37	0.91	27.11	64.86	0.09	0.02	3.44	0.14	1.13	0.00	1.93
114	0.53	0.37	20.43	73.51	0.02	0.15	1.89	1.54	0.64	0.00	0.92
147	0.89	0.20	21.14	74.28	0.12	0.06	1.83	0.89	0.18	0.00	0.41
156	0.51	0.17	18.80	77.18	0.05	0.06	1.11	1.11	0.54	0.00	0.46
128	1.43	0.45	17.15	42.32	16.12	0.00	1.97	20.20	0.00	0.00	0.36
129	0.75	0.46	17.82	43.34	15.35	0.00	2.68	19.05	0.00	0.00	0.55
130	1.27	0.43	17.56	42.88	15.83	0.00	1.90	19.64	0.02	0.00	0.47
132	1.47	0.50	14.46	41.16	18.35	0.02	1.81	22.06	0.02	0.00	0.15
133	1.17	0.56	15.95	36.72	19.47	0.00	2.21	23.78	0.00	0.00	0.14
134	1.71	0.06	22.60	68.90	0.00	0.04	5.13	0.54	0.30	0.00	0.72
135	1.50	0.07	21.95	70.07	0.00	0.05	5.52	0.32	0.00	0.00	0.52

Table 11.23: Averaged glaze composition for non-Chinese samples: earthenware, bone china and porcelain

Sample	Oxide concentrations (wt%)											
	Na_2O	MgO	Al_2O_3	SiO_2	P_2O_5	SO_3	K_2O	CaO	TiO_2	MnO	FeO	PbO
114	0.79	0.20	7.62	41.92	0.04	0.00	1.39	2.00	0.27	3.33	1.71	40.73
147	1.75	0.31	11.21	65.51	0.21	0.40	2.73	7.92	0.02	0.00	0.32	9.61
156	1.71	0.14	12.75	64.92	0.00	0.09	2.97	5.96	0.07	0.00	0.31	11.08
128	1.49	0.31	12.53	54.76	1.18	0.12	2.94	10.70	0.00	0.00	0.42	15.55
129	0.81	0.32	12.16	52.29	3.69	0.00	3.87	10.12	0.00	0.00	0.26	16.48
130	1.32	0.29	12.80	53.38	3.28	0.60	2.80	13.21	0.00	0.00	0.38	11.94
132	2.08	0.36	11.55	58.13	0.13	0.00	3.02	9.89	0.00	0.00	0.24	14.60
133	1.31	0.26	10.83	60.73	0.06	0.58	3.50	8.04	0.00	0.00	0.18	14.51
134	1.74	0.13	16.05	73.49	0.00	0.02	5.37	2.49	0.11	0.00	0.60	0.00
135	2.09	0.12	15.84	71.25	0.00	0.00	6.45	3.80	0.00	0.00	0.45	0.00

British/Australian earthenware

Sample 114 (KCC1998-1169), from KCC Hut 1, was wheel-thrown, has a shiny Rockingham glaze and is possibly a teapot or jug. It may have been made in either Britain or Australia. Inclusions of zircon, rutile, iron oxide and monazite (Ce, La and Nd composition) are present in the ceramic body. The glaze contains manganese and iron oxides and an average of 40.73 per cent lead oxide. The glaze composition suggests a suitable firing range of less than 1100°C.

British earthenware

Sample 147 (KCC2001-1661) is from a clear-glazed high-fired earthenware plate with relief-moulded decoration from KCC Hut 5, possibly dating to 1870-1885. The glaze contains an average 9.61 per cent lead oxide. Sample 156 (KC2001-1539) is from a high-fired earthenware flown-blue transfer-printed bowl, from KCC Hut 53. The glaze contains an average of 11.08 per cent lead oxide. The glaze compositions for these vessels suggest a suitable firing range of 1100-1200°C.

British bone china

Five samples were examined: 128 (KAP97-1172) from a residence in Kiandra Township; 129 (JCC-785.5) from JCC Temple Complex; 130 (KCC2003-1100) from the surface collection at KCC; 132 (KAP96-176.2) from Ah Chee's Kiandra Store; and 133 (KGQIV-1106.2) from the Gold Commissioner's residence at Kiandra. The last of these is known to pre-date 1860. All the bodies/glazes are slightly different; none are similar to the older dated sample. These glazes all contain lead oxide as the flux. Table 11.22 highlights the difference between bone china and other ceramic bodies; the high percentages of phosphorus and calcium oxides result from the addition of bone to the usual clay minerals.

Japanese porcelain

Samples 134 (KAP95-152.7) and 135 (KAP95-169.4), two sherds of Japanese origin (they are base-marked 'NIPPON/hand painted') were chosen for comparison with the Chinese sherds. The sherds were recovered during the 1995 ANU field school at the Kiandra Hotel, located near the Chinese stores in the township. The vessels were made in a mould with liquid clay (slip cast).

Both bodies are translucent. No inclusions, apart from silica, were seen in the body of Sample 134 while a small zircon, along with silica, was seen in Sample 135. Both have a pale-green glaze with an enamel over-glaze decoration. The glaze is a darker colour on the exterior and where it has pooled in the foot rings, and lighter, almost white, on the underside of the base. The chemical compositions suggest that both these glazes are suitable to be fired over 1300 °C. The main fluxes in the glazes are potassium and sodium oxides.

SAMPLES OF UNKNOWN ORIGIN

Sherds from seven vessels of unknown origin were examined in order to determine their provenance. Table 11.24 lists the sherds, the vessels from which they came and their apparent porosity. The apparent porosity of these samples shows a range of bodies from high-fired (clear-glazed cup) to low-fired (unidentifiable vessel). The averaged body and glaze compositions for these samples can be seen in Tables 11.25 and 11.26.

Table 11.24: Apparent porosity of samples with unknown origin

Sample	Vessel	Decoration	App.Por.
113	Bottle/jar	Brown glaze	0.18
115	Unidentifiable	Polychrome glaze	10.42
119	Large storage jar	Light-brown glaze	0.16
121	Ovoid jar	Mustard-brown glaze	9.64
116	Lamp dish	Clear glaze	0.27
117	Cup	Clear glaze	0.04

Sample 113 (KCC1998-1055) is a stoneware sherd from the shoulder of a possible ink bottle from KCC Hut 25 (Figure 9.12). It is wheel-thrown and has an angular shoulder of about 30°. Inclusions of zircon, rutile and iron oxide are present in the body. It has a shiny mid-brown glaze which totally covers the exterior but is patchy on the interior. The main fluxes in the glaze are potassium and sodium oxides.

Sample 115 (JCC-762.3) is an earthenware sherd from an unidentifiable vessel from JCC Hut 2. The body fabric looks similar to that of a Chinese wide-mouthed jar, particularly the base. Zircon, rutile and iron oxide inclusions are present in the body. The exterior has a multi-coloured glaze including areas of cream, brown, green and blue (Figure 7.9). The interior is unglazed, no traces of glaze being found on any of the interiors (22 sherds from this vessel). Higher percentages of magnesium, phosphorous, calcium and manganese oxides are present in the glaze than the body but a much smaller percentage of alumina. The main fluxes of the glaze are lead and calcium oxides.

Sample 119 (KAP96-299) is a stoneware sherd from a large storage jar from Ah Chee's store at Kiandra. Inclusions of zircon, rutile, iron oxide and monazite (Ce, La and Nd composition) are present in the body. The exterior appears to be glazed on the upper section of the jar only, not to the foot of the vessel. This glaze (119-1) is a honey colour containing manganese and iron oxides. The interior glaze (119-2) is a shiny clear lead glaze. The main flux in these glazes is calcium oxide.

Table 11.25: Averaged body composition of samples of unknown origin

Sample	Oxide concentrations (wt%)										
	Na_2O	MgO	Al_2O_3	SiO_2	P_2O_5	SO_3	K_2O	CaO	TiO_2	MnO	FeO
113	0.44	0.36	22.52	71.70	0.00	0.13	2.23	0.23	1.33	0.00	1.06
115	0.15	0.42	19.59	75.37	0.04	0.19	1.59	0.22	0.84	0.00	1.59
119	0.19	0.55	24.80	69.21	0.22	0.08	2.39	0.15	0.93	0.00	1.48
121	0.20	0.62	21.76	71.19	0.00	0.12	2.91	0.16	0.83	0.00	2.20
157	0.22	0.45	20.96	72.48	0.00	0.21	2.96	0.03	0.74	0.00	1.94
116	1.81	0.20	18.53	74.52	0.00	0.04	3.60	0.39	0.00	0.02	0.89
117	1.25	0.19	18.45	74.77	0.00	0.05	4.10	0.29	0.22	0.00	0.68

Table 11.26: Averaged glaze composition of samples of unknown origin

Sample	Oxide concentrations (wt%)											
	Na_2O	MgO	Al_2O_3	SiO_2	P_2O_5	SO_3	K_2O	CaO	TiO_2	MnO	FeO	PbO
113	3.18	0.52	17.79	65.00	0.00	0.04	6.71	1.99	1.43	0.00	3.34	0.00
115	1.88	2.06	2.64	60.77	2.01	0.43	3.21	16.43	0.17	0.40	1.16	8.84
119-1	0.35	0.48	17.96	63.64	0.24	0.09	1.93	11.72	0.53	1.49	1.57	0.00
119-2	0.55	0.43	17.11	65.12	0.13	0.05	3.09	9.69	1.01	0.00	0.86	1.96
121-1	0.30	4.02	15.82	53.66	2.83	0.00	3.02	13.11	1.03	1.35	4.86	0.00
121-2	0.35	2.72	18.65	54.96	1.28	0.00	2.34	10.73	1.47	0.53	6.96	0.00
157-1	0.32	4.28	15.57	51.76	3.22	0.08	2.80	14.92	1.00	1.55	4.50	0.00
157-2	0.35	2.68	18.63	54.33	1.14	0.00	2.12	12.30	1.37	0.49	6.59	0.00
116	2.29	0.42	16.58	71.63	0.09	0.04	4.03	3.51	0.02	0.08	1.31	0.00
117	1.91	2.61	13.91	73.92	0.00	0.02	5.46	1.91	0.00	0.00	0.26	0.00

Sample 121 (KCC2001-1140), a stoneware sherd from a possible ovoid jar, is from KCC Hut 4. Inclusions of zircon, rutile, ilmenite, iron oxide and monazite (Ce, La and Nd composition) are present in the body. The sample features a shiny exterior mustard-coloured glaze with a crystalline surface (121-1) and an interior flat dark-brown glaze (121-2). The main flux in both glazes is calcium oxide.

Sample 157 (KCC2001-1226) is from another possible ovoid jar, from KCC Hut 3. The jar features at least three incised lines around the circumference, possibly on the upper part of the vessel. Inclusions of zircon, rutile, ilmenite and iron oxide are present in the body. This sample has a similar, although slightly darker exterior glaze (157-1) to Sample 121 and a dark-brown glaze on the interior (157-2). The main flux in the glazes is calcium oxide.

Sample 116 (KCC2001-1087) is from a porcelain lamp dish from KCC Hut 4 (Figure 9.8). It is wheel-thrown with a grey body and has a very thin clear-glaze with a green tinge, on the interior only. The main flux of the glaze is potassium oxide.

Sample 117 (KAP96-22) is from a porcelain slip-cast cup from Ah Yan's store at Kiandra. It has a clear glaze which looks grey as a result of the colour of the clay body. The main flux of the glaze is potassium oxide.

The suggested range of suitable glaze firing temperatures, using the unity formula, is: Sample 115 – about 1000°C; Samples 121 and 157, 1100-1200°C; Samples 113 and 119, 1200 - 1300°C; Samples 116 and 117, around 1300°C (Hamer & Hamer 1997:379).

STATISTICAL ANALYSIS

Statistical analysis was undertaken using SPSS 19 (2012). A comparison of body compositions using principal component analysis (PCA) shows that the British bone china is distinct from the all other samples (Figure 11.9). The porcelain and refined earthenware group in the lower section of the graph and the coarse earthenware and stoneware towards the right hand side.

If the bone china is excluded and the rest graphed the distinction between the refined and coarse wares becomes more apparent, the porcelain and refined earthenware to the left and the coarse earthenware and stoneware to the right (Figure 11.10). These two groups are examined separately below. Eleven elements were used for PCA analysis.

Food/beverage containers and utilitarian vessels

Ceramic body comparison

Principal component analysis was used initially to group samples (Figure 11.11). A clear distinction is observed between the known Chinese samples and the British/Australian samples in the graph of Components 1 and 2.

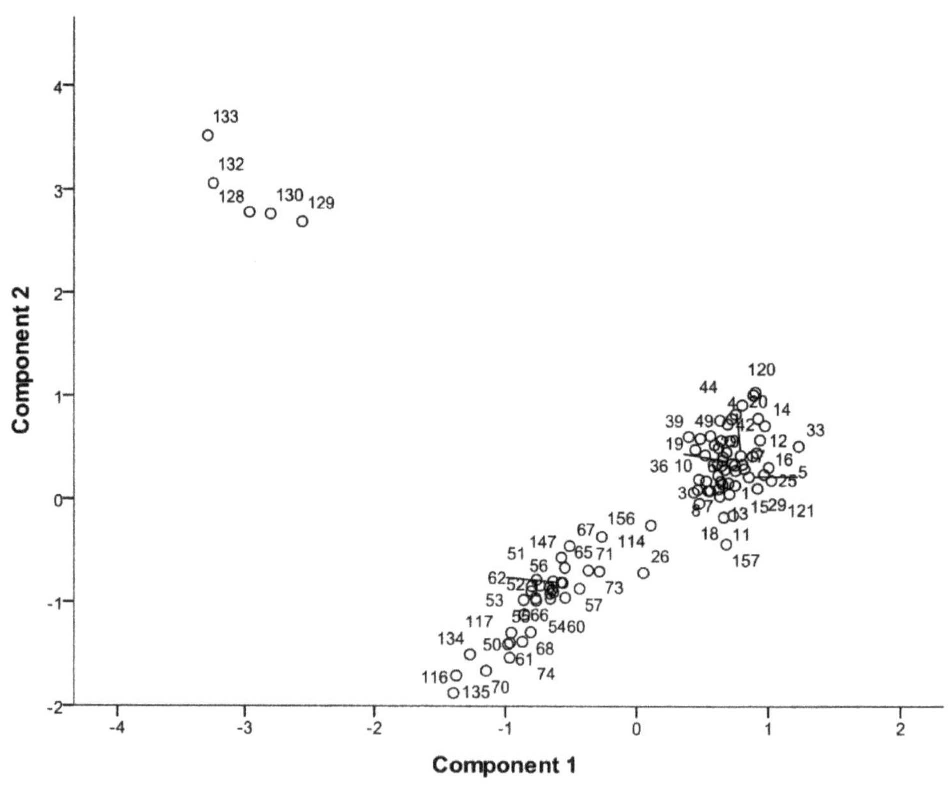

Figure 11.9: PCA scatter plot - all ceramic bodies

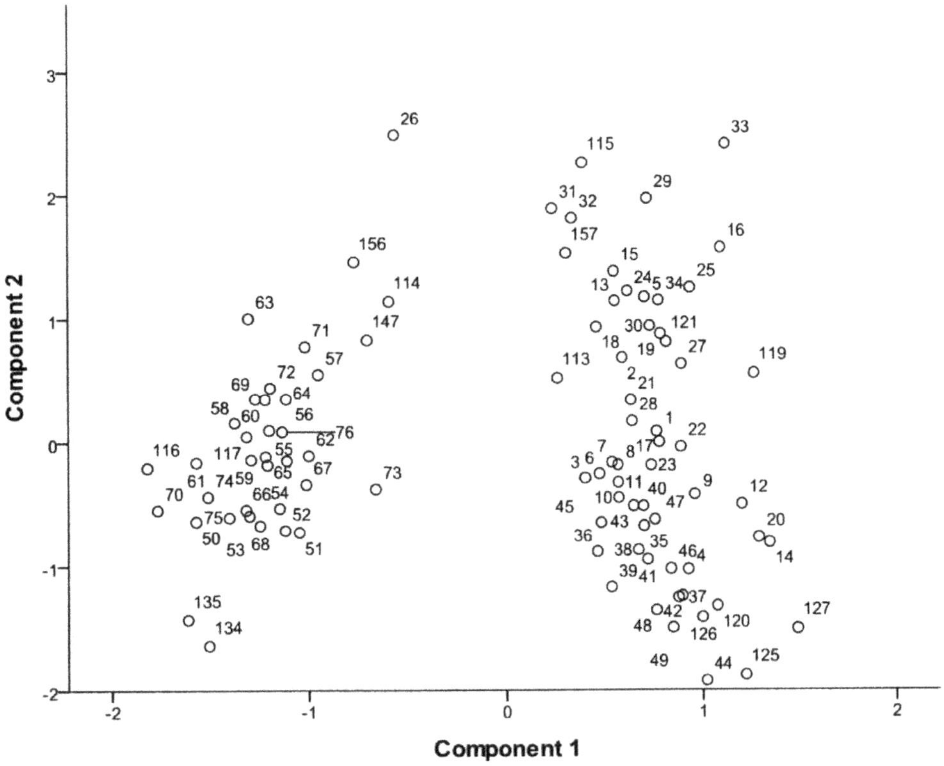

Figure 11.10: PCA scatter plot - porcelain, stoneware and earthenware ceramic bodies

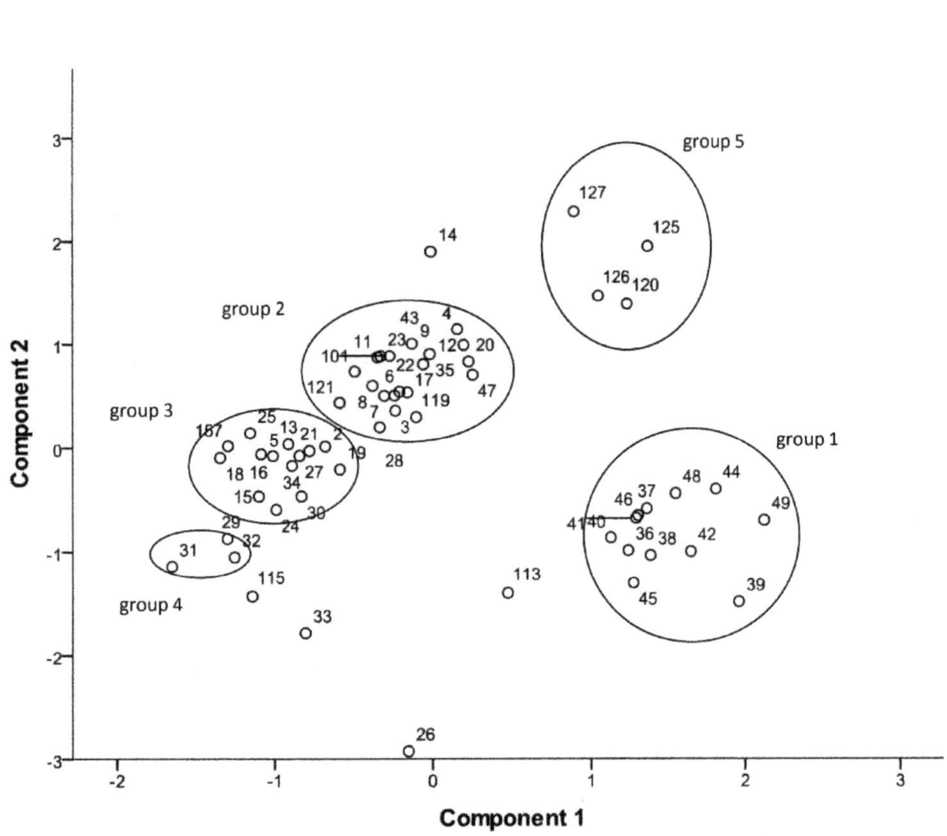

Figure 11.11: PCA scatter plot - ceramic body composition of all food/beverage containers and utilitarian vessels

Statistically, there are four components with Eigen values greater than one, together explaining 71.4 per cent of the total variance. Component 1 explains 27.4 per cent of the variance and Component 2, 21.2 per cent variance. The majority of the liquor bottles (Group 1) are together on the lower right of the graph and the British/Australian storage jar/ blacking bottles (Group 5) are together at the top right. The remaining groups contain Chinese-made vessels and those of unknown origin. Samples 14, 26 and 33 are not as closely aligned as others.

Discriminant analysis was used to test the probability of group membership. Figure 11.12 shows the group centroids when Functions 1 and 2 are graphed. Function 1 accounts for 59.9 per cent of the variance and Function 2, 33.0 per cent, totalling 92.9 per cent. Group 1 (MNV=12) comprises the majority of liquor bottles, all with 100 per cent probability of group membership. Group 2 (MNV=18), a mixture of globular jars, wide-mouthed jars and lids, also has 100 per cent probability of group membership. Group 3 (MNV=13), includes wide-mouthed jars and lids and Group 4 (MNV=3), wide-mouthed jar lids. The majority of these groups, 14 of 17, have over 99 per cent probability of group membership. Two samples have 93 and 94 per cent, while Sample 24 has 55.5 per cent probability of membership to Group 3 and 44.5 per cent to Group 4. Group 5 (MNV=4) contains the British/Australian stoneware samples, all with 100 per cent membership probability. The ungrouped cases include outliers (Samples 14, 26 and 33) and the vessels of unknown origin. Sample 14 is aligned with Group 2 and Sample 33 with Group 4. Sample 26, a Chinese wide-mouthed jar lid, is given 100 per cent probability of belonging to Group 5, the British/Australian stoneware (see later discussion). The samples of unknown origin 113, 115, 119, 121 and 157 are aligned with Groups 1, 4, 2, 3 and 2 respectively.

Reducing the number of elements used from 11 to eight (taking out P, S and Mn, all less than 0.2 per cent) produces a similar outcome (Figure 11.13). The only difference is a change of group for Sample 121, of unknown origin, from Group 3 to Group 2.

The averaged body composition of each group is detailed in Table 11.27 by oxide weight percentages and their standard deviation.

Group 1

This group comprises liquor bottles, 36, 37, 38, 39, 40, 41, 42, 44, 45, 46, 48 and 49. These are from the JCC, KCC, Ah Yan's Kiandra store and Narrandera. In comparison to the other Chinese-made samples, Group 1 ceramic body has the highest alumina / lowest silica content, the highest sodium and titanium oxides and lowest iron oxide content. The averaged apparent porosity of Group 1 is 7.69 per cent with a standard deviation of 1.31.

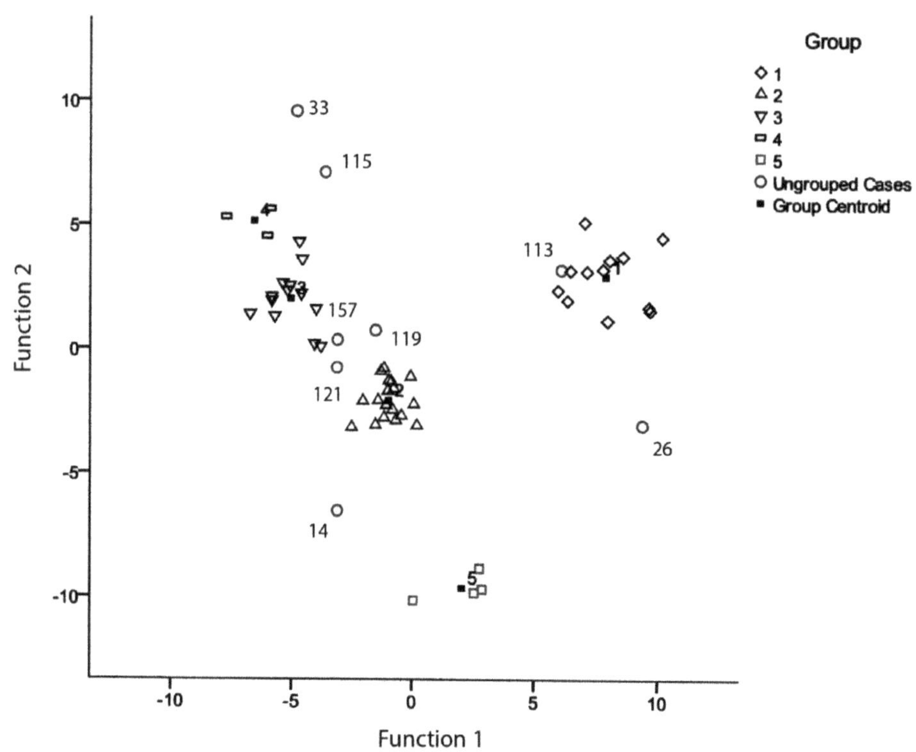

Figure 11.12: Discriminant analysis plot of the ceramic body of all food/beverage containers and utilitarian vessels examined using 11 elements

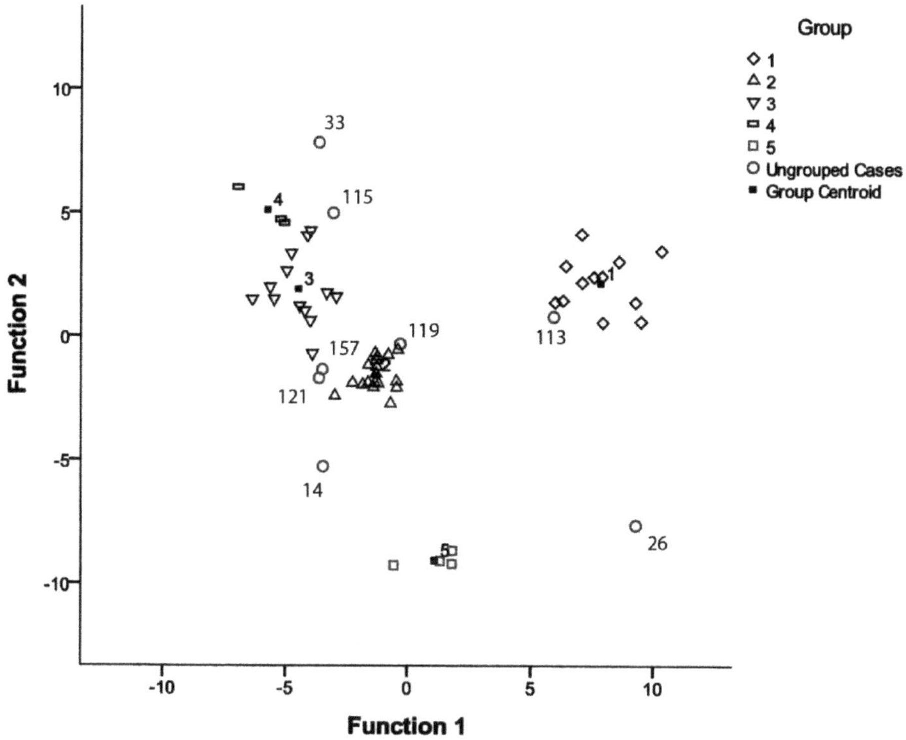

Figure 11.13: Discriminant analysis plot of the ceramic body of all food/beverage containers and utilitarian vessels examined using eight elements

Table 11.27: Averaged body composition of each group

Group		Oxide concentrations (wt%)										
		Na_2O	MgO	Al_2O_3	SiO_2	P_2O_5	SO_3	K_2O	CaO	TiO_2	MnO	FeO
1	Average	0.33	0.39	27.22	67.21	0.02	0.03	1.87	0.39	1.28	0.00	1.27
	St. Dev.	0.04	0.04	1.53	1.63	0.02	0.02	0.16	0.15	0.09	0.00	0.16
2	Average	0.21	0.50	23.72	69.57	0.02	0.04	2.78	0.13	0.90	0.00	2.12
	St. Dev.	0.02	0.07	1.33	1.41	0.02	0.03	0.22	0.07	0.14	0.00	0.23
3	Average	0.19	0.52	20.37	73.54	0.04	0.07	2.02	0.12	0.99	0.00	2.15
	St. Dev.	0.03	0.05	0.86	0.95	0.03	0.04	0.21	0.04	0.14	0.00	0.27
4	Average	0.17	0.44	17.25	77.48	0.04	0.05	1.55	0.10	0.94	0.00	1.98
	St. Dev.	0.02	0.02	1.45	2.01	0.03	0.06	0.22	0.04	0.17	0.00	0.28
5	Average	0.45	0.90	26.49	65.60	0.03	0.03	3.30	0.24	1.16	0.00	1.82
	St. Dev.	0.08	0.03	0.83	0.97	0.04	0.01	0.14	0.08	0.04	0.00	0.09

Group 2
This group contains liquor bottles, 43 and 47 from Kiandra Temple ancillary building and East Eucumbene Site 977; globular jars 1, 3, 4, 6, 7, 8, 9, 10 and 11, from JCC, FPCC, KCC, and Eucumbene Crossing Site 143; wide-mouthed jars 12, 17, 20, 22 and 23 from JCC, KCC, Eucumbene Crossing Site 143 and Ah Yan's Kiandra store; wide-mouthed jar lid 28 from JCC; and 35, an unidentified jar from ACC. The ceramic body of this group has the highest potassium and lowest titanium oxides of the Chinese-made samples. The average apparent porosity for this group is 5.81 per cent with a standard deviation of 1.97.

Group 3
This group consists of globular jars, Samples 2 and 5 from JCC; wide-mouthed jars 13, 15, 16, 18, 19 and 21 from JCC, KCC and Eucumbene Crossing Site 143; and wide-mouthed jar lids 24, 25, 27, 30 and 34, from JCC, KCC and Eucumbene Crossing Site 143. This group has the highest percentage of iron oxide. The average

apparent porosity for this group is 8.50 per cent with a standard deviation of 2.99.

Group 4

This group contains unglazed wide-mouthed jar lids, Samples 29, 31 and 32, all from KCC. This group has the highest percentage of silica and the lowest percentage of aluminium and potassium oxides of all the groups. The averaged apparent porosity of Group 4 is 10.68 per cent with a standard deviation of 1.08.

Group 5

This group includes the British/Australian Samples 120, 125, 126 and 127. This body group is characterised by higher percentages of sodium, magnesium and potassium oxides than the previous four groups. The averaged apparent porosity of Group 5 is 0.17 per cent with a standard deviation of 0.11.

The discriminant analysis outlined above is based on the absolute values of all the elements listed. Two other ways of treating the data to group the body samples were tried. First, calculations were made using all the elements with values transformed to log base 10, and secondly, transforming the data to a log-ratio using silica. All samples stayed within their allotted group.

Glaze composition

A PCA scatter plot illustrates the difference between the glazes (Figure 11.14). The Chinese brown-glazed wares are in two distinct groups, one comprising entirely of bottles, the other globular jars, wide-mouthed jars and bottles. Samples 119 and 120 have two readings, one each for internal and external glazes. When the glaze compositions of the globular and wide-mouthed jars are graphed alone they separate according to vessel type (Figure 11.15)

Discriminant analysis was used to test the probability of group membership of the brown-glazed wares. The majority of glaze samples grouped according to vessel type (Figure 11.16). Glaze A is found on all the liquor bottles except Samples 43 and 47, which have Glaze B along with all the globular jars and Sample 12, a wide-mouthed jar. All the other wide-mouthed jars have Glaze C. All samples have 100 per cent probability of group membership.

The characteristics of the glaze composition of each group are detailed in Table 11.28 by the averaged oxide weight percentages and their standard deviation.

Glaze A

This brown glaze has the highest percentages of alumina and silica, and sodium and potassium oxides; and the lowest magnesium, calcium, titanium and iron oxides in comparison to the other Chinese brown glazes. It also has no, or only trace-amounts of, manganese oxide. Based on the firing temperature suggested by Hamer and Hamer (1997:379), this glaze is suitable to be fired in the range 1200-1300°C.

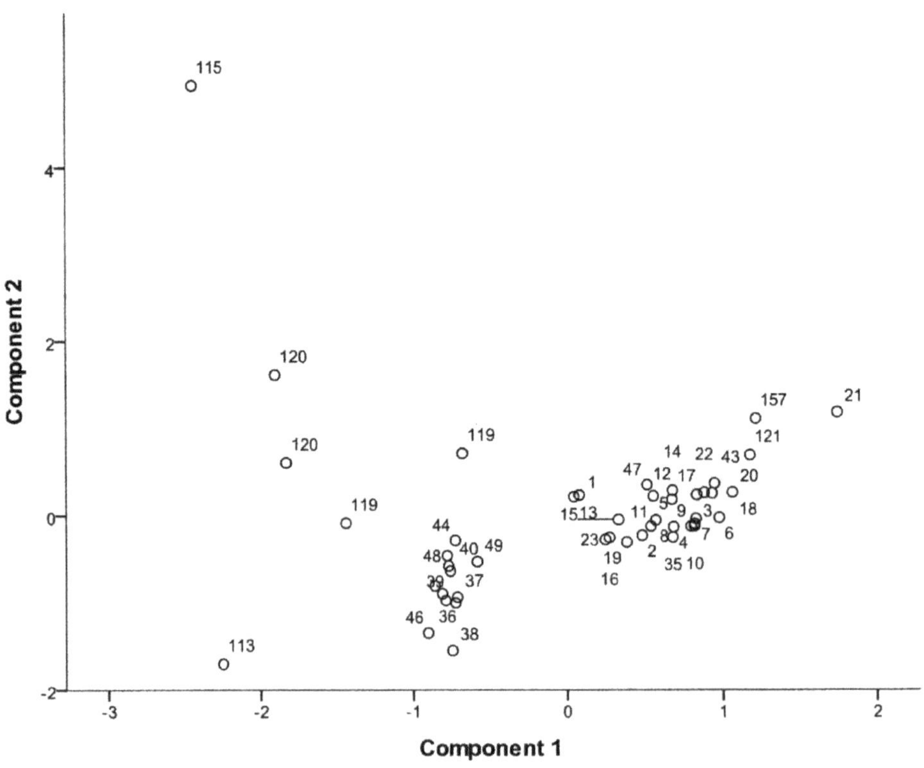

Figure 11.14: PCA scatter plot of food/beverage container and utilitarian glazes

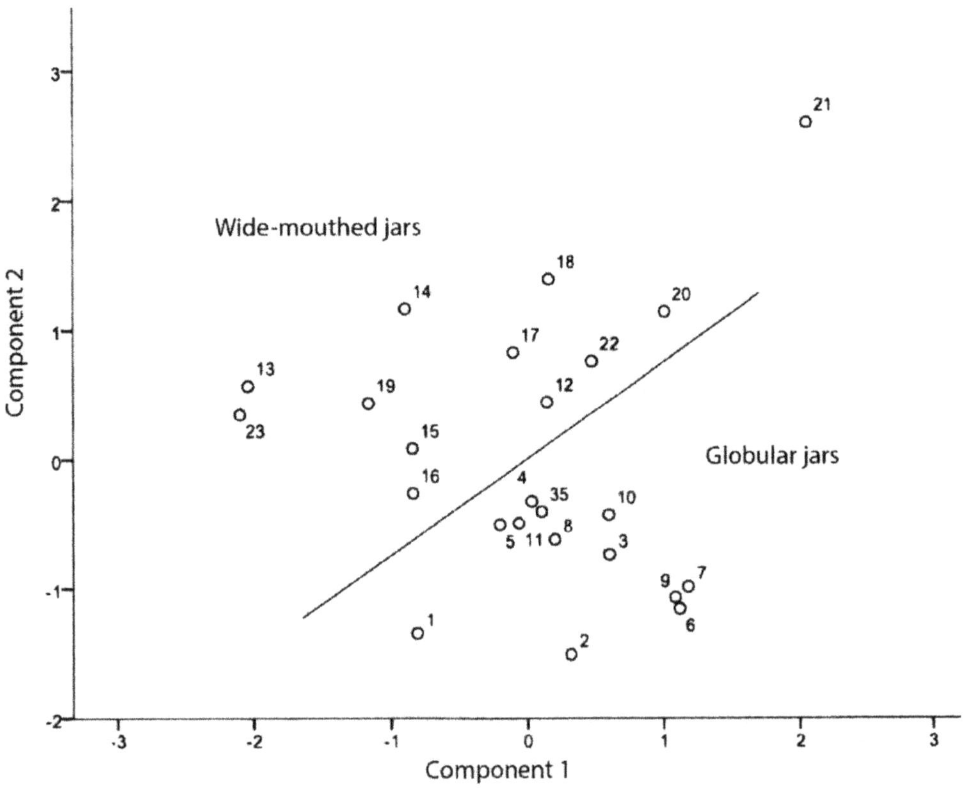

Figure 11.15: PCA scatter plot of brown-glazed globular and wide-mouthed jars

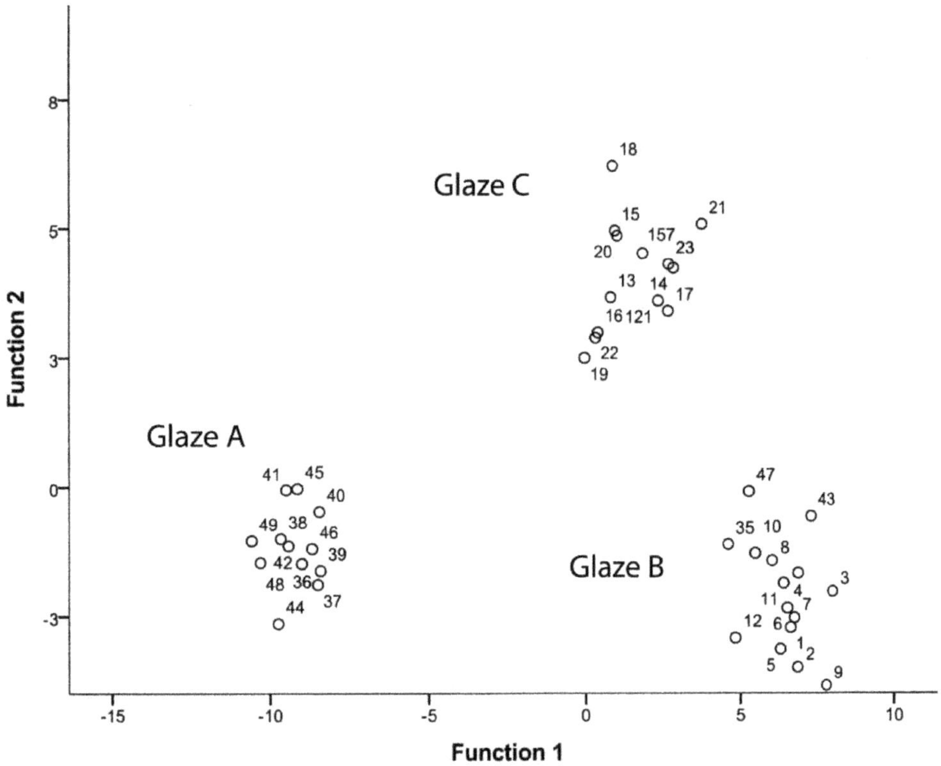

Figure 11.16: Discriminant analysis plot of the Chinese brown glazes

Table 11.28: Averaged composition of Chinese brown glazes

Glaze		Oxide concentrations (wt%)										
		Na_2O	MgO	Al_2O_3	SiO_2	P_2O_5	SO_3	K_2O	CaO	TiO_2	MnO	FeO
A	Average	0.71	2.15	17.90	65.66	0.11	0.00	2.84	5.55	0.91	0.01	4.13
	St. Dev.	0.15	0.45	2.44	0.63	0.05	0.00	0.23	1.23	0.10	0.02	0.68
B	Average	0.40	2.59	17.54	57.43	1.50	0.00	2.71	9.15	1.38	0.42	6.82
	St. Dev.	0.04	0.22	0.97	1.16	0.38	0.00	0.11	0.59	0.15	0.10	0.45
C	Average	0.39	2.93	16.94	59.21	2.07	0.00	2.70	8.07	1.22	0.72	5.74
	St. Dev.	0.05	0.48	1.13	3.28	0.69	0.00	0.21	1.26	0.08	0.21	0.51

Glaze B
This glaze is found on all the globular jars and on Samples 12, a wide-mouthed jar, 35, an unidentifiable jar, and bottles 43 and 47. It has the highest calcium and iron oxides of the brown glazes. Based on the firing temperatures suggested by Hamer and Hamer (1997), this glaze is suitable to be fired in the range 1100-1300°C.

Glaze C
This glaze is found on 11 of the 12 wide-mouthed jars. It has the highest percentages of magnesium, phosphorous and manganese oxides and the lowest percentage of alumina. Based on the firing temperatures suggested by Hamer and Hamer (1997:379), this glaze is suitable to be fired in the range 1100-1300°C.

Tea and tableware

Ceramic body comparison

A PCA scatter plot groups the celadon-glazed vessels separately to the other samples (Figure 11.17). Component 1 explains 32.4 per cent of the variance and Component 2, 21.2 per cent. Statistically, there are four components with Eigen values greater than one, in total explaining 79.8 per cent of the variance. Only the graphing of Components 1 and 2 separate the Chinese, Japanese and British-made wares.

Table 11.29 details the differences between the ceramic bodies. The Japanese-made wares have higher percentages of sodium and potassium oxides and a lower percentage of calcium oxide than the Chinese-made samples. The earthenware bodies have lower percentages of sodium and potassium oxide than the porcelain groups.

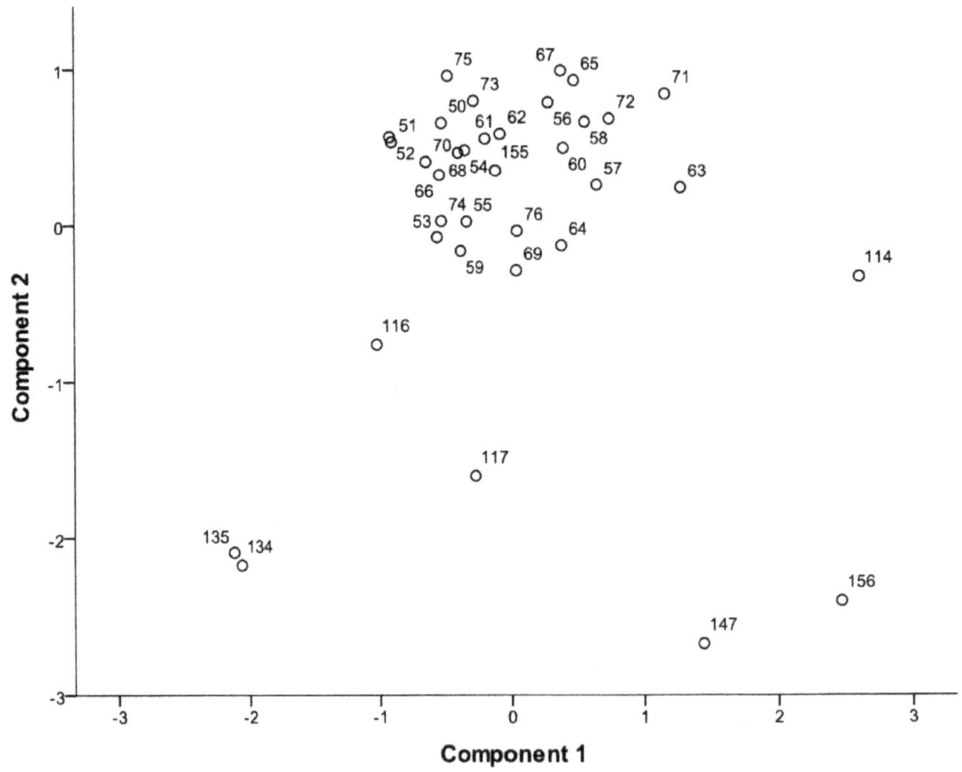

Figure 11.17: PCA scatter plot of the ceramic body composition of porcelain and refined-earthenware bodies

Table 11.29: Averaged body composition of porcelain and refined earthenware

Origin		Oxide concentrations (wt%)										
		Na_2O	MgO	Al_2O_3	SiO_2	P_2O_5	SO_3	K_2O	CaO	TiO_2	MnO	FeO
China	Average	1.09	0.28	19.94	73.02	0.00	0.05	3.38	1.04	0.02	0.00	1.16
	St. Dev.	0.22	0.04	1.30	1.58	0.01	0.03	0.36	0.25	0.02	0.01	0.14
Japan	Average	1.61	0.07	22.28	69.49	0.00	0.05	5.33	0.43	0.15	0.00	0.62
	St. Dev.	0.15	0.01	0.46	0.83	0.00	0.01	0.28	0.16	0.21	0.00	0.14
Britain	Average	0.64	0.25	20.12	74.99	0.06	0.09	1.61	1.18	0.45	0.00	0.60
	St. Dev.	0.21	0.11	1.20	1.93	0.05	0.05	0.43	0.33	0.24	0.00	0.28

Discriminant analysis suggests that Sample 116 belongs to the Chinese group and Sample 117 to the Japanese group. Sample 116 is part of a lamp dish, thought to be Chinese in origin. Sample 117 is from a cup, thought to be Japanese.

Glaze composition

A PCA scatter plot of celadon-glazed samples only, does not show any particular grouping within the samples (Figure 11.18). Component 1 explains 28.3 per cent of the variance and Component 2, 19.4 per cent. The scattered nature suggests diverse body compositions, perhaps caused by kiln origin but it could also have resulted from slightly differing proportions of the two clays and sand used.

Similarly, a PCA scatter plot shows no grouping of the celadon glazes (Figure 11.19). Component 1 explains 31.8 per cent of the variance and Component 2, 16.7 per cent. No groups are produced by graphing components which explain less of the variance.

The glazes on these vessels are similar, but the standard deviation of CaO is large in comparison to other oxides (Table 11.30). This variation in the calcium oxide content may have resulted from incomplete mixing of the raw materials used in the glaze (Tite et al. 1984:147).

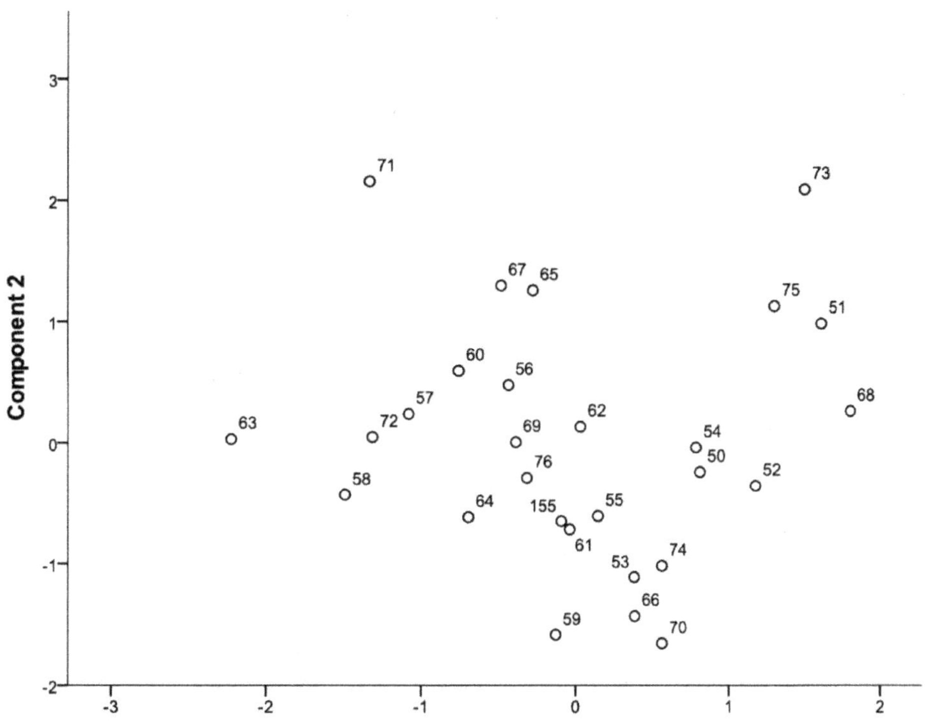

Figure 11.18: PCA scatter plot of the celadon-glazed ceramic bodies

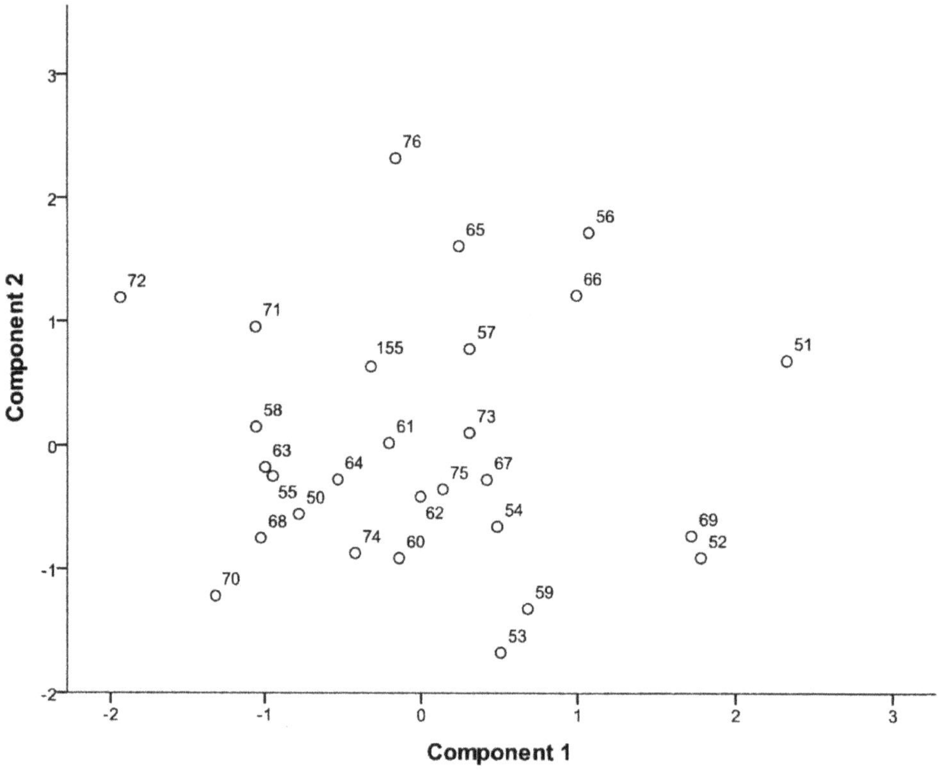

Figure 11.19: PCA scatter plot of the celadon glazes

Table 11.30: Averaged celadon glaze composition

Celadon	Oxide concentrations (wt%)										
	Na_2O	MgO	Al_2O_3	SiO_2	P_2O_5	SO_3	K_2O	CaO	TiO_2	MnO	FeO
Average	1.37	0.45	14.69	68.35	0.00	0.02	3.10	11.13	0.02	0.00	0.86
St. Dev.	0.34	0.09	0.82	1.41	0.01	0.02	0.40	2.07	0.05	0.10	0.09

DISCUSSION

Both macroscopic and microscopic analyses were carried out on 98 ceramic samples including: 49 Chinese-made food/beverage containers/lids; 28 celadon glazed vessels; 14 non-Chinese samples; and seven samples of unknown origin.

One aim of the technical analyses was to investigate whether the ceramic bodies tested could be grouped to suggest provenance. Pollard and Hatcher (1994:57) stated, with regards to chemical analysis, that 'it is possible to view the results in isolation from the actual Chinese sources, and simply use a chemical approach to answer the question "how many sources are represented in this collection of stylistically similar material?"'. One of the problems in ceramic provenance studies is that a kiln site may obtain materials from many sources while more than one manufacturing site may obtain materials from the same source. However, it is assumed that the potters within a kiln site would prepare the clay in a similar manner and that the refinement of the clay and the addition of temper would be consistent, allowing chemical profiles to be compared. Even when regions have been considered to be 'relatively homogenous' in their 'gross geological characteristics', differences have been detected between clay deposits (Bishop et al. 1982:294). Although slight variations in the fabric composition would be expected because of natural variation in clay deposits, it is likely that there would be more variation between sites than within a site.

In addition to the provenance studies, conclusions were drawn regarding the relationship between observed physical and chemical characteristics, and comparisons made to previous research on celadon vessels from overseas Chinese sites.

Macroscopic examination

When the body, glaze, profile and making methods of each sample were observed macroscopically it was found that the Chinese-made brown-glazed and celadon-glazed vessels generally conformed to previous descriptions of these wares.

Most of the jars/lids vary within vessel type. The globular jar samples could not be physically grouped as there was not enough of each sample to categorise profiles and the colour of their bodies and glazes varied. The wide-mouthed jars were tentatively grouped, according to their rolled-rim profile being either open or closed, as the rim diameter to neck depth ratio varied considerably amongst the samples. Once again there were differing body and glaze colours. The wide-mouthed jar lids were the most variable in body colour and texture, with only two of the samples having any similarity and that was in the rim profile. The liquor bottles had considerable colour variation in the bodies and glazes, but two samples were distinctly different to the others, one by the thickness of the rim and the other by the thickness of the foot ring.

There was not enough of each celadon-glazed sample present to establish a profile of each vessel. The body fabric of these vessels was similar but they had varying shades of green glaze that did not suggest any grouping. The main difference in these samples was the rim profile which tentatively suggested three groups according to the flat or rounded rim edge and the length of the thickened band of clay at the rim.

Microscopic examination

Electron microprobe analysis was used to determine the chemical composition of 98 samples from the study area.

Area scanned/number of readings per sample

In this study, the body area scanned (using x200 magnification) was about 600 x 400 μm² or 0.24 mm² and ten readings were taken. The area analysed is larger and the number of readings greater, than in many other similar studies. As an experiment, a smaller area on three samples was scanned, again taking ten readings, but using x10000 magnification, to compare the results to those of the larger area. It was found that the standard deviation of the average of each oxide was much larger when scanning the smaller area and thus not as reliable as when scanning the larger area. The glaze readings varied in number because of the thinness of the glaze in places. Readings were taken from both the exterior and interior.

There is more variation in the readings between samples than within samples. Generally the standard deviation of the averaged oxides is less for porcelain bodies than for the lower-fired brown-glazed and unglazed bodies. This could be because the clay has not been mixed as well or that the ingredients have not melted together as much in the lower-fired bodies. Titanium and iron oxide inclusions, particularly, caused more outlier readings in the globular and wide-mouthed jars.

Elements detected

Only seven elements, all recorded in combination with oxygen, were consistently present in amounts greater than 0.5 per cent for all types of clay and glaze. These are Na, Mg, Al, Si, K, Ca and Fe. Titanium is also present in the stoneware and earthenware bodies. Small amounts of phosphorus and sulphur are present in some but not all readings. Some samples contain trace amounts of vanadium, nickel, yttrium or gallium, but these were generally only present in one of the ten readings taken for those samples and so were not included in the final averaged composition. Manganese is present in some glazes. In total, 11 elements were consistently used to distinguish between the ceramic bodies and glazes.

The unmelted inclusions present in the samples were not used as provenance determinants. This is because the samples tested contain similar inclusions, i.e. those associated with clay or sand (zircon, rutile, ilmenite, iron oxide and rare earth minerals) and it was beyond the scope of this research to distinguish between similar mineral inclusions. The rare earth minerals, monazite and xenotime, for example, are known to be found in China, Britain and Australia. The zircons and rare earth minerals were very small, all less than 50 μm and most less than 20 μm, so were not present in sufficient amounts to be recorded as part of the total composition during the analysis.

Relationship between physical and chemical properties

As outlined in the summary of macroscopic examination, very few groups could be determined visually amongst the examined wares which included globular jars, wide-mouthed jars and lids, liquor bottles and celadon-glazed vessels. The main obstacle was the lack of vessel profiles because of small sherds. A number of groups were proposed by statistical analysis of the chemical compositions.

Visually the globular jars exhibited no difference in their body apart from the colour resulting from firing conditions, even though microscopic examination has determined two body groups. Chemically, the glaze of all samples is similar, although visually it is different in colour and texture.

Ten of the twelve wide-mouthed jars were able to be separated into two groups according to the shape of their rim. Chemically, two body groups and one outlier were proposed with eight of the ten samples visually separated being assigned to the correct chemical group. The glaze colour is variable between samples, although chemically eleven of the samples have a similar glaze. The sample with the different glaze does not visually stand out as being different to the other samples.

The unglazed wide-mouthed jar lids had the most variable body colour and texture. Statistically, three groups and two outliers were suggested. One of the outliers, Sample 26, has a red clay body and is very different visually to the other lids. The other outlier, Sample 33, has the same profile as a grouped sample, and a high probability of belonging to that group, although it appears graphically to be an outlier in Figures 11.12 and 11.13.

Physical differences were observed in the form and body texture of two of the liquor bottles, Samples 43 and 47, in comparison to the other 12 samples. This also agrees with the chemical analysis, which suggested two body groups. Once again, the glaze is visually quite variable, although 12 of the samples have a chemically similar glaze.

The celadon-glazed vessels were physically grouped according to rim type but no chemical grouping was suggested. The bodies are similar visually and chemically. The glazes are similar in thickness and texture but vary slightly in colour. Chemically they are similar. The most variation occurs in percentages of calcium oxide which could have been a result of incomplete mixing of glaze ingredients.

These findings highlight the visual variability obtained when using clay and glaze of the same chemical composition. These wares have been fired in long kilns which have uneven temperatures and uneven pockets of oxidation and reduction, affecting both the colour of the body and glaze, and the texture of the glaze. Hoh (1933), in his paper on Chinese making and firing methods, described placing particular vessel types in different parts of the kiln, in order to fire wares to different temperatures within a single firing. This is evidenced in this study. There are a range of apparent porosity percentages within each body grouping, proposed by the discriminant analysis. This suggests that, although the wares have similar body compositions, they have not all been fired to the same temperature or for the same length of time.

Interpreting statistical groups

Contemporary reports from the early twentieth century, regarding the making of Chinese ceramic vessels stated that they were made of two clays and sand (Laird 1918; Hoh 1933). Later microscopic research confirmed that two or more materials were used to produce Chinese ceramic wares (Yap & Hua 1994). Using two clays can produce a body that has better qualities than either of the clays alone. The addition of non-plastic inclusions such as sand to naturally occurring clay produces a body that can more easily withstand the processes of drying and firing and will be less likely to crack (Grimshaw 1971:416). The sand also allows the body to withstand the shock of rewetting when raw glazing, as it stops cracks going right through body.

It is known that different grades of porcelain were made at kiln-sites in China. Pere d'Entrecolles, a Jesuit missionary, travelled to China in the early 1700s, detailing some aspects of porcelain production. He stated that different proportions of two materials, porcelain-stone and kaolin, were used depending on the quality of the porcelain required (Burton 1906). Similar mixtures were used into the twentieth century for production of porcelain in Jingdezhen (Tite et al. 1984).

In this research, no attempts were made to distinguish between the clay and the temper, only the overall composition was considered as readings were taken across each sample. Chemical profiles, the 'weighted average of all the mineralogical components' of the fabrics (Bishop et al. 1982:294), were obtained. Thus, the chemical profile of each sample includes all elements in the two clay types and any added sand. Statistical analysis confirmed that the variance within each sample was less than between each sample.

Chinese-made food/beverage containers

Four body groups, designated Groups 1, 2, 3, and 4, and three glaze groups, designated A, B, and C, are suggested through statistical analysis.

Group 1 samples, all bottles, only have Glaze A. Group 2 samples have Glazes B, C or are unglazed. Similarly, Group 3 samples have Glazes B, C or are unglazed. Group 4 samples are all unglazed. This produces eight combinations of the body and glaze, nominally 1A, 2B, 2C, 2D, 3B, 3C, 3D and 4D.

Twelve of the 14 bottles have body Group 1 and Glaze A, while the other two bottles have body Group 2 and Glaze B. The Group 1 bottles may have come from a different geographical area to those of Group 2 or may have been made at a later date than the other two bottles. Chace (1976:517) posited that bottles made with the use of a mould were 'introduced late in the century' because bottles which were 'entirely hand thrown' were found at a site dating to 1860s, while a combination of thrown and moulded bottles were found at a later site dating to the 1880s. Olsen (1978:27) stated that the glaze on older bottles did 'not extend to the bottom of the container'. The glaze on the bottle belonging to Group 2 does not extend to the base.

The Group 1 bottles are from: JCC Hut 2, dating from mid-nineteenth to the early-twentieth century; KCC, 1860 to late-nineteenth century; Yan's store at Kiandra, 1882-1916; and Narrandera, dating from about 1870s. Group 2 bottles are from KCC Temple Ancillary Building, 1860 to late-nineteenth century; and East Eucumbene Site 977, 1870-1880s. The difference between these groups may be regional rather than temporal. The body of Group 1 bottles has almost half the iron oxide content of those in Group 2 and has less silica and more alumina as a percentage of the total composition. Whilst this could indicate a change in clay source over time, the glaze is also different. Both groups have calcium oxide as their main glaze flux, but the source of the calcium appears to have been different. The glaze on the Group 2 bottles has high levels of magnesium, phosphorous and manganese oxides indicating the source of the calcium oxide may have been wood ash (Wood 1999). Glaze A, on Group 1 body, has no phosphorous or manganese oxides suggesting the source of the calcium is more likely to be limestone (Wood 1999).

Groups 2 and 3 are interrelated. If B and C represent glazes used at a single kiln site, and these remained in use over time, then the grouping of the samples in this study may show a change of clay source over time. In 1933, Hoh reported that the clay source used in Shiwan, Guangdong, the largest pottery centre in southern China, was exhausted by about 1870 but that added sand still came from the local region. It is thought that most of the vessels used by overseas Chinese came from Guangdong province (Chace 1976). It is known that the Chinese used a combination of two clays for a vessel's body (Hoh 1933) and that the body clay or local river mud was commonly used in the glaze of a vessel (Wood 1999). It is possible that the source of one of the clays was exhausted before the other; while the clay used in the glaze remained the same. This scenario could account for glaze groups B and C which have body types 2 and 3 within each group. This could be possible if the wares came from kiln sites which shared an imported source of clay. The glaze could then be a combination of one of these clays and local clay. The kiln sites producing these samples could be from the same region given that they used the same resources.

Tables 11.31 and 11.32 detail the examined samples, the site from which they came and the date of those sites. It is possible that body Group 2 may be older than body Group 3, although there is some overlap in dates of the sites where the samples were found.

The unglazed wide-mouthed jar lids have three body types. Two of the body groups, 2 and 3, are also those of wide-mouthed jars. Sample 26, a wide-mouthed jar lid from JCC, was given 100 per cent probability of belonging to Group 5, the British/Australian stoneware. This suggests this sample was different to all the other Chinese wares and that its composition was closest to the British/Australian group rather than actually being part of it. The low-fired sherd came from a wide-mouthed jar lid which has an unusually high percentage of sodium oxide, possibly as a result of post-depositional change. Low-fired wares, such as this lid, can be affected by ground water saturation (Bishop et al. 1982).

Table 11.31: Samples with Group 2 body

Group	No.	Vessel	Site	Date
2B	1	Globular jar	FPCC Hut 4	1860s-1890s
2B	3	Globular jar	JCC Hut 2	Mid19th-early20th
2B	4	Globular jar	JCC surface	Mid19th-early20th
2B	6	Globular jar	JCC Hut 5	Mid 19th cent
2B	7	Globular jar	KCC Hut 3	Mid-late 19th cent
2B	8	Globular jar	KCC Hut 42	Mid-late 19th cent
2B	9	Globular jar	KCC Hut 2	Mid-late 19th cent
2B	10	Globular jar	KCC2003-2	Mid-late 19th cent
2B	11	Globular jar	Eucumbene Crossing 143	1860-1870s
2B	12	Wide-mouthed jar	JCC Hut 4	Mid19th-early20th
2B	35	Unidentifiable	ACC Hut 3	1860s-1880s
2B	43	Bottle	KCC Temple Ancillary Building	1860-late 19th cent
2B	47	Bottle	East Eucumbene 977	1870s-1880s
2C	17	Wide-mouthed jar	KCC Hut 1	1860-late 19th cent
2C	20	Wide-mouthed jar	Ah Yan Store	1882-1916
2C	22	Wide-mouthed jar	Eucumbene Crossing 143	1860s-1870s
2C	23	Wide-mouthed jar	JCC Store	Mid19th-early20th

Table 11.32: Samples with Group 3 body

Group	No.		Site	Date
3B	2	Globular jar	JCC Hut 2	Mid19th-early20th
3B	5	Globular jar	JCC Store	Mid19th-early20th
3C	13	Wide-mouthed jar	JCC Store	Mid19th-early20th
3C	15	Wide-mouthed jar	JCC Hut 2	Mid19th-early20th
3C	16	Wide-mouthed jar	JCC Hut 2	Mid19th-early20th
3C	18	Wide-mouthed jar	KCC2003-2	1860-late 19th cent
3C	19	Wide-mouthed jar	KCC Hut 3	1860-late 19th cent
3C	21	Wide-mouthed jar	East Eucumbene 977	1870s-1880s

Chinese tea and tableware

The body and glaze of all the Winter Green vessels is similar therefore no distinct groups were discerned. The averaged percentages of silica, 73.02 per cent, and alumina, 19.94 per cent, in the body of these samples are within the percentages determined for southern Chinese ceramic bodies; 69-83 per cent silica and 10-24 per cent alumina (Hall et al. 1973:66). Vast amounts of porcelain stone, the main ingredient in porcelain, have been found in the southern Chinese provinces of Zhejiang, Jiangxi, Fujian and Jiangsu (Yanyi 1987).

In 1993, Stenger proposed that some of the celadon-glazed bowls commonly found on overseas Chinese sites in America may have been manufactured in Japan. Her conclusion was based on composition comparisons between celadon-glazed wares recovered from Oregon, Washington, Idaho and Nevada and those of previously documented wares from both the nineteenth and twentieth centuries from China and Japan. She found that these four categories had different ratios of the main glaze elements. Her paper used the comparative terms: barely observable, trace, little, some and much, rather than numerical amounts (Table 11.33).

The major difference between the glazes of nineteenth (1870-1890) and twentieth (1910-1930) century of both China and Japan is the addition of manganese in the twentieth century. The major difference between the Chinese and Japanese glazes overall is the amount of calcium and more particularly the ratio of calcium to potassium. Stenger (1993:330) found that only one of her sites had celadon-glazed wares that were 'elementally characteristic of Chinese origin'.

Table 11.34 gives the oxide percentages of elements from this study for comparison to Stenger's study. The majority of celadon-glazed samples contain no manganese oxide suggesting that they were made in the nineteenth century. Only four samples, 56, 65, 72 and 76, contain any manganese oxide; 0.02-0.03 per cent of their total glaze composition. Overall, the celadon-glazed samples contain a large percentage of calcium oxide relative to the percentage of potassium oxide, which, according to Stenger, is characteristic of Japanese porcelain.

The Japanese-made samples are marked 'NIPPON', dating them about 1880-1920. Although there is no manganese in these glazes, suggesting a nineteenth century date, the amount of potassium is greater than calcium, suggesting a twentieth century date. The actual date for these samples is probably between the dates of Stenger's tested samples, perhaps showing a transitional period.

Table 11.34: Oxide percentages of elements in green-glazed wares from this study

Sample	CaO	P_2O_5	K_2O	FeO	MnO
Celadon	11.13	0.00	3.10	0.86	0.00
Japanese	3.15	0.00	5.91	0.53	0.00

Visually, the glaze on the Japanese-made samples conforms to the description given by Stenger; their exterior being green and interior being white. The glazes are thinly applied. Some of the Chinese-made bowls have whiter interiors than others but they are generally very pale green whilst the exterior is a darker green. These glazes have been thickly applied on the exterior in comparison to the interior.

While the celadon glaze of the porcelain from this study may suggest a Japanese origin according to Stenger's research, the ceramic body does not. A comparison of the body composition to samples from other Chinese sites is discussed below.

Interpreting origin of samples

Statistical analysis suggests that the stoneware and earthenware utilitarian vessels of unknown origin may be Chinese-made because their body is similar to that of the Chinese-made samples from this study. One unknown porcelain sample has a high probability of being Japanese-made because its body composition is similar to those marked 'NIPPON' and the other a high probability of being Chinese.

In order to test whether these assumptions are correct, data from other researchers were graphed along with those from this study using discriminant analysis (Figure 11.20).

Table 11.33: Relative amounts of elements in green-glazed wares from Stenger's study (after Stenger 1993:328)

Origin	Calcium	Phosphorous	Potassium	Iron	Manganese
19th century					
China	some	barely observable	some	trace	none
Japan	much	trace - little	little	barely observable	none
20th century					
China	some	none	some	trace	some
Japan	little	trace	little-some	little	little

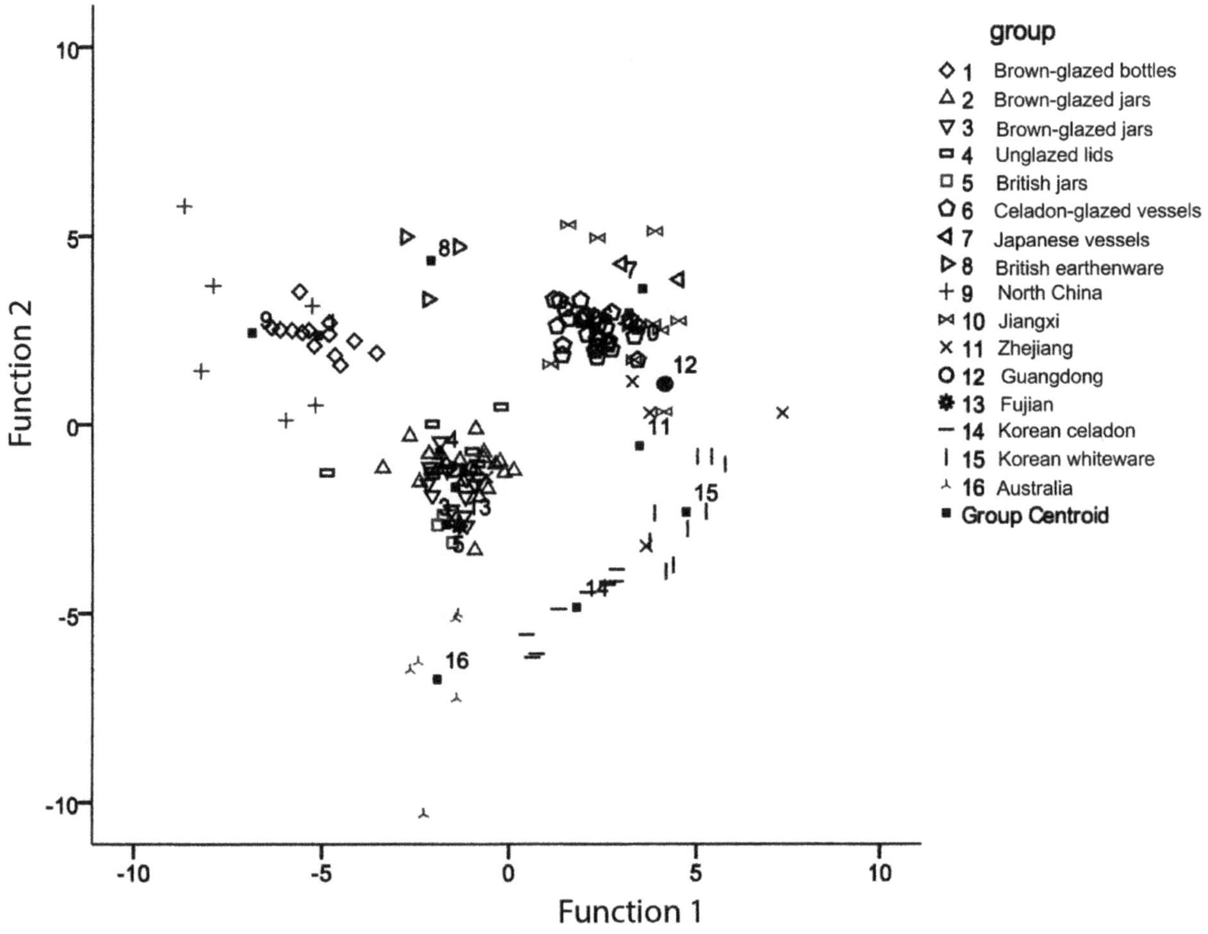

Figure 11.20: Discriminant analysis plot of ceramic body groups (Groups 1-8 from this study; 9-16 data from other researchers)

Eight elements were used: Na, Mg, Al, Si, K, Ca, Ti and Fe. As previously shown, the use of eight elements, in comparison to the original 11, does not greatly affect the grouping of samples. Much of the data from other researchers are from older wares, which would have used local materials for the production of ceramic vessels. Groups 1-8 are from this study. The data from other researchers include: Group 9 – northern China, Group 10 – Jiangxi, Group 11 – Zhejiang, Group 12 – Guangdong, Group 13 – Fujian (all data from Hall et al. 1973:77); Group 14 – Korean celadon, Group 15 – Korean white ware (data from Koh Choo et al. 2004:251-252); and Group 16 – Australian terracotta (data from Kelloway 2008:129). Unfortunately, there are no comparable data regarding the composition of Japanese ceramic bodies or Australian stoneware.

Interpretations are based on general trends in the data in addition to absolute values. Limited data are available for comparison and this means that samples have been assigned to the group having the most similar characteristics, not always the same characteristics.

Food/beverage containers

Nine liquor bottles in Group 1 (MNV=12) have 100 per cent probability of membership. Three bottles have 0.1-1.2 per cent probability of belonging to the North China group, the characteristics of which are not shared by any other group. This suggests that Group 1 bottles do not come from southern China or if they do, they were made using imported clay.

Groups 2, 3 and 4 all have similar body compositions. The probability of group membership consistently overlaps within these groups and with Group 13, Fujian. There is only one sample from Fujian, which is black-glazed stoneware (Hall et al. 1973:61). Sample 26, a wide-mouthed jar lid, previously grouped with the British vessels, is now in Group 4 (93.4 per cent) after the introduction of more data.

Sample 14, a previous outlier, has 61.9 per cent probability of belonging to Group 5, the British stoneware, and small percentages to Groups 2, 3 and 13. This wide-mouthed jar is from Southwest Eucumbene Site 740, known from documents to have been occupied in the early 1860s. It is possible that this jar came from a different source to the majority of brown-glazed jars, perhaps being brought into the site by a Chinese miner.

The probability of group membership of the stoneware samples from Fujian and Zhejiang (Table 11.35) show a link to the brown-glazed wares recovered from southeast NSW.

Table 11.35: Probability of group membership for stoneware samples of known southern Chinese origin (original data from Hall et al. 1973)

Sample	Origin	Body	Glaze	Grp2	Grp3	Grp4	Grp13
1217	Fujian	Stoneware	Black	2.4	9.0	0.0	88.6
1219	Zhejiang	Stoneware	Celadon	11.5	79.3	6.6	2.5

In summary, compositional analysis suggests the brown-glazed wares from Chinese sites in southeast NSW are from two main regions. One region was South China, possibly Fujian or Zhejiang (although no stoneware data were available for comparison from Guangdong). The vessels from this region are from a number of kiln sites, as evidenced by the slightly differing body compositions. The other region is possibly further north.

Tea and tableware

Nineteen of the 28 celadon-glazed samples have greater than 99 per cent probability of group membership to Group 6. Six samples have 93-99 per cent probability. The other three samples have probability percentages of 73.5, 79.9 and 88.9. All samples with less than 100 per cent probability have small probabilities of belonging to Groups 10, 11 or 12, that is, Jiangxi, Zhejiang or Guangdong. The only other group associated with Group 6 is Group 7 (Japanese porcelain). Only five of the 28 samples have any possibility, 1-4 per cent, of belonging to Group 7.

The probability of group membership of the porcelain samples of southern Chinese origin is shown in Table 11.36. There are similarities between the southern Chinese Groups 10, 11 and 12, Japanese Group 7, and Korean Groups 14 and 15. One of the Jiangxi samples has 100 per cent probability of belonging to Group 6, the celadon from this study. Previous research has noted the similarity in the body of Korean porcelain to that of the Chinese celadon-glazed porcelain from Jingdezhen (Ham et al. 2002).

Based on the statistical analysis, the celadon from this study is from southern China, perhaps Jiangxi.

Samples of unknown origin

Although the body of Sample 113 belongs statistically to Group 1, the glaze is completely different from the brown-glazed bottles. Its main flux is potassium rather than calcium oxide. This is a characteristic of Japanese iron glazes which have a higher or equal amount of potassium/sodium oxides than calcium oxide (Wood 1978). Similarly modern Western brown (temmoku) glazes have potassium and/or sodium as their main flux, not calcium (Wood 1978). It was thought that Sample 113 may have been from a Chinese ink bottle, but this has not been substantiated, therefore the provenance remains unknown.

Sample 115 has 87.4 per cent probability of membership to Group 4 and small percentages to Groups 3 and 13. Based on these characteristics, this unidentifiable vessel is from southern China. In addition, its glaze has calcium oxide as the main flux.

Table 11.36: Probability of group membership for porcelain samples of known southern Chinese origin (original data from Hall et al. 1973)

Sample	Origin	Grp2	Grp6	Grp7	Grp10	Grp11	Grp12	Grp14	Grp15
1201	Jiangxi	0.0	100.0	0.0	0.0	0.0	0.0	0.0	0.0
1202	Zhejiang	0.0	0.0	0.0	0.0	100.0	0.0	0.0	0.0
1203	Zhejiang	0.0	0.0	0.0	0.0	5.2	0.1	94.7	0.0
1204	Jiangxi	0.0	19.1	0.0	21.4	0.0	59.5	0.0	0.0
1205	Jiangxi	0.0	0.0	5.4	94.3	0.0	0.3	0.0	0.0
1206	Jiangxi	0.0	0.0	0.2	99.8	0.0	0.0	0.0	0.0
1207	Jiangxi	0.0	0.0	0.0	100.0	0.0	0.0	0.0	0.0
1208	Jiangxi	0.0	4.2	7.7	77.2	0.0	10.9	0.0	0.0
1209	Jiangxi	0.0	0.8	9.6	62.8	0.2	26.7	0.0	0.0
1210	Jiangxi	0.0	0.4	53.6	38.7	0.0	7.3	0.0	0.0
1216	Guangdong	0.0	0.1	0.0	4.4	0.5	95.0	0.0	0.0
1220	Zhejiang	0.0	1.3	0.0	0.4	56.1	32.0	0.0	10.1
1221	Zhejiang	0.0	4.4	0.0	0.0	95.6	0.1	0.0	0.0
1222	Jiangxi	0.0	0.0	0.0	0.2	0.3	99.5	0.0	0.0
1223	Jiangxi	0.1	0.0	97.6	2.1	0.0	0.2	0.0	0.0

Statistically the body of Sample 119 has 82.2 per cent probability of belonging to Group 2 with small percentages to Groups 1, 3, 4, and 13. This is the only vessel having this combination, suggesting a slightly different origin to the other vessels in Group 2. This jar is from Ah Chee's Kiandra store, dating it to the late-nineteenth or early-twentieth centuries. This vessel type was not found at the Chinese camps.

Sample 121, from a possible ovoid jar, is from KCC Hut 35, thought to be one of the earlier occupied huts. Statistical analysis suggests that this jar has 58.7 per cent probability of belonging to Group 2, with small percentages to Groups 3, 4, 5 and 13. The only other jar with any possibly of belonging to the British group as well as the Chinese groups is Sample 14, which also is from a site occupied in the early gold rush. Both these jars are probably from southern China, but perhaps from a different source to the other jars.

Sample 157, a possible ovoid jar, belonged between Groups 2 (44.6 per cent) and 3 (52.4 per cent) with small percentages to Groups 4 and 13. Based on these characteristics, this jar is from southern China.

Sample 116 is from a porcelain lamp dish, thought to be Chinese-made. Statistically, the sherd has 83.1 per cent probability of belonging to Group 10, Jiangxi, and a small probability of belonging to Groups 6, celadon from sites in southeast NSW, group 7, Japan, or group 12, Guangdong. Based on these results, it is likely to have been made in southern China. However, the glaze has higher percentages of potassium then calcium, a trait of Japanese glazes.

Similarly, Sample 117 has 73.5 per cent probability of belonging to Group 10, Jiangxi, and a small probability of belonging to Groups 6, 7 or 12. It is thought that this cup is from Japan as it was recovered from Ah Yan's Kiandra store and would have dated to the period of early Japanese imports. The glaze of this sample has potassium and sodium oxides as the flux. Wood (1999:30) stated that glazes using potassium and sodium oxides as their main flux were 'extremely rare in Chinese ceramic history' but 'quite common in Japan'.

SUMMARY

This research has discovered more about the Chinese-made ceramic body and glazes recovered from nineteenth century Australian sites. The source of the ceramics has been investigated along with the relationship between the observed physical and chemical differences.

This study confirms that most of the Chinese-made vessels are from southern China. A distinct difference is observed in the groups of brown-glazed vessels. Two of the three brown-glazed groups proposed by the discriminant analysis share similarities of the body and glaze which suggests a common source of materials. The wares from these groups are from southern China, possibly Fujian, although no comparative data were available from Guangdong. They have a similar body composition to a stoneware vessel from Fujian previously examined by Hall, Schweizer and Toller (1973:77). In this study, Glaze B was found on globular jars, a wide-mouthed jar and bottles, perhaps suggesting that kiln-sites used the same brown glaze for all storage vessels they made. A change over time may have been seen within these two groups, if the glaze remained the same and the body composition altered slightly, coinciding with the exhaustion of clay resources at some kiln-sites in the latter part of the nineteenth century. The other brown-glazed group, which consists entirely of bottles, is from a different source, perhaps northern China. The amount of silica, 67 per cent, lies between the amounts for northern and southern Chinese vessels determined by Hall, Schweizer and Toller (1973). The alumina is within the range for northern China. The statistical analysis of the celadon-glazed samples showed no distinct groups. All fall within the range of element proportions for porcelain vessels from southern China.

Physical differences between samples tested do not always relate to their chemical differences. It is also clear that differences in glaze appearance or body colour do not indicate provenance. This study has been hampered by the small size of the sherds and the lack of vessel profiles; however this assemblage is probably similar to many other Chinese sites in Australia, in that respect. In most cases the chemical groups could not be visually determined. The exception was the brown-glazed bottles. The two groups with different chemical compositions have physical dissimilarities in the thickness of the rim and foot ring and in the exterior area covered by glaze. Within the brown-glazed vessels, the apparent porosity relates more to the vessel type than the body composition, supporting Hoh's (1933) report that the position of wares in a kiln depends on the heat required for a particular vessel. Firing conditions, including the temperature and oxygen supply, vary considerably within the long kilns used by the Chinese potters and this is seen in the differing body and glaze colours within each chemical composition group.

This study has also added to the discussion on whether the ubiquitous celadon-glazed wares found on overseas Chinese sites were made in China or Japan. The element-ratios in the celadon glaze conform to the proportions cited by Stenger (1993) for nineteenth century Japanese-made wares. However, the body composition of the celadon-glazed wares from this study is different to tested Japanese-made samples which probably date to about 1900. Their composition is similar to that of a celadon-glazed vessel known to be from Jiangxi province. In conclusion, the body composition of the celadon-glazed wares suggests that they are more likely to have come from China. Their glaze is also more thickly applied in comparison to the Japanese samples tested. Statistical analyses of porcelain from China, Japan and Korea have shown some similarities in their characteristics.

CHAPTER 12

Summary of vessels in this study

All the ceramic data from each site in this study is collated and compared to other overseas Chinese sites within the broad functional categories of food/beverage containers, tea and tableware, and smoking.

FOOD/BEVERAGE CONTAINERS

A MNV of 270 storage containers have been identified in this study; 252 of Chinese origin, 13 non-Chinese and five of unknown provenance. Sixty-two of the Chinese containers cannot be identified to type; 36 are able to be described as small-medium jars and 26 as medium-large jars.

Only seven of the 76 sites have both Chinese and non-Chinese containers. These include: JCC Huts 1 and 2; Kiandra Township Stores; KCC Hut 2; and sites at Southwest Eucumbene and New Chum Hill. Thirty-two assemblages have both small and large containers, 22 have small containers only and nine, large containers only. East Eucumbene Site 635 has a non-Chinese made container only. Thirteen sites without any containers are at FPCC (2), ACC (5), JCC Grid 9, KCC Well and outlying Kiandra camps (4).

Chinese

Barrel jars and lids

A MNV of 26 jars and 23 lids are from 18 sites. There are ten jars and ten lids from the KCC2003-2 assemblage, some from an area at the lower end of a drainage line, which may have been used as a dump. The only other sites with multiple barrel jars are the Kiandra Township stores and the outlying camp at East Eucumbene, Site 143, covering two huts. Fourteen sites have single jars (11) and/or lids (7).

In Australia and New Zealand, barrel jars are most commonly recovered from Chinatowns or areas associated with commercial activities and this trend continues in southeast NSW where multiple barrel jars were found at the Kiandra Township stores.

Globular jars

There are twenty-two of these jars from 19 sites. The only sites with multiple jars are JCC Hut 2 (MNV=2) and Upper Adelong surface scatter (MNV=3).

Small numbers of these jars have been recovered from Chinatowns in Cooktown, Queensland (Rains 2005:279) and from Narrandera, NSW (Esposito 2012:653). They have also been found in New Zealand at Cromwell Chinatown and at Arrowtown (Ritchie 1986:700) and one jar from Sam Chew Lain's house in the Lawrence Chinese Camp (Davies 2012:57).

Wide-mouthed jars and lids

These are the most common brown-glazed vessels in this study and are found at a range of sites. A MNV of 33 jars and 23 lids have been recovered from 21 sites; eight sites with more than one vessel.

Although wide-mouthed jars are the most common jars identified in this study, they are not commonly reported in the literature from other areas in Australia. This may be because they have been included as general 'brown-glazed ware' or 'jars'. It may also be because the lower section of these jars is similar to that of spouted jars. Nevertheless, these jars are common on New Zealand and American sites.

Spouted jars

In southeast NSW these are from eight sites (MNV=13). Six sites have single jars (including JCC Store, KCC Ancillary Building, KCC Oven and three huts at KCC), KCC2003-2 has two, and JCC Hut 2 has five jars.

Small numbers of spouted jars have been found at other sites in NSW, but they appear to be more common elsewhere in Australia and overseas, for example, there were a minimum of 19 from Chinaman's Point in Victoria (Bowen 2007:170) and nine from Atherton Chinatown (Dunk 2010:56).

Straight-sided jars and lids

None of these jars have been identified in the assemblages. It is possible that they may have been present but no diagnostic sherds were found. However, there are two lids: one from Ah Chee's Store; and the other from FPCC Hut 5.

The jars are also uncommon elsewhere in Australia and in New Zealand, where they have mainly been recovered from Chinatown sites rather than hut sites (Ritchie 1986:245).

Unidentified storage jar lid

A lid, diameter 90 mm, from an unidentified vessel is from KCC Hut 3. It has dark-brown glaze over a relief-moulded pattern on the top exterior, which is flat, and an unglazed interior and flange (Figure 9.16, lower left).

A similar lid was identified at Sam Chew Lain's house in the Lawrence Chinese Camp (Davies 2012:79), although no others have been seen in the literature.

Liquor bottles

There is a MNV of 25 bottles from a range of sites in this study. There are six bottles from JCC Hut 2, five from the Kiandra Township Stores (Ah Chee MNV=2; Ah Yan MNV=3), two from KCC Hut 4 and two from the KCC2003-2 assemblage. Ten sites have single bottles.

Liquor bottles appear to be more common elsewhere than in southeast NSW. Olsen (1978) suggested that the observed differences in bottles are temporal, but this research has confirmed a different source of clay for bottles from southeast NSW with different visual characteristics, one from southern China and the other possibly from northern China.

Ginger jars and lids

There is a MNV of 13 jars and four lids from 11 sites. Eight jars have an exterior green-glaze and interior brown-glaze and are round in shape. Two of the lids are from green-glazed jars. These come from JCC Hut 2 (MNV=2 jars, 2 lids), JCC Store (2 jars), Ah Yan Store (2 jars), KCC2003-2 (1 jar) and Eucumbene Crossing, Site 143 (1 jar). Five jars, from JCC Hut 4, JCC surface scatter, KCC Hut 46, Ah Chee's Store and Ah Yan's Store, are decorated with blue-underglaze and a clear glaze. Two lids from KCC Oven and Hut 1 have a clear glaze only and are different in form. Both lids have a flat top with an adjoining flange. The lid from KCC Hut 1 is only glazed on the exterior top and the flange joins at an angle of about 70°, slanting outwards. The lid from the Oven is glazed over the whole exterior and the flange is at an angle of about 90°.

Ginger jars are commonly found on Chinese sites. In this study green-glazed ginger jars were slightly more common than the blue-underglaze decorated ones. In New Zealand Ritchie (1986:263) found that the green ones far outnumbered the blue ones, 51 to 6.

Ovoid jars

There is only a MNV of two jars, both from KCC, one from Hut 3 and one from Hut 35. Although the sherds are small, the colouring, and the patterning on one sherd, suggests that they may belong to ovoid jars.

It has previously been suggested that ovoid jars were only found in northern Australia. McCrae (2001) noted that the jars from Cooktown sites were not detailed in reports from America, Canada or New Zealand. He stated that the Cooktown examples are similar to those found at sites in Northern Territory and northwest Western Australia. Rains (2005:285) found a similar jar in Brisbane's Chinatown which contained 'salted Chinese radish greens' and the storekeeper related that the vessel had been imported from Guangzhou. These jars are the most common type recovered from the dump site in Cooktown with a MNV of 16 (Rains 2005:415). They are also commonly found in Bendigo, Victoria (personal communications, Dennis O'Hoy, 21 February 2012).

Spouted liquor storage jar

There is a MNV of two possible spouted liquor storage jars; one from the Bone Pit at KCC and the other from KCC Hut 46.

These have not been seen previously in archaeological literature. The ones in this study have been identified by the glaze and shape of one sherd, which matches that of a spouted liquor storage jar (personal communications, Dennis O'Hoy, 21 February 2012). Its exterior glaze is different to the usual brown-glazed jars but the interior glaze appears similar. It is possible that these jars have been present in other assemblages without the difference being recognised and the sherds simply recorded as brown-glazed wares. Chandler (2005:32) recorded the presence of a globular jar with a small spout in the collection that she studied. She noted that she was 'unable to locate a similar vessel in the literature' (Chandler 2005:32).

Number of large/small containers at sites

In this study, the highest MNV of large jars are from the stores in Kiandra Township. Few other sites have a greater number of large than small jars. The surface collection at KCC (KCC2003-2) contains 21 large containers and 12 small containers, some of which are from a probable rubbish dump at the end of a drainage line. KCC Hut 3 (four large, three small) is the only other site to have a greater number of large than small containers and a total MNV over five.

Jembaicumbene Hut 2 has the most food/beverage containers of any site in this study. Its 43 containers are predominantly small food jars; only five are large containers, three Chinese-made and two of non-Chinese origin. The JCC store has the highest number of storage containers after Hut 2, while at KCC the most containers are from Hut 4, a two-roomed hut, and Hut 3, possibly occupied by a merchant.

Previous research has suggested that a greater proportion of large containers at a site may indicate a commercial establishment. For example, in Cooktown, a dump associated with commercial activity was dominated by large containers but these containers were not so prevalent closer to residential areas (McCrae 2001). Similarly, in Market Street Chinatown, San Jose, California, it was found that commercial sites had a greater percentage of large containers (Brabyn 2007).

Summary

The most common brown-glazed vessel in this study is a wide-mouthed jar (MNV=33) followed by a barrel jar (MNV=26). These are followed by liquor bottles

(MNV=25), globular jars (MNV=22), spouted jars (MNV=13) and ginger jars (MNV=13). Generally, the most commonly recorded food/beverage containers from Chinese sites are spouted jars, liquor bottles and ginger jars.

Non-Chinese

There are far fewer non-Chinese than Chinese containers at the examined sites. Large non-Chinese made ceramic jars were only found at sites which were occupied later – JCC Huts 1 and 2, and Ah Yan's Kiandra Township store, although ceramic bottles have been found at earlier sites.

There are only 13 non-Chinese made containers in the assemblages, six large, six small and one of indeterminate size. The large jars include: a Bristol-glazed jar from JCC Hut 1; two Bristol-glazed jars from JCC Hut 2; and two salt-glazed jars and a Bristol-glaze demijohn from Ah Yan's Store. The small containers are: a blob-top salt-glazed bottle and a white earthenware food jar from JCC Hut 2; and Bristol-glazed bottles from Ah Chee's Store, East Eucumbene Site 635, and Southwest Eucumbene Sites 698 and 740. The other container, from KCC Hut 2, is either a jar or bottle with a two-tone Bristol glaze. These vessels may have been made in Britain or Australia.

There are few comparative data from Australasian sites. A salt-glazed stoneware jar was recovered from Chinaman's Point, Victoria (Bowen 2007:200). A number of white earthenware food jars and paste pots, stoneware bottles and a Bristol-glazed crock were found by Ritchie (1986:298, 302), predominantly from the Chinatown settlements.

Unknown origin

There are two jars and two lids of unknown origin. The jars are from the Kiandra Township stores. One, from Ah Chee's Store, has a light-brown exterior glaze, on the upper section of the body only, and an interior clear-glaze. There is a similar interior glaze on a jar sherd from Ah Yan's Store but its exterior is unglazed. The lids are from JCC Hut 2 and include a brown-glazed lid knob and part of an unglazed stoneware lid.

TEA AND TABLEWARE

Overall, tea and tableware examined in this study is dominated by British-made wares. A MNV of 708 includes 121 Chinese-made vessels, 574 identified as non-Chinese, predominantly British, and 13 which are possibly Japanese. Only four sites have more Chinese than non-Chinese tea and tableware: JCC Store front; a storage building at the back of Ah Yan's Store at Kiandra; UACC Hut 1; and a hut in Giandarra Gully. The MNV is small in all cases; 5:4, 2:1, 1:0 and 1:0 respectively.

Numerous British tea and tableware vessels were found during Ritchie's (1986:297) analysis of his sites in Otago, New Zealand. Overall, he found 135 Chinese-made tea and tableware vessels and 256 non-Chinese vessels. The non-Chinese vessels dominate in the Chinatowns and at Ah Lum's store. Of the seven rock shelters, two have more Chinese ceramics, three, more non-Chinese, one with equal numbers and one with no ceramics. Four of his hut sites have more Chinese-made ceramics and one, more non-Chinese.

Chinese

The most common decorative patterns on the Chinese tea and tableware in this study are Winter Green, Four Seasons and Bamboo. These patterns along with other blue-underglaze patterns also dominate assemblages found elsewhere. For example, seventy-five per cent of vessels (MNV=2767) from Riverside Chinatown, California, are Winter Green, Four Seasons, Bamboo or Double Happiness (Mueller 1987:264).

Serving bowls and dishes

Eleven Chinese-made serving vessels were identified in this study, five serving bowls and six shallow dishes.

The serving bowls, all decorated with Four Seasons pattern and all from Kiandra, are from the Stores (2), KCC2003-2 (1), and KCC Huts 1 and 4.

The shallow dishes have varying decoration. A large shallow dish from Ah Yan's Store is in Winter Green. There are four medium shallow dishes: a Four Seasons dish from JCC Temple; an unidentifiable green-blue underglaze pattern on a dish with a stacking-ring from KCC2003-2; and two Rocks and Orchid dishes, one each from JCC Hut 2 and KCC Hut 2. A small shallow dish in Four Seasons pattern is from KCC Hut 5.

Most of the serving bowls and dishes were from later sites at JCC and KCC. A similar Rocks and Orchid dish to those found at JCC and KCC was recovered from The Rocks, Sydney, NSW (Lydon 1996). Serving bowls and dishes are more commonly found in Chinatowns than rural sites in America and New Zealand (Evans 1980:92; Ritchie 1986:210, 216). No serving dishes were found at Bowen's (2007) fish curing camp in Victoria.

Rice bowls

The most commonly recovered Chinese-made vessels in this study are rice bowls (MNV=49), found at 26 sites. These are predominantly decorated in celadon-glazed Winter Green (MNV=35). In addition, there are five bowls with Bamboo decoration, two with probable Bamboo decoration and two bowls with Double Happiness. Four further bowls have blue-underglaze decoration but there is insufficient pattern to be identified. Three of the bowls, a Bamboo bowl from KCC Temple Ancillary Building, a probable Bamboo bowl from KCC Hut 35, and an unidentifiable pattern from Ah Yan's Store have green-blue underglaze decoration rather than pure blue. There are no Four Seasons rice bowls.

A major difference in the assemblages of Australia and New Zealand compared to those of America is the absence of Four Seasons decorated rice bowls. Celadon-glazed rice bowls dominate the Australian and New Zealand assemblages in comparison to blue-underglaze decoration. In America the numbers of blue-underglaze decorated rice bowls are higher than those for Winter Green or Four Seasons. Ritchie (1986) reported that Bamboo rice bowls from New Zealand sites had differing hues. Two bowls from Chinatowns had a blue cast to the glaze and the other bowl from Firewood Creek, a rock shelter site, had a green cast to the glaze (Ritchie 1986:216).

Teacups and teapots

There are very few cups in this study. Three are from JCC Hut 2 and one each from the Chinese stores in Kiandra, all in Winter Green. Teapots are also uncommon with only five being identified. One, from JCC Hut 2, is decorated with blue-underglaze. The other four, three from KCC and one from JCC Temple, are identified as possible teapots. The sherds are from cylindrical vessels and have a clear glaze with a green cast on the exterior only. As porcelain has a porosity of close to zero, glaze is not necessary to enable the pot to hold liquid. It may also be that the interior of the vessels was unevenly glazed as the recovered sherds are small.

Relatively few teacups were found at Ritchie's (1986:210, 216) New Zealand sites. In contrast, the majority of Winter Green vessels from a San Francisco dump were teacups (Pastron et al. 1981:659).

Liquor cups and pots

Liquor cups (MNV=16) are from 11 sites. Twelve, decorated in Winter Green, are from JCC Hut 2 (MNV=5), JCC Hut 4 (1), JCC Store (1), KCC Hut 1 (1), KCC Hut 4 (1), KCC Hut 35 (1) and Ah Chee's Store (2). Four cups, decorated in Four Seasons, are from KCC Hut 5, KCC Hut 62, KCC Oven and KCC2003-2. A single liquor pot from JCC surface scatter, located between Huts 3 and 4, is decorated in blue-underglaze Sweet Pea.

In New Zealand, as in Australia, Winter Green liquor cups are more common than Four Seasons (Ritchie 1986). Conversely, sites in America have more Four Seasons than Winter Green liquor cups (Evans 1980; Pastron et al. 1981). Although suggestions have been made concerning the possibility of ritual associations with liquor cups (Mueller 1987:304), no liquor cups have been found in the temple sites in this study. They come from hut sites and stores in Jembaicumbene and Kiandra and one from KCC Oven.

Spoons

There is a minimum of 12 spoons in this study, eight in Winter Green and four in Four Seasons. The Four Seasons spoons all come from JCC, two from Hut 2, and one each from Hut 1 and the surface collection. The Winter Green spoons are from KCC Huts 2, 5 and 28, KCC2003-2, Ah Chee's Store (2), Ah Yan's Store and the Upper Adelong surface collection.

Spoons are commonly found at overseas Chinese sites. As with other forms, Winter Green spoons are more common than Four Seasons spoons in New Zealand (Ritchie 1986:210, 216) while the reverse is true in San Francisco (Pastron et al. 1981).

Unidentifiable vessels

Twenty-two vessels could not be identified to vessel type. Seventeen may have been cups or bowls but have no diagnostic parts. Sixteen of them are decorated in Winter Green and one has a clear-glaze only on the sherd. Four vessels are either bowls or serving bowls. One has Four Seasons pattern and the other three have a clear-glaze only on the sherds. The final unidentifiable sherd is from a lid and is decorated in blue-underglaze.

Makers' marks

There are four blue-underglaze marks on the Winter Green rice bowls, and three on wine/liquor cups, all different. These share common characteristics with other marks reported in the literature although none are identical.

Summary

The common Chinese vessel/decorative types found at other sites have also been found in this study, but differences have been noted in the type of vessel forms and decoration most commonly found.

Non-Chinese

The majority of non-Chinese made wares in this study are British white-earthenware. Porcelain vessels, probably of Japanese origin, and British bone china, are from sites in Jembaicumbene and Kiandra. There are few comparative data for non-Chinese vessel types from Australasian Chinese sites.

Serving vessels

A MNV of 15 vessels manufactured for food preparation or serving for a Western-style diet are present in this study. These vessels come from a variety of sites: JCC Hut 2 (tureen lid), JCC Store (tureen), JCC Temple Complex (mixing bowl), JCC Oven (pan/dish), JCC surface collection (pudding bowl, platter); KCC Hut 3 (platter), KCC Temple ancillary building (jug), Ah Chee's Store (serving plate/dish), Ah Yan's Store (mixing/pudding bowl, jug), FPCC Oven (platter); ACC Hut 1 (platter) and ACC Hut 3 (pudding bowl, platter).

Few serving vessels were found at Ritchie's (1986:298, 302) sites. There were two platters and a serving dish from Chinatowns. The only food preparation and serving vessels in an assemblage from Narrandera Chinatown are a mixing bowl and jug (Esposito 2012:653).

Bowls

A MNV of 69 bowls have been identified. Nine of the bowls are the size of Chinese serving bowls, 160-180 mm, 31 are the size of rice bowls, 130-150 mm, four are 120 mm and 25 are of indeterminable size (Table 12.1). They were found at most sites, the highest number being from JCC Hut 2.

Bowls, of these sizes, used in the second half of the nineteenth century appear to have been made for serving, for example open vegetable bowls, or as waste bowls which could be table or tea ware. A perusal of collectors' books and web sites shows they were made in a number of sizes from about 120 mm to 180 mm.

Ritchie (1986:298) found 14 bowls in total. Six were from the Chinatowns, three from Ah Lum's store and five from other sites. Two of these are described as serving bowls, three as 'steep-side bowls', three as 'small dishes/bowls', one as a small bowl and two as sugar bowls (Ritchie 1986:302-308). The steep-sided bowls may be similar to those from this study measuring 130-150 mm. No non-Chinese bowls were identified at Chinaman's Point in Victoria (Bowen 2007:200).

Plates

The most common non-Chinese vessel type is the plate, in a range of sizes (Table 12.2). Dinner plates (>250 mm) are more common than smaller sized plates, although 45 plates could not be sized.

Details of American research, regarding the use of plate profiles to date British ceramics, are outlined in Chapter 5. Profiles from sherds in this study have been examined to see whether they conform to Moir's (1997) diagram. Comparisons are also made to ceramic vessels from the Gold Commissioner's residence at Kiandra (KGQ), a site known to have been occupied 1860-1862.

A number of plates from JCC, UACC, ACC and KCC correspond to Moir's 1825, 1845 and 1860 profiles. Plates known to have a later manufacture date, for example the ivory bodied plate with jigger marks from JCC Grid 9 and the blue-banded plate from UACC, do not fit with any of his profile drawings. This was expected as the later sequence of plates in his diagram is based on American-made as well as British-made plates. The profiles of some plates and saucers from JCC, ACC, KCC and East Eucumbene were found to be similar to those of wares from KGQ.

Table 12.1: Sizes of non-Chinese bowls in this study

Size	JCC					KT	KCC														KV				FP		ACC		UA	Total
(mm)	1	2	3	St	SS	AY	SS	1	2	3	4	5	21	25	35	53	62	O	T	AB	EE	EC	SE	NC	2	5	2	3	SS	
180						1					1													1					1	4
160		2														1				1									1	5
150		2		1	1		1						1		1			2		1	1	1						1	1	14
140		5			2				1					1					1				1		1			1	1	14
130		2					1																							3
120		2										2																		4
Unk	1		1	1	1		4	1	1	1	2	3					1				3				1	1	1		2	25
Total	1	13	1	2	4	1	6	1	2	1	3	5	1	1	1	1	1	2	1	2	4	1	1	1	2	1	1	2	6	69

Key: JCC-Jembaicumbene: St-Store, SS-surface scatter; KT-AY: Kiandra Township-Ah Yan's store; KCC-Kiandra Chinese Camp: SS-KCC2003-2, O-Oven, T-Temple, AB-Ancillary building; KV: Kiandra Valley – EE-East Eucumbene, EC-Eucumbene Crossing, SE-Southwest Eucumbene, NC-New Chum Hill; FP-Flanagan's Point Chinese Camp; ACC-Adjungbilly Chinese Camp; UA SS-Upper Adelong surface scatter

Table 12.2: Size of plates in this study

Plate	Size	JCC							KT		KCC													KV				FP	ACC			UA	Total
	(mm)	1	2	3	St	TC	SS	G9	C	Y	SS	O	T	AB	2	3	4	5	35	36	42	53	62	EC	EE	S	NC	O	3	5	R	SS	
Dinner	>250	1	3						3	1	1	1	1					2							2		1	1	4	1		6	28
Medium	200-240		1			2		1	1	2	1					1								1	4		2				1		17
Small	150-190		1													1						1	1								1	1	6
Child's	150-160	1	1		1	1	3																						1				8
Soup	250		2																														2
Unknown			1	1		1	2		1	5	2		1	3	1	6	2	2	1	2	1			5		1	2	2	3				45
Total		2	9	1	1	4	5	1	5	8	4	1	2	3	1	8	2	4	1	2	1	1	1	6	6	1	5	3	8	1	2	7	106

Key: JCC-Jembaicumbene: St-Store, TC-Temple, SS-surface scatter, G9-Grid 9; KT: Kiandra Township – C-Chee's store, Y-Yan's store; KCC-Kiandra Chinese Camp: SS-KCC2003-2, O-Oven, T-Temple, AB-Ancillary building; KV: Kiandra Valley – EC-Eucumbene Crossing, EE-East Eucumbene, S-Southwest Eucumbene, NC-New Chum Hill; FP-Flanagan's Point; ACC-Adjungbilly: R-Race; UA SS-Upper Adelong surface

The old manufacturing date suggested by these profiles fits with dates of the decorative patterns; for example, Eton College is thought to be a pre-1850 pattern (Lawrence et al. 2009) and all recovered plates in this pattern have a double foot ring. This dating method may be useful for determining whether wares were bought new or second-hand. It was common in NSW for second-hand consignments of wares to be offered for sale through newspaper advertisements in the nineteenth century (Atkins 1991).

A variety of plate sizes are found at other sites. Nine plates from Chinaman's Point are 180-200 mm in diameter (Bowen 2007:199). Plates from Narrandera Chinatown include two 180 mm, one 200 mm, three dinner plates and three of indeterminate size (Esposito 2012:653). Ritchie (1986:298) described the plates in his collection as large (MNV=39) or side (MNV=6). The sizes of 14 were indeterminable.

Teacups, saucers and teapots

Overall, there are fewer cups (MNV=66) and mugs (MNV=3) than saucers (MNV=80). There are no obvious patterns relating to the numbers of cups and saucers compared to the overall numbers of Chinese/non-Chinese vessels.

Only four possible teapots were identified. Ah Chee's store has a red-earthenware teapot with a blue slip and clear glaze. KCC Hut 1 and New Chum Hill, Site 485, both have a teapot with a buff-earthenware body and a Rockingham glaze and New Chum Hill, Site 1069 has a red-earthenware teapot with a black glaze.

Numerous cups/mugs (MNV=57) and saucers (MNV=35) were found at a range of sites in New Zealand (Ritchie 1986:298).

Miscellaneous vessels

A possible water monkey is from KCC Hut 28 and two eggcups are from JCC Huts 1 and 2.

Unidentifiable vessels

Some vessels could only be identified as cup/bowls (MNV=128), plate/saucers (MNV=66), hollowware (MNV=12) or flatware (MNV=6). Other vessels with transfer-prints, most of which have only the base present, are identified as possible tea or tableware (MNV=27).

Bone china and porcelain

At JCC bone china saucers are found at Huts 2, 3 and 4, Grid 12, the Store and Temple Complex. A single bone china cup is from Hut 1. In addition, a bone china vase or jug is from the Temple Complex. A porcelain eggcup comes from Hut 2 and a saucer from Hut 3. The Temple Complex has two fine porcelain unidentifiable hollowware vessels.

At Kiandra, bone china cups and saucers are from KCC2003-2 and from both stores in the Township. A number of cups, saucers and a plate are also from New Chum Hill, Sites 485, 742 and 1069, which were occupied from the late 1870s. Possible Japanese-made porcelain comes from the Kiandra stores and from New Chum Hill, Sites 742 and 1069.

Bone china and porcelain has been found at other Chinese sites in Australia and New Zealand. A clear-glazed bone china cup and saucer were found at Chinaman's Point (Bowen 2007:199). Four bone china vessels came from Narrandera Chinatown, including a plate, a hollowware item and two saucers. Three porcelain cups were found, two with banded decoration (Esposito 2012:653). Twenty-six porcelain vessels were found in Otago, mostly from Chinatowns, including nine plates/dishes, four cups/mugs, nine saucers and four miscellaneous items (Ritchie 1986:300). A gilt-banded saucer, possibly of Japanese origin was found at Ah Lum's store in Arrowtown (Ritchie 198:286).

Sets and matching vessels

In general, vessels in each assemblage have different decoration. Exceptions at JCC include three bowls (one serving and two smaller) banded in blue from Hut 2 and a matching cup and plate or saucer in blue Willow from Hut 4. At Kiandra, the only site to have matching vessels is Hut 3 from the Chinese Camp, which has: blue Albion – a platter and two plates; blue Asiatic Pheasants –a cup, medium plate, dinner or soup plate; Athens – green cup and grey plate; Rhine – blue cup, green saucer and grey cup/bowl; blue Willow – two plates; and a matching cup and saucer in a flown-green pattern. At ACC, Hut 3 also has a few matching vessels: an oval dish and two dinner plates in blue Willow; a saucer in Asiatic Pheasants and a cup or bowl which is possibly Asiatic Pheasants, both in green; a cup in black Antique and a cup or bowl in blue Antique.

Although, Lawrence, Brooks and Lennon (2009:75) cautioned that sets of the most common transfer-printed patterns, Willow, Asiatic Pheasants and Rhine, may not necessarily indicate an intention to purchase sets but reflect the availability of those items, most Chinese sites in this study do not have multiple vessels in the same pattern. The sets at JCC Huts 2 and 4 could be reflective of the commonly available wares at the time, banded and Willow, respectively. In contrast, KCC Hut 3 and ACC Hut 3 have other patterns apart from the most common ones, i.e. Athens and Antique respectively. Lawrence and her colleagues (2009:75) argued that the presence of an Eton College set in the household of Reverend Bean in Gippsland, Victoria, showed this was an intentional purchase because that transfer pattern was not common in regional Victoria. In this study Eton College is the seventh most common pattern, in terms of vessels found, while Athens and Antique are equal ninth. Interestingly, Reverend Bean left his Gippsland home in 1859. If he chose to sell his household contents, they may have been purchased by a retailer who sold second-hand wares to those in Kiandra, where Eton College was common from 1860 and found at nine sites in Kiandra.

There is a distinct lack of sets at Chinese sites in New Zealand (Ritchie 1986; Davies 2012:90).

Makers' marks

A total of 19 makers' marks were found on British-made vessels but only four makers have been identified: Sampson Bridgwood & Son (Longton, 1870+); Bates, Elliot & Co (Burslem, 1870-1875); Wood & Son (Burslem, 1891-1907); and J & G Meakin (Hanley, post 1890). Another mark may have belonged to either Pinder, Bourne & Hope (Burslem, 1851-1862) or Pinder Bourne & Co (Burslem, 1862-1882). Most other marks are partial or illegible. A clear mark 'H. N. & Co.' was unidentifiable. Marks associated with Rhine and Fibre, have the pattern name only. The Rhine cartouche marks, coming from sites at Jembaicumbene and Kiandra, are all similar but not identical.

Vessels made by some of these makers have been found elsewhere in regional NSW. For example, a vessel made by 'H. N. & Co.' was recovered from Lake Innes near Port Macquarie (Brooks 2007), while vessels made by Wood & Sons and J & G Meakin have been found at Kinchega Homestead near Broken Hill (Allison & Cremin 2006:50). Five vessels marked Pinder Bourne & Co were recovered from Otago sites (Ritchie 1986:289).

SMOKING

Only 33 of the 76 assemblages examined contain opium or tobacco pipes. Eighteen sites have both types of pipes, while there are 11 sites with opium pipe bowls only and four with clay tobacco pipes only. Sites without any pipes are found in all regions. The sites with tobacco pipes only are: JCC Hut 4, FPCC Hut 2, and East Eucumbene Sites 636 and 637. This discussion focuses on the opium pipe bowls as not all details were available regarding clay tobacco pipes.

Opium pipe bowls in this study are from a variety of sites. The largest number are from JCC Hut 2 (MNV=9). Of the other sites, the highest numbers come from KCC Oven (MNV=4) and KCC Hut 2 (MNV=4). Three hut sites have a minimum number of three pipe bowls. The stores have a small number of pipe bowls: JCC Store has a MNV of two and the Kiandra Township stores, MNV of two (Ah Chee) and one (Ah Yan).

It has been assumed that the dark-orange and light-orange pipe bowls in this study correspond to Ritchie's (1986:373-375) 'slipped' or 'unslipped' pipe bowls. Microscopic examination shows that clay particles in the dark-orange slip are much finer than those in light-orange slip, making the dark slip appear shinier and the light slip closer to the terracotta body colour (personal communication, Dr Anthony Flynn, December 2011).

The minimum number of opium pipe bowls present in this collection is 58. The most common type is the light-orange slipped bowl with circular rim and ten faceted sides (Ritchie's 1986 type C3), MNV of 21. It is found in 17 assemblages. The second most common is the dark-orange slipped bowl with a circular rim and smooth sides with a single raised ridge; it has a MNV of seven (Ritchie's 1986 type C1), and is from six assemblages. Sixteen terracotta pipe bowls are unidentifiable to type, ten of them having light-orange slip, two, dark-orange slip and the others with no remaining decoration. Type C1 is also found in light-orange, MNV of three, and C3 in dark-orange, MNV of two. The only other pipe bowl with more than one recovered has an octagonal rim and sides and is slipped in light-orange (MNV=4).

Single examples were found of five pipe bowls: a grey-stoneware hexagonal bowl with a buff-orange slip; a brown-stoneware octagonal bowl with grey slip; a buff-earthenware octagonal bowl with grey slip; a terracotta-bodied octagonal bowl, with four straight and four concave sides (similar to Wylie & Fike 1993:281, type B4), with light-orange slip; and a terracotta-bodied wheel-thrown bowl.

There is scant literature detailing opium pipe bowl types from Australian sites. At least 32 were found at the fish curing site of Chinaman's Point, Victoria (Bowen 2012:127). Two of these bowls are stoneware and the rest earthenware. The decoration on the bowls incudes light-orange, dark-orange, and brown slips, and a 'highly fired black glaze' (Bowen 2012:129). Types include C1 and C3, as well as two circular types not found in southeast NSW, but individual numbers of each are not listed. A light-orange slipped opium pipe bowl with a circular rim and faceted sides, #0387, was recorded in the Cooktown Sewerage Scheme main data sheet (data courtesy Gordon Grimwade, 22 September 2011). A MNV of two opium pipe bowls, one circular and the other hexagonal or octagonal, were recovered from a rubbish dump area at Cooktown (Rains 2005:288). A MNV of two were found in the Atherton assemblage examined by Dunk (2010:52), one of which appears to be a light-orange slipped circular type C1. An illustration in Gaughwin (1995:231) shows five opium pipe bowls from Tasmania: one type C1, one C3 and three with circular smoking surfaces and smooth sides. One of the latter appears to be decorated with flowers.

The most common opium pipe bowl in this study, the light-orange slipped C3, is also the most commonly found bowl, in terms of the number of sites at which it was found, at Ritchie's (1986) sites. However, the most commonly found bowl type from Ritchie's (1986) Otago sites with regards to absolute numbers is the dark-orange slipped C1 (MNV=22) followed by the light-orange slipped C3 (MNV=19). He also found a number of light-orange slipped C1 (MNV=10) and dark-orange slipped C3 (MNV=14). Overall, Ritchie (1986:371) identified 26 types of opium pipe bowls from his sites. About 70 per cent of Ritchie's pipe bowls could confidently be identified to type, a similar proportion to that of this study with about 72 per cent identifiable. The most common type of opium pipe bowl at the Lawrence Chinese Camp, New Zealand, was Ritchie's C3 (Arun 2012:47-50). Other common types were Ritchie's C1 and octagonal varieties.

Wylie and Fike (1993:274-275) examined opium pipe bowls, primarily from private collections in western USA and found circular bowls to be the most common (30 types) followed by octagonal bowls (13 types). In total, they identified 50 bowl types and speculated that this may represent 'half to three-quarters of the total number of types present at all overseas Chinese sites' (Wylie & Fike 1993:274). Bowl type C3 (Wylie & Fike, type A1) rates third on their list of most common types while C1 (Wylie & Fike, type A6) has only a single specimen.

The findings from this research agree with those of Williams' (2004) study of San Jose's Market Street Chinatown, which found that opium was consumed in a variety of contexts, rather than being exclusively associated with brothels, clubs or the back of stores.

Makers' marks

There are nine makers' marks on opium pipe bowls. A stamped mark, with Chinese characters for water and east, is found on pipe bowls recovered from KCC Hut 2, Sawyers Hill, ACC Hut 3 and JCC Hut 2. This mark may be the same as ones found in Lawrence, New Zealand (Arun 2012). One mark, with the Chinese characters for pot/jar and mountain, was found on opium pipe bowls recovered from sites in Lawrence (personal communication, Lana Arun, 4 October 2012). This mark may be linked to the water/east mark.

In summary, the opium pipe bowls from southeast NSW are not as diverse in type as those from New Zealand or USA. There are some similarities: the most common types found in New Zealand were Ritchie's C3 and C1; and the most common types identified in USA were circular followed by octagonal.

NON-FOOD RELATED VESSELS

Chinese

There is a possible miniature pot, similar to that described by Ritchie (1986:223) in the Temple Complex at JCC.

Another unusual Chinese-made find is a lamp dish from KCC Hut 4. No similar items have been reported in archaeological literature from Australian sites.

Non-Chinese

There are seven paste or ointment pots and seven pot lids. The contents of only three of these can be identified, two Holloway's ointment pot lids, one from JCC surface collection and the other from KCC Oven, and a White Rose toothpaste lid from JCC Hut 2. The most common non-food related vessels are blacking bottles: two from JCC Hut 2; and single bottles from KCC Temple, Hut 42, KCC2003-2 surface collection and Upper Adelong road and surface scatter. There are three salt-glazed ink bottles, two from Ah Yan's Store and one from KCC Hut 3.

Nine Holloway's ointment pots were found at Otago sites (Ritchie 1986:288). Five ink bottles were from Arrowtown and two penny inks from other sites (Ritchie 1986:304, 308).

CHAPTER 13

The use and supply of ceramic vessels

This research utilises several ceramic collections from Chinese mining camps and stores in southeast NSW, allowing a comparison between settlements, and between individual sites within them. The analyses enable a detailed examination of the active role of ceramic vessels in everyday life, not only within the domestic sphere but in various communal aspects of food preparation, consumption and feasting. For example, they allow an understanding of the broader social and economic characteristics of the Chinese population, through the assessment of the differing roles of specific buildings, such as temples, communal ovens and stores, as well as the variation over time in the domestic use of vessels. In addition, the analyses provide an insight into the Chinese supply of ceramics and foodstuffs and demonstrate how Western-style ceramics became appropriate substitutes, as those systems broke down.

COMMUNITY ASPECTS

Ceramic vessels play a part in the creation and maintenance of social networks and identity. In nineteenth century China, market networks for selling and buying vessels were also social networks (Skinner 1964). Villages, which operated markets, and clans within them, were seen as groups through which the central system of Chinese Imperial control could be administered, as the vastness of China meant that it was necessary to employ sub-administration at a local level (Hsiao 1960). The local aspects of this administrative system extended into NSW, and elsewhere, and as groups of miners arrived, the headmen set up a hierarchical system of settlements (Smith 2006).

The Chinese arrived in southeast NSW in groups from south-eastern China. For example, at the Kiandra Chinese Camp, 150 men worked under a headman (The Sydney Morning Herald, 11 May 1861:7) who controlled the provisions to those miners, including food, opium and ceramic vessels. The Qing government had exercised ideological control over the population 'which stressed social duties and human relationships' (Hsiao 1960:184). The provision of familiar peasant-ware vessels to the miners reinforced their Chinese identity and links to their homeland. The administration also set up temples and communal ovens to enable social gatherings, again reinforcing a sense of community.

Temples

Temples were a focus of the Chinese community. They were often dedicated to one primary deity and several secondary deities (Niemeier 1995) and were meeting places, not only for worship, but also where national ideologies could be reinforced (Gao 2012). Temple complexes usually comprised a number of buildings and had a caretaker. The temple was a modest building with a raised floor and veranda. Outbuildings included a kitchen and storeroom (Grimwade 1995, Wegars 2003; Smith 2006).

The exterior of the buildings were ornately decorated. For example, the 'large recessed entrance' of the temple at Emerald Hill in Victoria was 'decorated with some large painted lanterns, bright scarlet streamers, large gilt tablets with inscribed characters, and a small glass mirror over the centre of the doorway' (The Argus, 13 December 1866:5). The building had a central hall which opened onto wings, one of which was used as assembly rooms and apartments for those living on the premises, and the other included a large hall which housed a 'very large number - apparently several thousands - of small white wooden tablets', each a 'memorial of an individual deceased' (The Argus, 13 December 1866:5).

Other religious artefacts and ceremonies have been described. For example, the interior of an unspecified temple in Victoria was described by a reporter in 1865:

> Joss was seated on a throne of state, and surrounded by all the curious tinsel and gilding obtainable. Artificial flowers of English manufacture, the silver shreds from the top of boxes of French prunes, the gold and silver paper tacked under pearl buttons, a few of the button's themselves...some stray pieces of coloured glass, several yards of various coloured ribbons, many bunches of coloured paper, and a background of crimson crape made up the surroundings of the Joss (The Sydney Morning Herald, 6 October 1865:2).

Worshippers would burn incense upon entering the building (The Sydney Morning Herald, 6 October 1865:2) and burn 'pieces of paper impressed with Chinese characters' and discharge fire crackers (Australian Town and Country Journal, 20 July 1872:18).

Temples may also have been a place of learning. A newspaper report from Daylesford, Victoria, stated:

> In one Chinese tent I saw no less than six abaci, that is, counting tables, anciently used in calculations. The Celestials there harbor a description of persons who act as tutors, and who are sometimes provided for in some office in the Joss House (The Mercury, 27 May 1864:3).

Opium smoking may also have occurred in temples. In 1866, 'The hon. G.M. Isaacs, Solicitor-General of New South Wales...visited Sandhurst...and the Chinese Camp, where he was shown the joss house, and we believe initiated in the mysteries of opium smoking' (Bendigo Advertiser, 7 August 1866:2).

Ceramic vessels were used to offer food or to hold flowers or incense. A reporter described the fare offered at the opening of the temple at Ironbark, Victoria.

> In the middle of the floor, spread on some Chinese matting, were various edibles, in the shape of a cooked fowl, rice, tea, and what appeared to be to our unenlightened observation, a saucer full of some description of pickles...Repairing inside again, we found a table had been placed in front of the altar, and on which were placed a second collection of offerings' (Bendigo Advertiser, 23 July 1859:2).

Visitors to a temple in Canton, China, were served with 'tea and sweetmeats' (The Sydney Morning Herald, 26 April 1867:6).

The artefacts found at temple sites are therefore temple fittings, ritual paraphernalia and those associated with other communal activities. There may also be domestic vessels associated with the caretaker. In terms of specific artefacts, incense burners, candlesticks, vases and lanterns have been found (Wegars 2003:71). However, much of the ritual paraphernalia from inside the buildings may not have survived in the archaeological record. The temple ceramic pattern may not equate with ritual artefacts.

Few temple ceramic assemblages have been analysed in Australia. The Atherton Temple assemblage is dominated by Chinese tea and tableware (Grimwade 1995:312). Croydon Temple in North Queensland, has low densities of artefacts (Grimwade 2003:56) although it does have a Sweet Pea patterned vessel, possibly a teapot or liquor pot, and some Four Seasons vessels. The Mongarlowe Temple, in southeast NSW, also has tea and tableware among its excavated ceramic assemblage (Smith 2006:413). Tableware and brown-glazed storage jars were seen at the Tingha Temple Complex, including a rice bowl, tea or liquor cup, and a plate (Yit 2005:65). The Inverell Pioneer Museum has a collection of vessels which were associated with the Temple and these include celadon-glazed plates, bowls, spoons and liquor cups (Yit 2005:66).

Two ceramic assemblages have been examined in this study, from Temple Complexes at Jembaicumbene (JCC) and Kiandra (KCC) Chinese Camps. Both were built in the 1860s and appear to have been in use for the length of the settlements. These sites are dominated by wares used for food/beverage consumption. Western-style saucers and small plates, similar in size and shape to traditional Chinese shallow serving dishes, dominate the JCC Temple assemblage. A number of 'unusual' vessels from JCC Temple Complex include a possible Chinese-made miniature pot, similar to one described by Ritchie (1986:224), and a cream-glazed bone china vase or jug. In addition, children's china with English script has been found, along with writing slates, suggesting that one of the functions of that temple complex may have been as a place where English was taught. These artefacts may have been used by adults or children. Education was seen as a way to improve oneself and succeed and Chinese children were often told by their parents to 'blend in' to their local society (Low Choy 2012) but it was also important for adults to learn English. Speaking English was not about assimilation, but about empowerment, being able to communicate (Martinez 2011). The assemblage from the Temple Complex at KCC contains no vessels that specifically identify it as a temple site, but it is also dominated by food/beverage consumption vessels. In addition, a blacking bottle found at the Temple may have been used as a vase.

Communal Ovens

Overseas Chinese communal ovens were only recognised at camps relatively recently (Bell 1995; Smith 2006; Grimwade 2008). Communal ovens are generally circular in shape, often measuring about 1 m in width and height, and are partially sunk into the ground, although some regional variation has been noted (Bell 1995:214). Although there is no mention of them in contemporary documents, oral evidence relates to their use for cooking meat, particularly for festive occasions (Bell 1995; Grimwade 2008). They may also have been used more often when camps were occupied predominantly by single males. Chinese workers' camps in California, occupied by single males, show evidence of communal eating (Greenwood 1980). Sharing food on a daily basis reinforces the sense of community and belonging to a group. Artefacts associated with the ovens include hooks and wire to secure a carcass as it cooks, steel plates to keep in the heat while roasting and various knives and meat cleavers (Grimwade 2008). Little is documented about ceramic artefacts associated with oven sites.

Ceramics from four communal ovens have been examined in this study and these assemblages vary between sites. Two camps, Kiandra and Flanagan's Point (FPCC), were occupied predominantly by single males. The oven assemblage from the major camp at Kiandra is dominated by tea and tableware. Cups and bowls are the most common vessels along with a serving bowl, small plate and saucers. The majority of brown-glazed wares are small food/beverage containers and there are also at least six opium/tobacco pipes indicating that this site was used communally for eating, drinking and smoking. In addition to the ceramics, a hook from a sulky and a metal bar, which had been discarded after an unsuccessful attempt at forging, have been reused at the KCC Oven (Myles 2004:87). At the smaller camp of FPCC, the majority of serving vessels were found at the oven, while individual consumption vessels, cups and bowls, were found in the huts. Eating around the oven provided a sense of community and the ceramic vessels were the link between cooking, serving and consuming food.

The demography at Jembaicumbene Chinese Camp changed from single males to families and its oven has few associated artefacts, suggesting less use of the oven than at KCC and FPCC. At JCC, a major camp, the oven has the lowest MNV per excavated area for that settlement suggesting that, in later years of occupation, it was only used for festive occasions, rather than for communal eating on a daily basis. Similarly, there may have been families at Adjungbilly Chinese Camp (ACC), which has no ceramic finds from its oven. Other artefacts associated with the ovens at these camps include a large metal rod and numerous pieces of charcoal and animal bone at JCC oven and an iron cooking pot at ACC (Smith 2006:89, 154).

Stores

During the 1850s and 1860s the Chinese miners came to New South Wales and Victoria in their social groups and were supplied with food, opium and other necessities by their headmen through Chinese networks. These provisions were distributed from the headman's tent. By the 1870s, this system had broken down and the independent Chinese store took over the role of supplying miners. They competed with other local stores by offering lower prices (McGowan 2010:27).

Chinese stores were a focus for social activity for their countrymen (Wilton 2004). One such store is found in Bolong, NSW. It was built in the 1880s and was closed up, with the stock still on the shelves, when its owner Wong Sat died in 1916 (Meacham 2003). It was the subject of an exhibition at the Powerhouse Museum in Sydney in 2003. Wong's store was the focal point of the Bolong community, only rivalled by churches in the area. It was also important as a financial institution, extending credit to customers (Meacham 2003). Many Chinese stores offered a range of services including finding work, providing accommodation for employees and helping with travel documents (Wilton 2004; McGowan 2010).

In terms of ceramics, researchers have suggested that a greater proportion of large containers, at a site, indicates a commercial establishment. For example, in Cooktown, a dump associated with commercial activity was dominated by large containers but these containers were not so prevalent closer to residential areas (McCrae 2001). Similarly, in Market Street Chinatown, San Jose, California, it was found that commercial sites have a greater percentage of large containers (Brabyn 2007).

While the majority of occupants of the camps in this study remain anonymous, the names of the Chinese store owners have been documented. Three of the sites in this study are known to be stores, one at JCC and two in Kiandra Township. The JCC Store was occupied by Ah How from at least 1866 until 1875 when he sold or leased it to You Watt. The end date of this site is unknown but it was probably occupied into the early-twentieth century along with the nearby huts. Tom Ah Yan owned one Kiandra store from 1882 and lived there with his wife and children. George Ah Chee owned the other from 1895 until 1902 when he sold it to Henry Ping Kee. Both were destroyed by fire in 1916.

The main difference between the stores and other sites in this study, regarding ceramics, is the larger number of food/beverage storage containers at the former. All store assemblages contain a mix of Chinese and non-Chinese vessels. The front and back of JCC Store have differing assemblages. Although the MNV is small, the front, which may have been the sales area, has more Chinese-made wares. However, the rear of the store has better quality Chinese tableware and a wider range of non-Chinese tableware. The Chinese Stores in Kiandra Township have more non-Chinese vessels, mirroring the trend of later sites at KCC. In this study, the highest MNV of large jars are from those stores. Although stores are often thought of as a social centre for the Chinese community and a place for opium smoking, few opium pipe bowls or tobacco pipes were found at the Jembaicumbene and Kiandra stores, in comparison to other vessels. In addition to the known stores, this research has identified Hut 4 as a possible store at the Kiandra Chinese Camp. It has the highest MNV of storage containers at that camp.

Domestic aspects

Ceramic vessels are a part of everyday life, and are used to store, prepare, serve and consume food and drink. They become inextricably linked to the creation and maintenance of identity as 'how and what we eat is one of the fundamental ways we define ourselves as social beings and members of a given group' (Bray 2003:3). When the Chinese miners first came to Australia, they were supplied with Chinese food, rice bowls and opium pipes, reinforcing the link to their homeland. Evidence of these is found in the assemblages of the early hut sites in southeast NSW. Later sites in this region do not have the same proportion of Chinese-made wares. This discussion on domestic aspects explores the type of vessels found at hut sites and how they were used.

The most common group of vessels recovered from these sites are the brown-glazed storage containers. Every hut site within the major camps and in most of the smaller camps has these vessels, including all common types, such as barrel jars and lids, globular jars, wide-mouthed jars and lids, spouted jars, liquor bottles and ginger jars and lids. Some of these may have been reused. For example, carbon deposits on two bases, either spouted or wide-mouthed jars from huts at KCC, show that these vessels could have been used to heat water or liquor, or to cook food. The large quantity of containers confirms the addition of traditional Chinese food to the miners' diet. In contrast, there are few non-Chinese containers. Most are ginger beer bottles, but there are also a few large storage jars and a small food jar from later sites.

Tea and tableware is dominated by British vessels, predominantly bowls, plates, cups and saucers. The most common decorative type is a transfer-print. The most common Chinese tea and tableware vessel is the celadon-glazed rice bowl. Other vessel types include serving dishes and bowls, teacups, spoons, teapots and liquor cups. These are decorated with Four Seasons, Bamboo, Double Happiness, Rocks and Orchid and Sweet Pea. Surprisingly, few Chinese teacups are identified in the collections, for example, there are none at KCC or the smaller camps. The only identified teacups are from JCC and the Kiandra Stores. However, this may be because a traditional Chinese way of brewing tea was directly in bowls (Pettigrew 2003) and the earlier residents upheld this tradition.

The use of non-Chinese tea and tableware by Chinese has often been discussed in terms of acculturation, but in southeast NSW it appears that Western-style vessels were used as a substitute for Chinese vessels, which were no longer available. Table 13.1 lists Chinese forms, their sizes and suggested Western substitutes. If the Western vessels were used as substitutes for similar unavailable Chinese ones then plates and saucers, traditionally made for consumption use, would have become serving vessels. Non-Chinese bowls, similar in size to the Chinese ones, were traditionally made as serving vessels or as waste bowls for food scraps or tea leaves. These would then become consumption vessels for the Chinese.

Table 13.1: Chinese tea and tableware forms and Western substitutes

Chinese Form	Size (mm)	Western Substitute
Serving vessels		
Plate-large	c250	Dinner plate; soup plate; platter
Plate-medium	c150	Small plate/saucer
Shallow dish - large	c170-225	Small/medium plate
Shallow dish - medium	c90-140	Saucer
Shallow dish - small	c78-108	Saucer
Serving bowl - large	185-200	Serving dish/tureen
Serving bowl - medium	160-180	Bowl - large
Serving bowl - small	90-120	Cup/Bowl
Consumption vessels		
Rice bowl - standard	130-150	Bowl
Rice bowl - small	110-130	Cup/Bowl
Teacup - standard	75-85	Cup
Teacup - small	67	Cup

An 'ordinary' Chinese family used 'a minimum of serving and eating vessels' which included 'individual rice bowls, main dish platters, small sauce-and-dip dishes, cups, soup bowls (and) soup spoons' (Anderson & Anderson 1977:365-366). Based on a study of pertinent literature regarding Chinese settlement in America and contemporary illustrations of late-nineteenth/early-twentieth century table settings, Mueller (1987:261) proposed a hypothetical table setting for four people, consisting of 20 items. The assemblage from JCC Hut 2 can be considered to be representative of what had been used by those occupants at the time the hut burnt down. It has 21 possible serving vessels and 34 consumption ones (Table 13.2), that is, 38 per cent serving to 62 per cent consumption. This compares well with the proportional composition of Mueller's (1987) Chinese table setting, which can be interpreted as 40 per cent serving and 60 per cent consumption vessels. The implications of this are that these Chinese people were using the Western vessels in a Chinese way.

Table 13.2: JCC Hut 2 tea and tableware

Vessel	Size (mm)	No.	%
Serving vessels			
serving bowl	160-180	2	3.6
plate-dinner	>250	3	5.5
soup plate	250	2	3.6
plate-medium	200-240	1	1.8
plate-small	150-190	1	1.8
shallow dish	145	1	1.8
serving dish	175	1	1.8
saucer	160-180	4	7.3
plate-unknown size	unknown	1	1.8
plate/saucer	unknown	2	3.6
small bowl	120	2	3.6
teapot/lid		1	1.8
Consumption vessels			
rice bowl	130-150	14	25.5
spoon		2	3.6
teacup	60-100	7	12.7
cup/bowl	unknown	3	5.5
liquor cup	45-50	5	9.1
plate-child's	150-170	1	1.8
mug-child's	60	1	1.8
eggcup	unknown	1	1.8
Total		55	

To test this assumption the assemblage of JCC Hut 2 is compared to a contemporary assemblage of a family of European descent (Table 13.3). Hennessy's Hut, a farmstead located near Bungendore, about 45 km from the JCC Hut, was occupied by Richard and Catherine Hennessy and their six children from 1875 to 1904.

Table 13.3: MNV from JCC Hut 2 and Hennessy's Hut

Vessel type	JCC Hut 2		Hennessy's Hut	
	No.	%	No.	%
Food containers	45	40.2	5	3.1
Food consumptior	55	49.1	137	84.0
Non-food vessels	12	10.7	21	12.9
Total	112		163	

Note: this table does not include opium or tobacco pipes

The main difference between these assemblages is the amount of food/beverage storage containers at the Chinese site. The European site has a few ceramic containers; the majority are glass (Esposito 2008). If the percentages of tea and tableware are compared in terms of Western-style serving and consumption vessels, the proportions are similar for both sites (Table 13.4).

If the vessel types are examined, the differences between the assemblages become apparent (Table 13.5). The number of bowls is greater in the Chinese hut while the European one has a greater number of plates. The cups and saucers in the European hut are almost equal in number while in the other, the number of cups is double that of the saucers. The Chinese were not using the same proportions of vessel types. In addition, the majority of bowls from JCC Hut 2 have rim diameters of 130-150 mm, the size of a Chinese rice bowl. This confirms that the occupants were maintaining their Chinese diet and using the Western-style vessels as a substitute for similar Chinese vessels that were not available.

Table 13.4: Percentages of serving and consumption vessels from JCC Hut 2 and Hennessy's Hut in terms of a Western diet

Vessel type	JCC Hut 2		Hennessy's Hut	
	No.	%	No.	%
Serving	5	9.1	13	9.5
Consumption	50	90.9	124	90.5
Total	55		137	

Table 13.5: Tea and tableware vessels from JCC Hut 2 and Hennessy's Hut

Vessel	JCC Hut 2		Hennessy's Hut	
	No.	%	No.	%
Serving vessels				
serving bowl	2	3.6	1	0.7
serving dish	1	1.8	1	0.7
tureen/lid	1	1.8	4	2.9
platter	0	0.0	2	1.5
gravy boat	0	0.0	1	0.7
jug	0	0.0	2	1.5
teapot	1	1.8	2	1.5
Consumption vessels				
bowls	16	29.1	4	2.9
plates	9	16.4	40	29.2
cups	8	14.5	29	21.2
saucers	4	7.3	28	20.4
cup/bowl	3	5.5	9	6.6
plate/saucer	2	3.6	10	7.3
eggcup	1	1.8	4	2.9
spoon	2	3.6	0	0.0
liquor cup	5	9.1	0	0.0
Total	55		137	

The proportions of serving and consumption vessels can provide an indication of the type of diet, that is, Chinese or Western, as this ratio relates directly to it. This analysis of assemblages suggests that MNV would need to be sufficient, perhaps greater than 20, to obviate sample size bias. Tables 13.6 and 13.7 show the results of tea and tableware vessels ordered in accordance with a Chinese diet, that is, the plates and saucers are considered to be serving vessels. Few of the assemblages from this study have MNV greater than 20. The assemblages from JCC Hut 2, KCC Hut 5, the surface scatter across huts at KCC (KCC2003-2) and the surface scatter across huts at Upper Adelong (UACC) are close to the 40:60 per cent ratio, while KCC Hut 3 and ACC Hut 3 have different proportions of serving to consumption vessels to the other sites from this study.

KCC Hut 3 is thought to have been occupied by a Chinese man and a woman of European descent. Similar percentages to that hut have also been found at ACC Hut 3 and Dolly's Creek C75 (data from Lawrence Cheney 1995). It is possible that both of these huts were also occupied by a Chinese man and a woman of European descent. The Dolly's Creek hut, C75, is known to have had a female occupant (Lawrence 1999:133). Although it is thought that all living there were of European decent, it is possible that the male was Chinese. A celadon-glazed cup or bowl was recovered from that site and it is noted there were numerous Chinese in the Dolly's Creek area in Victoria. The other assemblages, including Hennessy's Hut, the Gold Commissioner's residence at Kiandra (KGQ), and three houses, Quebec Street, McKay Cottage and Farrow Cottage, all in Port Adelaide, South Australia, (data from Briggs 2005) are known to have belonged to people with a Western-style diet and the percentages of serving and consumption vessels, when treated in terms of a Chinese diet, are very different to those of the Chinese sites.

A comparison of vessel forms, from Chinese and non-Chinese sites in southeast NSW, reflects the choice of vessels for different foodways (Figure 13.1). The vessels are shown as a percentage of total vessels, the main difference being the larger proportion of bowls at the Chinese sites. Overall, the number of hollowware vessels is higher than flatware at Chinese sites, while flatware vessels dominate at non-Chinese sites. The choice of vessel type relates to the food consumed.

Table 13.6: Tea and tableware ordered in terms of serving/consumption in a Chinese diet – sites from this study

Site	JCC Hut 2	KCC Hut5	KCC03-2	UA scatter	KCC Hut3	ACC Hut3
MNV	**55**	**27**	**40**	**31**	**29**	**39**
Serving	38.2	37.0	40.0	41.9	48.3	48.7
Consumption	61.8	63.0	60.0	58.1	51.7	51.3

Table 13.7: Tea and tableware ordered in terms of serving/consumption in a Chinese diet – sites outside this study

Site	Hennessy's Hut	KGQ	Quebec St	McKay Cottage	Farrow Cottage	Dolly's Creek C75
MNV	**132**	**51**	**204**	**157**	**493**	**42**
Serving	66.4	68.6	71.6	65.0	62.3	45.2
Consumption	33.6	31.4	28.4	35.0	37.7	54.8

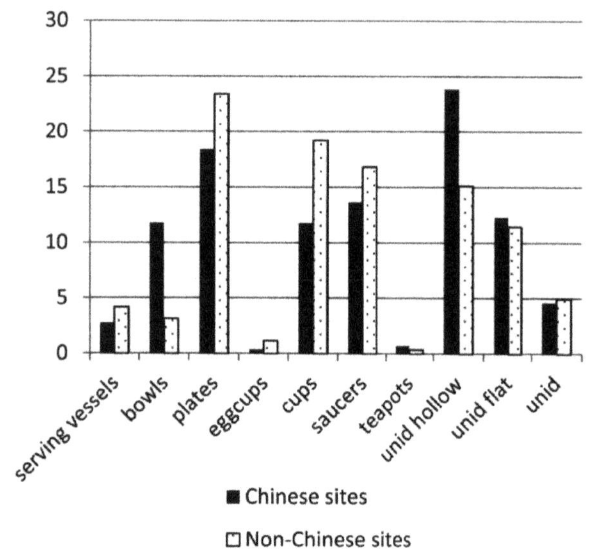

Figure 13.1: A comparison of vessel forms from Chinese and non-Chinese sites in southeast NSW as a percentage of total vessels

Two of the sites compared within KCC are Hut 5 and the surface collection, KCC2003-2, covering six hut sites. Both sites confirm an adherence to a Chinese-style diet, but the former has predominantly non-Chinese vessels and the latter more Chinese vessels. Hut 5 is closer to the Temple Complex and therefore was probably occupied at a later date. This suggests that although the supply source for tea and tableware vessels may have changed, the occupants of the camp were still conforming to Chinese foodways.

In general, the huts at JCC and KCC located on the outer parts of the settlement have more smoking paraphernalia than tea and tableware. These would have been occupied at the start of the rush, as it has been established that the camps contracted over time towards the centre, containing the temple and communal oven (Smith 2006). Similarly, some of the smaller work camps, occupied in the 1860s, have high proportions of opium pipe bowls.

Early/short-term miners' huts in southeast NSW, whether Chinese or European, have fewer tea and tableware vessels associated with them (Gant-Thompson 2008). A dominance of clay tobacco pipes has been associated with short-term occupation in European miners' huts at Township Hill in Kiandra, where 185 structures have been recorded by students from ANU, and few were found to have ceramic remains apart from clay pipes. While the European miners probably used metal cups and plates, the use of ceramics for eating and drinking has always been important to Chinese.

A comparison of the MNV of opium and tobacco pipes to all tea and tableware is presented for the site assemblages (Table 13.8). This is based on the study by Wylie and Fike (1993) who used a comparison of opium pipe bowls to Chinese tea and tableware to ascertain the relative proportions of opium use. They made the assumption that 'culinary behaviour' would remain 'relatively uniform' over time and that increases or decreases in opium use could be identified (Wylie & Fike 1993:292). They established that opium use was greater in Chinese rural work camps in America than in comparative urban sites.

If higher proportions of opium/tobacco use are associated with male workers' camps then the assemblages from sites with greater proportions could be representative of the households of single males. Sites with a greater or equal number of opium pipe bowls to tableware are JCC Hut 5 and the Southwest Eucumbene Site 740. At the former, opium and tobacco pipes account for almost half of the total assemblage. The ephemeral nature of JCC Hut 5 remains and the scarcity of artefacts suggest that this hut was one of the original huts occupied short-term by single miners. Site 740 is an outlying workers' camp at Sawyers Hill, known from documents to have been occupied at the start of the Kiandra rush. Five of the six sites with 31-50 per cent pipes/tableware have more Chinese than non-Chinese vessels in their assemblage. FPCC Hut 2 has no Chinese-made ceramics.

Table 13.8: MNV of pipes compared to MNV tea and tableware

Pipes/T&T	JCC	FPCC	KCC	Kiandra Valley	Kiandra Township
0-10%	Hut 3, Oven, Temple Complex	Huts 1, 3, 4, 5, Oven	Huts 3, 5, 21, 26, 37, 47, 53, 55, 57, 59, 62	Eucumbene Crossing 127, 143, 268; East Eucumbene 977; New Chum Hill; Giandarra Gully	Ah Chee, Ah Yan
11-20%	Huts 1, 2, 4, Store		Temple and Ancillary building, Huts 4, 35, 36	East Eucumbene 636, 637	
21-30%			Huts 1, 2, 42, Oven	East Eucumbene 635	
31-50%		Hut 2	Huts 25, 28, 43	Eucumbene Crossing 125; Southwest Eucumbene 698	
Over 50%	Hut 5			Southwest Eucumbene 740	

In general, the hut sites at JCC and KCC with the highest proportion of opium pipe bowls and tobacco pipes are further from the Temple Complex than the huts with less. At KCC, the Oven also has a high proportion of smoking paraphernalia. All sites, at both ACC and UACC, as well as the surface collections from JCC and KCC, (not included in Table 13.8) have results of less than 10 per cent. Although these data suggest that opium smoking may have reduced over time as fewer pipes were recovered from later sites, relative proportions may be accounted for by the increase in tea and tableware as household composition changed.

Some huts occupied in the late-nineteenth century have assemblages which suggest occupation by couples or families rather than single males. For example, JCC Hut 2, located near the centre of the settlement, and occupied from the mid-nineteenth to the early-twentieth century, has a greater range of vessels including children's items. Children's vessels do not necessarily mean that there were children at the site, but given the length of occupancy, the documented presence of women at the camp and the importance of family to the Chinese, it is reasonable to conclude that children were present. The only other camp, apart from JCC, to have children's ceramics is ACC. While these examples are extreme ends of the spectrum, from short-term single males to long-term families, the hut assemblages in this study vary. There is a continuum of assemblages, reflecting different occupation periods and different living arrangements.

A distinctly different assemblage, to the others in this study, is from KCC Hut 3, which includes matching vessels, a trait not commonly seen in the Chinese occupied huts. It is suggested that this is indicative of the presence of a European rather than Chinese woman. The cost of matching ceramic vessels decreased through mass-production and after the mid-nineteenth century they could be afforded by the working-class, which emulated the upper and middle-class Victorian ideals of gentility. Choice of goods demonstrated an understanding of social standards (Young 2003:153).

Although many successful relationships were formed between Chinese men and European women, they faced prejudices from the white Australian community and the Chinese communities in Australia and China (Bagnall 2006:299). This is reflected in the location of KCC Hut 3, which would have been on the edge of the Chinese camp after it had contracted. The only other hut to have matching vessels is ACC Hut 3 and this, too, would have been towards the outskirts of the camp. Similarly, at Jembaicumbene, the assemblages of Huts 3 and 4, while small in number, suggest a female presence, perhaps European and they would also have been located on the outskirts of the camp after it had contracted.

ECONOMIC ASPECTS

Chinese networks

This research has revealed a changing composition in the numbers of Chinese/non-Chinese vessels in assemblages over time as the earlier sites within JCC and KCC have a higher proportion of Chinese-made wares. In addition, different vessels have been found in different regions, although the sites are all located within 150 km of each other. For example, Four Seasons decorated serving bowls and liquor cups are found at Kiandra but not at Jembaicumbene, while Four Seasons decorated spoons are only found at Jembaicumbene.

At present, little is known about the supply of Chinese ceramics to Australia in the nineteenth century. In the 1860s-1870s, supplies were controlled by merchants linked to clan groups; however this changed in the 1870s-1880s, when commercial networks of English-speaking Chinese began working with foreign firms (Chan 1977; Motono 2000). Crawford (1877:18) reported that 'The bulk of the Chinese business in Australia...is in the hands of two or three great houses'. One of these may have been Lowe Kong Meng (1831-1888), the son of a Chinese merchant operating in Penang. He travelled to Victoria in 1853 and set up a business importing goods from China (Macgregor 2012). Kong Meng was fluent in English, Cantonese, French and Malay and had numerous contacts with British, Chinese and French merchants. He became more involved in providing for Western consumers as the Chinese population declined in Victoria.

The location of a site, and the size of its population, appears to determine what vessels are present. The migration of Chinese in the nineteenth and twentieth centuries was one of the 'world's largest population movements' (Voss & Allen 2008:5). During the second half of the nineteenth century, more than 2.5 million people left China for Southeast Asia, the Americas, Australasia and Africa, of which about 380,000 migrated to mainland USA (Voss & Allen 2008:6). The most diverse range of nineteenth century exported Chinese-made vessels in America has been found in San Francisco as that city was the 'demographic, economic and social' centre of the overseas Chinese during the second half of the nineteenth century (Pastron et al. 1981:465). It was the main port for imports as well as having a different demographic composition to other areas in California, including upper class Chinese, middle-class merchants and females. Pastron, Gross and Garaventa (1981:466-467) stated that Chinese labourers in America 'were generally isolated and segregated from the mainstream of Anglo American society' and therefore they did not have access to a full range of goods available to the rest of Americans. They depended on Chinese merchants who were supplying a 'virtually captive market' (Pastron et al. 1981:467).

It might be argued that the numbers of Chinese in Australia were not sufficient for a supply network of the size operating in Asia or America as the variety of form and decorative types seen in the former does not appear to be present throughout Australia. Numerous blue-underglaze patterns have been found in Southeast Asia (Willetts 1981) and on shipwrecks bound for there. The cargo of the Tek Sing, a Chinese ship which set sail from Amoy (Xiamen) in 1822 heading for Southeast Asia but sinking in the South China Sea, contained hundreds of different blue-underglaze patterns (Pickford et al. 2000). Many blue-underglaze and polychrome enamel patterns have been found at sites in urban America.

Some vessels may have arrived with Chinese miners. Blue-underglaze rice bowls were provided to the miners for use on their voyage to America (Mueller 1987:308). Similarly, Chinese miners going to work in Malaysia took with them 'chinaware for personal use' including rice bowls in the common blue-underglaze patterns, Bamboo and Double Happiness (Willetts 1981:11). Crawford (1877:27) observed that Chinese arrived in Australia with their 'own bamboo-pole brought from China', carrying 'a few humble necessities' and it is possible that these included the cheap blue-underglaze rice bowls.

Melbourne and Sydney would have been the equivalent Australian cities to San Francisco, in terms of arrivals, but little is known about the ceramics used by nineteenth century Chinese residents, beyond the research of Muir (2008) and Lydon (1996). Excavations in Melbourne's Chinatown have shown that there is a relative lack of Chinese ceramics in household assemblages (Muir 2008:123). In Sydney, some less common blue-underglaze patterns have been found in excavations, for example, Peach & Fungus, Chrysanthemum & Conch and Om (Lydon 1996: Appendix 1). The Chinese miners initially arrived in Melbourne and Sydney to travel to the goldfields in those colonies. The residents in Kiandra were linked to both cities, while those in Jembaicumbene arrived via Sydney. Census figures from 1861, 1871 and 1881 show a reduction in numbers of Chinese in Victoria and NSW as restrictive laws stopped the influx of migrants (Choi 1975:21). In Victoria there were 24,732 Chinese inhabitants in 1861, 17,826 in 1871 and 10,205 in 1881, while for the same years in NSW, the population numbered 12,988, 7,220 and 10,205.

About 20,000 Chinese migrated to Queensland in the 1870s (Crawford 1877:2). In the late-nineteenth century, Cooktown and Cairns were described as being the 'stations on the highway of commerce between Australia and Java, Singapore and China' (Crawford 1877:25). By 1880, Cooktown merchants imported a wider range of goods than Brisbane (Grimwade et al. 2007) as they supplied the numerous Chinese on the Palmer goldfields (Rains 2005). Vessels previously undocumented in Australia have been recovered in Cairns, including two blue-underglaze patterns, a celadon-glazed plate of an undocumented shape and a blue-underglaze 'cube-like' item 'of an unidentified configuration and function' (Grimwade & Rowney 2004:72). The 'diversity of material' from Pine Creek Chinatown, a large settlement in Northern Territory, went 'well beyond that from comparable published Chinese sites either in Australia or overseas' (McCarthy 1988:139). In addition to the usual brown-glazed jars, ginger jars, celadon and Four Seasons, decorative patterns included 'tranquil landscapes, dragons and plum blossom vistas' (McCarthy 1995:200).

The strength of the Chinese networks for new immigrants is demonstrated in northern Queensland. Excavated sites in Cairns Chinatown, dating from late-nineteenth century to early-twentieth century, have 'few, if any, examples of English or European derived fine earthenware or porcelain tableware pieces' (Grimwade & Rowney 2004:66). Rains' (2005:303) evaluation of the Chinese supply network in Cooktown concluded that the 'most overtly "Chinese" period of consumption' was between 1874, the arrival of the Chinese, and 1890. After this time the high levels of distinction seen between the Chinese and non-Chinese networks were not as evident. By the end of the 1880s, Cooktown was in decline, mainly because of the depletion of alluvial gold at Palmer River. There was an exodus of Chinese from Cooktown but some Chinese merchants set up general stores for non-Chinese consumers, requiring them to have access to Western market sources (Rains 2005).

In contrast, in NSW in the 1870s, the existing Chinese population was moving to different parts of the colony for work, for example to Narrandera and Tingha. Although there are Chinese ceramic vessels at these locations, the number and variety does not seem to be as great as in northern Australia at the same time. The Narrandera Chinese lived in a camp of '30 or 40 houses, a joss house, a hospital...stores, restaurants, banks, and gambling and opium dens' (Gammage 1986:141). Their leader was Sam Yet, who had moved from Victoria in 1871. He was a storekeeper and financier who organised and supplied work gangs for the European timber industry. By 1883, there were 303 Chinese in the camp but numbers had dropped to 124 by 1891 (Gammage 1986). Generally, the types of vessels in the Narrandera Chinatown assemblage are similar to those from the later sites at Kiandra, although there are three blue-underglaze rice bowls in Double Happiness (Esposito 2012:653). Small numbers of Chinese arrived in Tingha in the 1870s and numbers peaked at 2,268 in 1888, declining to a few hundred in following years (Yit 2005:13). The Tingha ceramic assemblage contains a mix of Chinese and non-Chinese vessels including a hexagonal green-glazed ginger jar (Yit 2005), a vessel type not found at the sites in this study.

The sites in this study are inland, away from the main supply ports and the numbers of Chinese on these goldfields were small. At Jembaicumbene, there were 400-500 in 1861 but from 1877 onwards, there were less than 50 (Smith 2006:287). The highest number recorded at Flanagan's Point was 118 in 1877 (Smith 2006:429). Although there were about 1,000 Chinese at Upper Adelong in the early 1870s, there were less than 100 from 1877 onwards, a similar number to those at Adjungbilly (Smith 2006:512, 538). At Kiandra, there were about 700 in 1860 but by 1883 there were only 125 (Smith 2006:607). While the examined sites do not have as wide a variety of vessel types as those in America there are similarities in most types of storage containers and opium pipe bowls. A notable difference is seen in the tea and tableware. Blue-underglaze decorated wares are the most common rice bowls in America, but celadon-glazed wares are more common in this study. In addition, there have been no Four Seasons rice bowls recovered in Australia or New Zealand but they are common in America.

The supply source of vessels to southeast NSW may have diminished with declining population numbers. Hut sites further from the Temple Complex, at both JCC and KCC have more Chinese than non-Chinese ceramics, while closer huts have the reverse. If the camps contracted over time, the inner hut sites would have been occupied at a later date suggesting that the sites with more Chinese-made vessels were earlier. Although only five huts were excavated at JCC, the outermost Hut 5 has more Chinese-made ceramics while the inner Huts 1, 3 and 4 have more non-Chinese ceramics. Hut 2, the unique site, is located in the centre of the camp but it has slightly more Chinese-made ceramics, predominantly storage jars. At KCC, there is a significant difference between the ceramics recovered from huts further from the Temple Complex than those close by.

The range of Chinese vessels in assemblages, by form and decoration, changes from the initial camps of the 1850s and 1860s to the occupations of the 1870s and 1880s. The earlier huts at KCC have a greater proportion of Chinese-made vessels, a larger variety of opium pipe bowls and contain all but one of the blue-underglaze rice bowls. The remaining Bamboo rice bowl is from the Temple Ancillary Building, but its underglaze is a green-blue rather than a clear blue, like those from the earlier huts. Two other forms are found with a green-blue underglaze: a dish with an interior stacking ring from the surface collection (KCC2003-2) and a block-printed rice bowl from Ah Yan's Store. There are a greater variety of celadon-glazed Winter Green forms at the store than from the earlier dated sites at the camps. Four Seasons vessels are also more common at the later sites, which may be a result of a reduction in their cost in the late 1870s (Sando & Felton 1993). Generally, Winter Green is found at major and minor camps; however Four Seasons is only at the main camps of Jembaicumbene and Kiandra and in Kiandra Township.

A comparison of common decorative patterns, both Chinese and British, has been undertaken between two groups of huts (Table 13.9 - data are from all sites examined with a MNV greater than five). It reveals that huts with more Chinese vessels have a greater variety of Chinese patterns. Similarly, huts with more non-Chinese vessels have a larger variety of British patterns.

Table 13.9: Comparison of presence of nameable patterns at sites with more Chinese vs more non-Chinese ceramics

Pattern name	Colour	Group 1 Huts	Group 2 Huts
Chinese			
Bamboo	blue	x	
Double Happiness	blue	x	
Four Seasons	polychrome	x	x
Rocks & Orchid	blue	x	
Winter Green	green	x	x
Non-Chinese			
Albion	blue	x	x
Alhambra	purple		x
Amoy	grey		x
Antique	black		x
Antique	blue		x
Aquatic	blue		x
Asiatic Pheasants	blue	x	x
Asiatic Pheasants	green		x
Athens	blue		x
Athens	green		x
Athens	grey		x
Cable	blue	x	x
Cable	purple	x	x
Cable A1	blue		x
Crystal	brown		x
Crystal	green		x
Fibre	black		x
Fibre	blue	x	x
Fibre	green		x
Forest	blue	x	x
Gem	blue		x
Rhine	blue	x	x
Rhine	brown		x
Rhine	green	x	x
Rhine	grey	x	x
Sydney	purple		x
Willow	blue	x	x

x = presence; Group 1 huts have more Chinese-made ceramics overall (MNV > 5); Group 2 huts have more non-Chinese made ceramics overall (MNV > 5)

It appears from this study that the miners were initially supplied with necessities such as foodstuffs, opium and rice bowls. In that period the network controlled what was used by the new migrants, reinforcing their sense of community and Chinese identity through the supply of familiar goods. As numbers declined and the system broke down, it may not have been as profitable for the networks to supply goods in the same quantities as before. As a consequence the Chinese miners in southeast NSW had to supplement their Chinese-made vessels with those of non-Chinese origin. This scenario would explain the greater numbers of blue-underglaze bowls and the greater variety of opium pipe bowls at earlier sites. The miners could also have bought some ceramic wares with them when they initially arrived, such as the blue-underglaze decorated bowls, which are found at sites occupied in the 1860s. For example, the only Double Happiness decorated bowl in this study from the Kiandra region is from a site near Sawyers Hill, in the Kiandra Valley, known to have been occupied in the 1860s. Chinese stores took over the supply of the camps from the 1870s and this could also account for changes in the composition of the assemblages. The Chinese made some purchases at local stores in Kiandra as early as 1861 (The Sydney Morning Herald, 11 May 1861:7), but it is not known if this included ceramic vessels. It was also noted that Chinese miners bought rice from a store at Rocky Plains, near Kiandra, in 1861 (Holme n. d.).

Origin of Chinese-made vessels

One of the aims of this research was to determine the number of sources of the Chinese ceramic vessels by using electron microprobe analysis to establish the elemental composition of the ceramic bodies and glazes. Clay, the main component of any ceramic vessel, has a particular elemental signature. Although slight variations in the fabric composition would be expected because of natural variation in clay deposits, it is likely that there would be more variation between sites than within a site. Although the exact number of sources is not known, analysis reveals that vessels were sourced from multiple kiln sites in China. Two distinct ceramic body types have been detected within the brown-glazed storage containers. In addition, the glaze compositions of five ginger jars are all different, suggesting either different sources, or time, of manufacture. Although the celadon-glazed wares appear to come from one region, the blue-underglaze and Four Seasons wares may have come from multiple sources.

Variation within the first brown-glazed group suggests that they came from more than one kiln site within the same region. In addition, two tested vessels have similar, but slightly different compositions to the majority of this group. Both of these vessels are from Kiandra sites occupied in the early 1860s. It is possible that early miners brought food jars with them from southern China. Crawford (1877) hinted that food may have been part of the personal possessions carried by each man arriving in Australia. The actual origin of this first group has not been established, but comparisons with other data show a similarity to ceramics from Fujian and Zhejiang. No stoneware data were available for comparison from Guangdong. Hetherington (1921:9-15) listed ten potteries in Fujian, two in Zhejiang and eight in Guangdong, which were operating in the nineteenth century. Hoh (1933) stated that about one third of the wares produced in Shiwan, Guangdong, were made for the overseas Chinese market.

The other distinct brown-glazed group is closer to the characteristics of north Chinese ceramics than any other tested data and is made up entirely of liquor bottles. It has previously been thought that brown-glazed storage jars on overseas Chinese sites are from southern Chinese provinces, with most coming from Guangdong (Quellmalz 1976; Hellmann & Yang 1997; Pastron et al. 1981). However, goods were transhipped around the coasts of China; therefore, a ceramic vessel, such as a bottle from Tientsin (Tianjin) in northern China, could have ended up on a ship to Australia from Hong Kong (personal communication, Paul Macgregor, 6 September 2012).

Compositional analysis reveals that all the celadon-glazed samples tested have a similar body and glaze. Statistical analysis shows a strong similarity with ceramic bodies known to have come from Jiangxi and, to a lesser extent, from Zhejiang and Guangdong. Previously, Stenger (1993) had suggested that celadon-glazed wares found on overseas Chinese sites in America may have been made in Japan but only five of the 28 celadon-glazed bodies in this study have any probability, between 1 and 4 per cent, of belonging to the tested Japanese group. Although the celadon glazes analysed in this study conform to Stenger's Japanese glaze characteristics, they are very different to Japanese glazes tested in this study. The ceramic bodies are more closely aligned to those of Chinese vessels than Japanese, confirming a Chinese origin. None of the Chinese makers' marks on tea and tableware are the same, although this may not be indicative of a different source. They may have been copied and degenerated over time (Pastron et al. 1981) or be illegible due to bad application (Olsen 1978).

The blue-underglaze and Four Seasons decorated vessels are from multiple kiln sources. Compositional analysis of the clear glaze on blue-underglaze decorated vessels shows that the rice bowls have a different glaze to that of a dish. Although an elemental analysis of the blue-underglaze itself was not undertaken, the variation in colour was noted. The majority of bowls have a clear blue colour, while a few have green-blue colouring. Research has shown that the source, and purity, of the cobalt used changed over time (Yap & Tang 1985; Yap 1986, 1988; Steele 1993; Yu & Miao 1996). Some Four Seasons vessels have a very white body and a clear glaze, while others have a greyer body and green cast to the glaze which appears similar to pale celadon. The Four Seasons glaze and the clear glaze on the blue-underglaze dish contain lead, but the other glazes do not.

The opium pipe bowls in this study are not as diverse in type as those from New Zealand or the USA. The most common types are made of terracotta and have circular rims, one with ten faceted sides and the other with smooth sides and a single raised ridge. These coincide with common types found in New Zealand. They are decorated with light or dark orange slip. Other bowl types include octagonal and hexagonal shapes. One bowl has been wheel-thrown but only a small sherd is present, therefore its exact profile is not known. The hexagonal bowl and one of the octagonal bowls are stoneware and all the other bowls are earthenware. At least five different marks are found on opium pipe bowls. One impressed stamp, containing the Chinese characters for water and east, is found on pipe bowls from JCC Hut 2, ACC Hut 3, KCC Hut 2 and Kiandra Valley Site 740. While compositional analysis shows that brown and celadon-glazed vessels from the same source were supplied throughout southeast NSW, the maker's mark on the opium pipe bowls, noted above, is the only mark which provides a link in the supply network across the whole region. It is known that the Kiandra Valley hut was occupied in the 1860s. The other huts may also have been occupied at that time, confirming the operation of a Chinese network across the whole region during that period.

Chinese access to Western markets

The examined collections also contain a large number of non-Chinese ceramics, the majority of which are British. Some late-nineteenth to early-twentieth century sites have Japanese or European porcelain. The only confirmed Australian-made vessel is from JCC Hut 2, a terracotta lead-glazed pan or dish. A vessel from KCC Hut 2 is possibly a terracotta water monkey, which may have been made in Australia. Some of the sponge-printed wares may have been made in Holland. Very few of the vessels have makers' marks. As common patterns were made by a large number of potteries unmarked vessels could not be attributed to a certain maker.

No previous research has been undertaken to ascertain whether the Chinese in Australia had the same access to British ceramics as the wider population. To date, knowledge of Western-style ceramic supply in nineteenth century Australia is scant (Brooks 2005; Lawrence et al. 2009). The most common decorative type in the nineteenth century was the transfer-print. Coysh and Henrywood (1989) have recorded hundreds of British transfer-printed patterns, stating that the number of patterns printed in the nineteenth century appears to be an inexhaustible subject. Similarly, in Australia, numerous transfer-printed patterns have been recorded (Casey & Lowe 2009). Lawrence (1999:129) recorded 43 'diverse and colourful' patterns from four huts at Dolly's Creek, Victoria.

A comparison of nameable transfer-printed patterns found at Chinese sites with those from non-Chinese sites in southeast New South Wales has determined common patterns are found at all sites. The non-Chinese sites include: the Kiandra Hotel, which was later a private residence, the Teachers' Residence and Dance Hall, a Township residence, The Gold Commissioner's Residence and outbuildings and the hospital, all in Kiandra; and the Hibernian Hotel and Hennessy's Hut near Bungendore. These sites were occupied during the period from 1860 to the early-twentieth century.

A MNV of 464 nameable transfer-printed vessels were identified. They comprise 35 different patterns, 11 of which are found in different colours. A MNV of 139 are from Chinese sites and 325 from non-Chinese (Table 13.10). The six most commonly found patterns in this study, Willow, Rhine, Asiatic Pheasants, Albion, Fibre and Cable, are among the most common transfer-printed patterns found at nineteenth century sites in Australia. They have been found at many sites in Sydney and Parramatta (Casey & Lowe 2009). The top four patterns have been found at sites in Adelaide (Briggs 2005), at Dolly's Creek in Victoria (Lawrence Cheney 1995), at Lake Innes (Brooks 2007) and Kinchega in NSW. Rhine and Asiatic Pheasants were recovered from Chinaman's Point in Victoria (Bowen 2007:199). Willow, the most common pattern in terms of the number of sites at which it is found, was the cheapest transfer-printed pattern in the early-nineteenth century and potters classed it as a separate category for price listing (Miller 1980, 1991).

Twenty of 35 nameable transfer-printed patterns, or 57.1 per cent, are found at both Chinese and non-Chinese sites, one pattern at Chinese sites only and 14 at non-Chinese sites only. The patterns found only at non-Chinese sites have small MNV or are from one site only. In addition to the nameable patterns there are 358 unidentifiable transfer-patterns. Only 19, or 5.3 per cent, of these were found at both Chinese and non-Chinese sites; 126 at Chinese sites only and 213 at non-Chinese sites only. Many patterns are found at one site only.

Tables 13.11 and 13.12 show the data broken down into regions. The Chinese sites include Braidwood (JCC & FPCC), Tumut (ACC & UACC) and Kiandra (KCC, Kiandra Valley and Township stores). The non-Chinese sites are Kiandra (as listed previously) and Bungendore (Hibernian Hotel & Hennessy's Hut). While eight nameable transfer-printed patterns were found at Chinese sites in all regions, there are no unidentifiable transfer-printed patterns. Kiandra has the highest number of patterns because it has the most number of sites, however, most patterns were localised to one region. The large camps, JCC and KCC, have a greater variety of patterns than the smaller camps. A similar scenario occurs at the non-Chinese sites with the majority of patterns found in one region only.

Although these patterns have been numbered separately, some may have been duplicated as many sherds are small and all motifs within a single pattern are not identifiable. Regardless of this, there are numerous transfer-printed patterns from these sites. This concurs with other Australian researchers who have found numerous transfer-printed patterns at their sites.

Other decorative types found, apart from transfer-prints, include flown, sponge, sprigged, banded, decal and relief-moulded patterns. Banded, decal and relief-moulded patterns were more common in the later nineteenth century; consequently these decorative types are only found at sites occupied in the late-nineteenth/early-twentieth centuries.

Figures 13.2 and 13.3 compare decorative types at Chinese and non-Chinese sites respectively. The graphs show the numbers of sites at which decorative types were found as a percentage of the total sites, that is, of 76 Chinese sites or 19 non-Chinese sites.

Although the percentages are higher at the non-Chinese sites, the trend of decorative types is similar in both graphs. The higher numbers of banded and relief-moulded decoration at non-Chinese sites reflects the occupation dates of those sites.

Table 13.10: The number of sites in southeast NSW at which nameable transfer-printed patterns were found

Pattern	MNV Chinese sites	MNV Non-Chinese sites	Total vessels	Ch find sites	Non-Ch find sites	Total find sites
Willow	28	66	94	21	18	39
Rhine	33	64	97	22	11	33
Asiatic Pheasants	11	31	42	9	11	20
Albion	10	17	27	9	10	19
Fibre	9	14	23	13	4	17
Cable	8	19	27	8	6	14
Eton College	3	16	19	2	8	10
Antique	7	4	11	6	2	8
Crystal	6	2	8	6	2	8
Athens	5	6	11	3	3	6
Gem	3	5	8	3	3	6
Cable A1	2	11	13	1	4	5
Kulat	1	5	6	1	4	5
Magenta	2	5	7	2	3	5
Forest	3	2	5	2	2	4
Alhambra	1	6	7	1	2	3
Amoy	2	3	5	2	1	3
Clyde (2)	0	8	8	0	3	3
Dresden	2	4	6	1	2	3
Aquatic	1	1	2	1	1	2
Melbourne	0	2	2	0	2	2
Sydney	1	3	4	1	1	2
Chinese Marine Series	0	1	1	0	1	1
Clyde (1)	0	1	1	0	1	1
Gothic	0	5	5	0	1	1
Gothic Scroll	0	2	2	0	1	1
Honeysuckle	0	2	2	0	1	1
Lichfield	0	1	1	0	1	1
Melrose	0	1	1	0	1	1
Peacock	0	3	3	0	1	1
Poppy	0	2	2	0	1	1
Trellis	0	4	4	0	1	1
Wattle	0	1	1	0	1	1
Wicker	1	0	1	1	0	1
Windsor	0	8	8	0	1	1
Total	139	325	464			

Note: data ordered by total number of find sites

Table 13.11: Transfer-printed patterns from Chinese sites

Transfer-printed patterns	All regions	Braidwood only	Tumut only	Kiandra only	Braidwood & Tumut	Braidwood & Kiandra	Tumut & Kiandra	Total
Nameable	8	1	2	8	0	0	2	21
Unidentifiable	0	35	19	87	1	2	1	145
Total	8	36	21	95	1	2	3	166

Table 13.12: Transfer-printed patterns from non-Chinese sites

Transfer-printed patterns	Kiandra only	Bungendore only	Kiandra & Bungendore	Total
Nameable	14	9	11	34
Unidentifiable	157	64	11	232
Total	171	73	22	266

The other notable similarities between British tea and tableware in this study are the plate profiles and the quality of the wares. A large number of plate profiles are similar to those dated by Moir (1997) 1825-1870s. This is consistent across the region when the same patterns are found at Chinese or non-Chinese sites. If Moir's dates are correct, it could indicate that consignments of second-hand wares were being brought by a supplier/s and sent to areas within southeast NSW. Alternatively, old stock from Britain was being dumped on the Australian market.

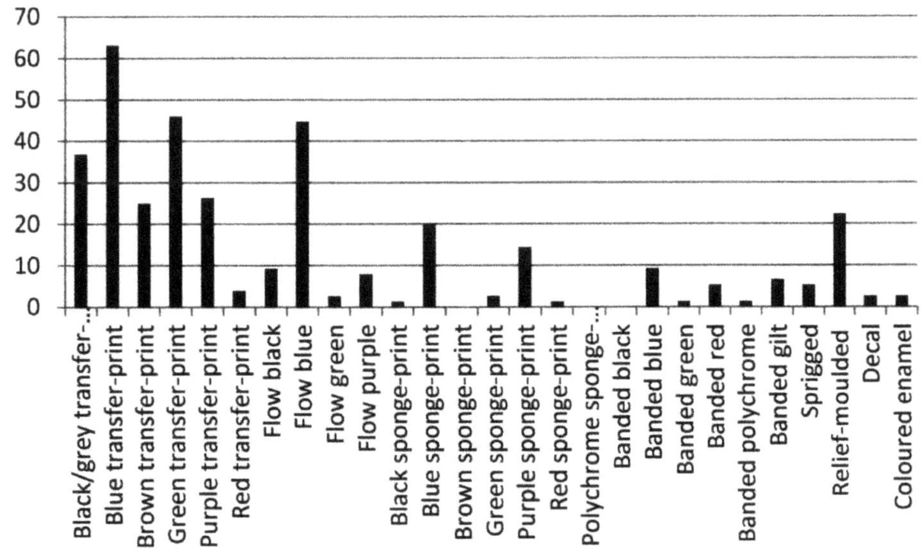

Figure 13.2: Decorative types at Chinese sites

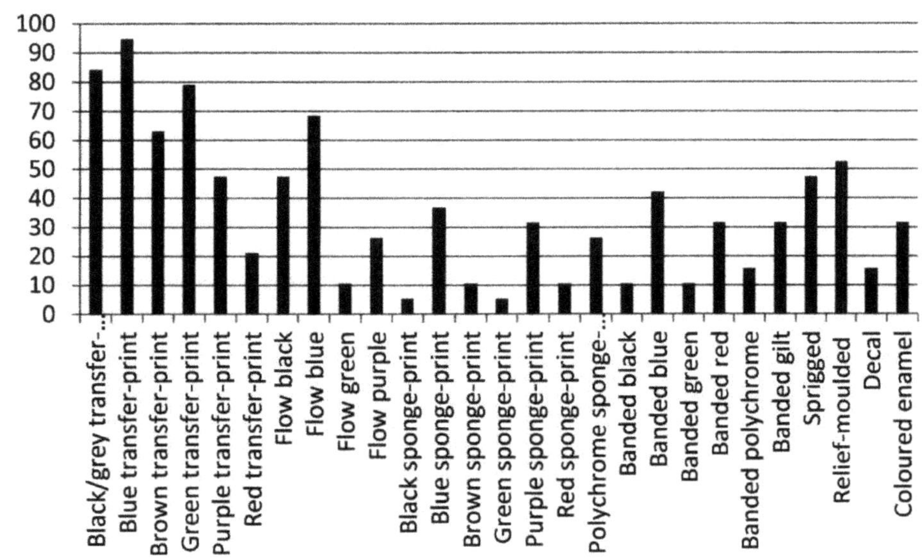

Figure 13.3: Decorative types at non-Chinese sites

Numerous sherds reveal manufacturing faults including parts of transfer-prints that are missing, patchy, smudged or wrongly joined; and unglazed patches or pooled glaze drips. It was thought that Chinese bought cheaper vessels because they were expected to be frugal in order to be able to send money back to China. However, the quality of wares is similar whether from Chinese or non-Chinese sites. While these faults are small and may not have been particularly noticeable, the purchase of 'second-grade and generally faulty goods' could result in considerable savings (Crook 2008:155). Nineteenth century quality management documents have shown a correlation between the quality of goods and their cost (Crook 2008). A mining community, needing quick supplies, would have been an ideal place for merchants to offload old or faulty stock.

The colour of vessels is similar at both Chinese and non-Chinese sites, showing the relative abundance of blue patterns, whether transfer-printed, sponge-printed, flown or banded, in comparison to other colours (Figure 13.4). However, the proportion of blue vessels is higher at the Chinese sites. This may reflect a cultural preference as a result of the lack of Chinese blue-underglaze decorated vessels or it could be that the blue vessels were cheaper because they were more common. Green is the second most common colour at the Chinese sites, while black/grey is the second choice for non-Chinese sites.

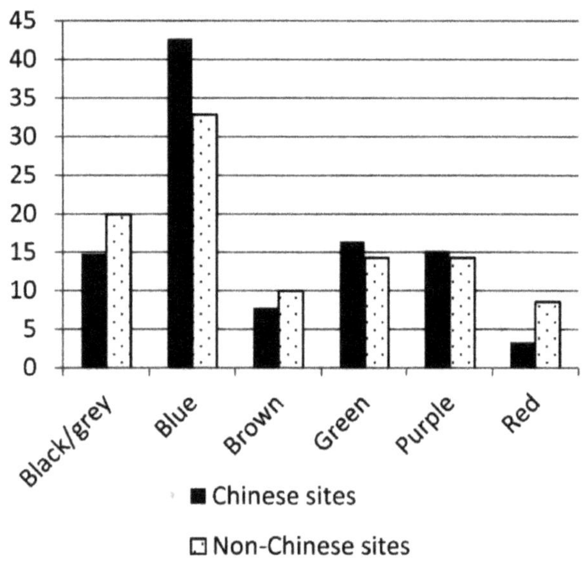

Figure 13.4: Percentage of vessels by colour

Bone china is generally only found at the later Chinese sites although it was present at the non-Chinese sites in the region from the 1860s. Elemental compositional analysis of a bone china sample from the Gold Commissioner's residence in Kiandra, occupied 1860-1862, shows it to be different to tested samples from Chinese sites.

It would appear that there are similarities in common decorative types, colour and quality of non-Chinese tea and tableware across all site types, suggesting that the Chinese had access to the same market as the non-Chinese inhabitants. Although the patterns and colours recovered from sites were determined largely by availability, vessel form was a deliberate choice as the Chinese maintained their traditional foodways.

Proportions of Chinese/non-Chinese vessels were seen as measures of acculturation when overseas Chinese sites were first examined in the 1970-1980s but researchers have now put forward a number of alternate explanations. In the 1850s-1860s merchants in Sacramento chose to use Western-style vessels to entertain American business associates because they were seen as 'familiar symbols of European culture' and their use gave the impression that the merchants were civilised (Praetzellis 1999:133). One of those merchants, Yee Ah Tye, moved from Sacramento to a small gold mining town to supply Chinese miners in the 1890s (Praetzellis & Praetzellis 2001:649). Although he considered himself American, he effectively prevented the miners from access to American goods by supplying their boarding house with Chinese-made wares. Only one British saucer was found, all the other ceramics were Chinese. At the same time in Australia, the residents of Melbourne's Chinatown used Western-style vessels as a way of reinforcing 'new cultural and social identities' (Muir 2008). In contrast to those studies, the ceramics from sites in this research indicate that Western-style vessels were used as a substitute for Chinese vessels that were no longer easily available.

CHAPTER 14

Conclusion

This research set out to explore the nature of Chinese gold miners' society in southeast NSW through their use of ceramic vessels. It has enabled conclusions to be drawn about the active role of vessels in everyday life, not only within the domestic sphere but also in communal aspects of food and feasting. On a broader scale, it has considered features of Chinese supply networks. It has used traditional and non-traditional methods of ceramic analysis to answer major questions which have expanded the archaeology of Chinese in Australia. The sites examined had previously been placed within a Chinese settlement pattern and were thought to have functioned as 'homogenous and segregated communities' until the end of the nineteenth century (Smith 2006:2). These sites include camps at Jembaicumbene, Flanagan's Point, Upper Adelong, Adjungbilly and Kiandra.

Although the archaeological study of overseas Chinese began in Australia in the 1980s, this is the first intra-site analysis of a Chinese camp. The comparison of individual hut sites has allowed an insight into social change at two major camps and has raised questions about the supply of Chinese-made vessels. This research has analysed all ceramics, not just those of Chinese origin. At many sites, Chinese ceramics are visually dominant as a result of the larger brown-glazed storage jar sherds and the durability of porcelain. The often-weathered earthenware, the most common non-Chinese made ceramic, is frequently overlooked. The inclusion of these has allowed inferences to be drawn about the Chinese supply network with regards to the changing proportions of Chinese/non-Chinese vessels over time. While this research has added to our understanding of the Chinese in Australia, it has also highlighted gaps in our present knowledge.

To date, little is known about the supply sources of Chinese ceramic wares to the overseas Chinese in Australia. The availability of ceramic collections from nineteenth century sites across southeast NSW has allowed the provenance of the Chinese vessels to be examined. It is generally thought that the brown-glazed wares, containing food and beverages, were made in the southern Chinese provinces of Guangdong, Fujian, Zhejiang and Jiangxi. This research, the first compositional analysis of post-1850 brown-glazed wares from overseas Chinese sites, has confirmed that the majority of wares were from multiple kilns in those provinces, but it has also revealed that some of the liquor bottles found in southeast NSW may have been made in northern China. A common belief is that the various shapes and sizes of these brown-glazed vessels have similar body fabrics and glazes, but this research has illustrated that this is not necessarily the case. Microscopic analysis has shown that visual similarities are not always related to chemical similarities. Compositional analysis has also established that the celadon-glazed vessels were made in southern China, possibly Jiangxi, in contrast to Stenger's (1993) suggestion that similar vessels found on American sites may have been made in Japan.

There is scant knowledge about the Chinese network's distribution of vessels throughout Australia in the nineteenth century, particularly with regards to the supply of tableware. Analysis has shown that differing Chinese decorative types and vessel forms are found across a range of settlement types. For example, the Four Seasons pattern is only found at the major camps and stores but not at the minor camps. In southeast NSW however, this decorative pattern has only been found on serving bowls at Kiandra while it is only on spoons at Jembaicumbene. At present, it is not known if this is a result of artefact recovery or if these areas were supplied differently. Some celadon-glazed forms, for example, large serving dishes and teacups, only appear at the later sites in the study area. Furthermore, little is known about the quantity of tableware supplied to minor camps. At Flanagan's Point, for example, no Chinese-made tableware sherds were recovered. There was evidence of fossicking at that site and it is possible that Chinese sherds may have been removed prior to the archaeological survey and excavations by Smith in 2002.

This research has revealed a change in the relative amounts of Chinese and non-Chinese vessels in the assemblages from southeast NSW. The earlier sites have a higher proportion of Chinese-made wares. Supply sources may have altered as the Chinese networks changed from being clan based, to become more commercial in the 1870s. This coincided with a change in the demography at some camps as many miners moved away and family homes were established. An analysis of the vessel types recovered has concluded that Western-style tea and tableware vessels were substituted for Chinese ones which may no longer have been available.

There is a higher proportion of non-Chinese tea and tableware at the later sites within the major camps and at the Chinese stores in Kiandra. A comparison of forms used by Chinese and non-Chinese households has confirmed that the former maintained their Chinese-style diet while using non-Chinese vessels, as their choice of vessel forms related directly to the food they consumed. Plates and saucers were used by the Chinese as serving vessels. Non-Chinese bowls, originally made as open vegetable serving bowls or waste bowls, were substituted for Chinese rice bowls as they were similar in size.

A comparison of the ceramic assemblages from the major and minor camps indicates that the nature of their dining may have been different. The major camps and later stores had serving bowls and dishes, but these vessels were not found at the minor camps. This appears to be the pattern throughout Australasia as these forms are generally found at Chinatowns and larger camps but not at smaller all-male camps. While Chinese serving vessels were only found at the major camps, non-Chinese plates and saucers, which could be used for the same purpose, were found at minor camps. At the minor camp of Flanagan's Point, for example, these vessels were found at the oven but not at hut sites. This suggests communal eating at that camp. Conversely, while this scenario may have occurred in the early phases of the major camps, the later occupants were more likely to have dined at home, their communal ovens only being used on festive occasions.

One of the aims of this research was to ascertain whether the Chinese had the same access to the ceramic market as non-Chinese residents. The availability of ceramic collections from a number of non-Chinese sites in southeast NSW has allowed a comparison between these assemblages and those of the Chinese. It appears that British decorative types found at Chinese sites are similar to those found at non-Chinese sites, suggesting their presence is more to do with availability than choice. Similar conclusions were drawn by Lawrence and her colleagues (2009:75) when considering the presence of matching sets reflecting the 'acquisition of the most readily available items' as some transfer patterns are consistently found at sites across Australia. Although the proportions of particular vessel forms differed between the Chinese and non-Chinese, the decoration on all forms was similar, along with the quality of the wares, confirming that the occupants of the examined sites had access to the same market.

The similarity of the examined assemblages was enforced by common plate profiles which aligned with Moir's (1997) study of those vessels over time. Although there was only a small sample of plates examined, this is the first research in Australia to conclude that plate profiles from sites within this country concur with those dated by Moir in the range of 1825 to 1870s. His study related to British plates from American sites and it is not known whether the dates given by Moir are accurate for Australia, but there appears to be a trend in plate shapes over time. More research is needed to ascertain that post-1870 plates in Australia are distinct from earlier plates. If Moir's dates are correct, then it suggests that older goods were dumped onto the Australian market, or alternatively, suppliers were buying consignments of second-hand wares to sell in regional NSW.

In summary, the analysis of ceramic artefacts has given an insight into the miners' lives and those of the later families at the settlements in this study. When the Chinese miners arrived in southeast NSW, they were under the control of a headman who supplied provisions such as food, rice bowls and opium. The ceramic vessels found at sites which were occupied in the early period of settlement reflect the control exercised by the headmen. The miners used familiar peasant-ware from China such as blue-underglaze rice bowls. The artefacts show that the temple and communal oven were places where social gathering reinforced a sense of community and identity. Over time, the camps underwent a change in their demography. The early Chinese supply network was succeeded by another which accessed the Western market, requiring those remaining at the camps to use Western-style vessels as a substitute for Chinese ones. Some of the later sites have children's ceramics, featuring English script, suggesting that parents wanted their children to be familiar with the language of their new country. That does not indicate assimilation but points to their ability to communicate with the non-Chinese community. The presence of European women at the Chinese camps is suggested through assemblages with matching crockery, not a common feature in the Chinese huts. Those huts would have been on the outskirts of the camps towards the end of their occupation, reflecting that interracial relationships did not fit wholly into the Chinese or non-Chinese community. Overall, this research has highlighted short and long-term occupation sites within these settlements. The camps were not homogenous or static, they changed over time.

Bibliography

Abbott, D.R., A.D. Lack and G. Moore. 2008. Chemical assays of temper and clay: modelling pottery production and exchange in the uplands north of the Phoenix Basin, Arizona, USA. *Archaeometry*, Vol. 50, No. 1, pp. 48-66.

Abrams, D. 1997. Tea leaf ironstone china. In D. Stoltzfus and J. Snyder. *White ironstone, a survey of its many forms*, p.13. Schiffer Publishing, Atglen.

Adan-Bayewitz, D., F. Asaro and R.D. Giauque. 1999. Determining pottery provenance: application of a new high-precision x-ray fluorescence method and comparison with instrumental neutron activation analysis. *Archaeometry*, Vol. 41, No. 1, pp. 1-24.

Adburgham, A. 1972. *Victorian shopping Harrods's catalogue 1895, a facsimile of the Harrods's Store 1895 issue of the price list*. David and Charles, Newton Abbot.

Allen, F.J. 1969. *Archaeology and the history of Port Essington*. Unpublished PhD Thesis. The Australian National University, Canberra.

Allison, P.M. and A. Cremin. 2006. Ceramics from the Old Kinchega homestead. *Australasian Historical Archaeology*, Vol. 24, pp. 55-64.

Anderson, E.N. and M.L. Anderson. 1977. Modern China: south. In K.C. Chang (ed.) *Food in Chinese culture*, pp. 317-382. Yale University Press, New Haven.

Arnold, D.E., H. Neff and R.L. Bishop. 1991. Compositional analysis and "sources" of pottery: an ethnoarchaeological approach. *American Anthropologist*, Vol. 93, pp. 70-90.

Arnold, K. 2006. *Old bottles*. Crown, Castleton.

Arun, L. 2012. *Opium pipe bowls from the Lawrence Chinese Camp*. Unpublished Postgraduate Diploma. University of Otago, Dunedin.

Atkins, M. 1991. *Not to be excelled for elegance or utility: a study of the availability of ceramics in Sydney 1803-1868*. Unpublished Honours Thesis. University of Sydney, Sydney.

Australian Town and Country Journal, 20 July 1872. The Chinese gold-diggers at Maitland Bar, p.18.

Bagnall, K. 2006. *Golden shadows on a white land, an exploration of the lives of white women who partnered Chinese men and their children in southern Australia, 1855-1915*. Unpublished PhD Thesis. University of Sydney, Sydney.

Banks, M.S. and J.M. Merrick. 1967. Further analysis of Chinese blue and white. *Archaeometry*, Vol. 10, pp. 101-103.

Barker, D. 2001. *The usual classes of useful articles*. Viewed 17 June 2009. <http://www.chipstone.org/publications/CIA/2001/Barker/BarkerIndex.html>

Barker, D. and T. Majewski. 2006. Ceramic studies in historical archaeology. In D. Hicks and M.C. Beaudry (eds.) *The Cambridge companion to historical archaeology*, pp. 13-33. Cambridge University Press, Cambridge.

Baxter, M.J. and C.M. Jackson. 2001. Variable selection in artefact compositional studies. *Archaeometry*, Vol. 43, pp. 253-268.

Beale, A. 2006. *Analysis of a ceramic assemblage from a 19th century goldfield in Northwest Queensland*. Unpublished Honours Thesis. Flinders University of South Australia, Adelaide.

Bell, P. 1995. Chinese ovens on mining settlement sites in Australia. In P. Macgregor (ed.) *Histories of the Chinese in Australasia and the South Pacific, Proceedings of an international public conference held at the Museum of Chinese Australian History, Melbourne, 8-10 October 1993*, pp. 213-229. Museum of Chinese Australian History, Melbourne.

Bell, P. 1996. Archaeology of the Chinese in Australia. *Australasian Historical Archaeology*, Vol. 14, pp. 13-18.

Bendigo Advertiser (Victoria: 1855-1918), 23 July 1859. Opening of the new Chinese Joss House at Ironbark, p.2. Viewed 3 November 2012. <http://nla.gov.au/nla.news-article87990820>

Bendigo Advertiser (Victoria: 1855-1918), 25 October 1859. The Chinese brickmakers, p.2. Viewed 29 October 2012. < http://nla.gov.au/nla.news-article87993053>

Bendigo Advertiser (Victoria: 1855-1918), 7 August 1866. Another ministerial visitor from New South Wales, p.2. Viewed 3 November 2012. < http://nla.gov.au/nla.news-article87962798>

Berthoud, M. 1990. *A Compendium of British cups*. Micawber Publications, Bridgnorth.

Birks, S. 2002. *North Staffordshire pottery marks, J and G Meakin (Ltd)*. Viewed July 2010. <http://www.thepotteries.org/mark/m/meakin_jg.html>

Birks, S. 2007. *Types and examples of pottery: bone china.* Viewed 14 December 2007. <http://www.thepotteries.org/types/bonechina.htm>

Birks, S. 2012. *1862 – Second investigation into child labour in the pottery industry.* Viewed 2 August 2012. <http://www.thepotteries.org/history/scriven_index.htm>

Birmingham, J. 1976. The archaeological contribution to nineteenth-century history: some Australian case studies. *World Archaeology*, Vol. 7, No. 3, pp. 306-317.

Birmingham, J. and K. Fahy. 1987 (1970). Old Australian pottery. In J. Birmingham and D. Bairstow (eds.) Papers in Australian historical archaeology, pp. 7-11. Reprinted from *The Newsletter of The Australian Society for Historical Archaeology, 1970-1982*. The Australian Society for Historical Archaeology Incorporated, Sydney.

Bishop, R., R. Rands and G. Holley. 1982. Ceramic compositional analysis in archaeological perspective. In M.B. Schiffer (ed.) *Advances in archaeological method and theory, Volume 5*, pp. 275-330. Academic Press, New York.

Bowen, A.M. 2007. *The Chinese involvement in Victoria's early fishing industry.* Unpublished PhD Thesis. La Trobe University, Bundoora.

Bowen, A.M. 2012. *Archaeology of the Chinese fishing industry in colonial Victoria.* Sydney University Press, Sydney.

Brabyn, J. 2007. *Stoneware in the Market Street Chinatown, a comparative analysis.* Viewed April 2009. <http://marketstreet.stanford.edu/2007/04/student-research-projects-2006-2007/>

Bray, T. 2003. The commensal politics of early states and empires. In T. Bray (ed.) *The archaeology and politics of food and feasting in early states and empires*, pp. 1-13. Plenum Publishing, New York.

Breward, C. 2004. Fashionable living. In M. Snodin and J. Styles (eds.) *Design and the decorative arts, Victorian Britain 1837-1901*, pp. 99-128. V and A Publications, London.

Briggs, S. 2005. *Portonian respectability: working-class attitudes to respectability in Port Adelaide through material culture 1840-1900.* Unpublished PhD Thesis. Flinders University of South Australia, Adelaide.

Brooks, A. 2005. *An archaeological guide to British ceramics in Australia 1788-1901.* The Australasian Society for Historical Archaeology, Sydney.

Brooks, A. 2007. The artefacts: ceramics. In G. Connah (ed.) *The same under a different sky? A country estate in nineteenth-century New South Wales*, pp. 183-194. BAR International Series 1625, Oxford.

Brooks, A. and G. Connah. 2007. A hierarchy of servitude: ceramics at Lake Innes Estate, New South Wales. *Antiquity*, Vol. 81, No. 3, pp. 133-147.

Brown, R. and S. Sjostrand. n.d. *Maritime archaeology and shipwreck ceramics in Malaysia.* Department of Museums and Antiquities in collaboration with Nanhai Marine Archaeology, Kuala Lumpur.

Buckrell Pos, T.M. 2004. *Tea and taste, the visual language of tea.* Schiffer Publishing, Atglen.

Burgh, R.F. 1959. Ceramic profiles in the Western Mound at Awatovi, Northeastern Arizona. *American Antiquity*, Vol. 25, No. 2, pp. 184-202.

Burton, J.H. and A.W. Simon. 1993. Acid extraction as a simple and inexpensive method for compositional characterization of archaeological ceramics. *American Antiquity*, Vol. 58, No. 1, pp. 45-59.

Burton, J.H. and A.W. Simon. 1996. A pot is not a rock: A Reply to Neff, Glascock, Bishop, and Blackman. *American Antiquity*, Vol. 61, No. 2, pp. 405-413.

Burton, W. 1906. *Porcelain, its art and manufacture.* B.T. Batsford, London. Viewed 26 July 2011. <http://www.ceramicstoday.com/articles/entrecolles_part1.htm>

Casey, M. 1999. Local pottery and dairying at the DMR Site, Brickfields, Sydney, New South Wales. *Australasian Historical Archaeology*, Vol. 17, pp. 3-37.

Casey and Lowe Pty Ltd. 2009. Unpublished pattern series list.

Chace, P. 1976. Overseas Chinese ceramics. In R.S. Greenwood (ed.) *The changing faces of Main Street,* Ventura Mission Project Archaeological Report, pp. 509-530. Redevelopment Agency, City of San Buenaventura, California.

Chan, H.D.M. 2001. Becoming Australasian but remaining Chinese: the future of the down under Chinese past. In H. Chan, A. Curthoys and N. Chiang (eds.) *The overseas Chinese in Australasia: History, settlement and interactions*, pp. 1-16. Interdisciplinary Group for Australian Studies, National Taiwan University, Taipei and Centre for the Study of the Chinese Southern Diaspora, The Australian National University, Canberra.

Chan, K.B. 1997. A family affair: migration, dispersal and the emergent identity of the Chinese cosmopolitan. *Diaspora*, Vol. 6, No. 2, pp. 195-214.

Chan, W.K.K. 1977. *Merchants, mandarins and modern enterprise in late Ch'ing China.* East Asian Research Center, Harvard University, Cambridge.

Chandler, J. 2005. *Chinese sojourners in Victoria, a collection of artefacts from the Upper Ovens Goldfields.* Unpublished Honours Thesis. La Trobe University, Bundoora.

Chang, K.C. 1977. Introduction. In K.C. Chang (ed.) *Food in Chinese culture*, pp. 1-22. Yale University Press, New Haven.

Cheng, L., X. Ding, S. Feng, H. Cheng, W. Zhang, and C. Fan. 2006. PIXE analysis of Chinese ancient greenish white porcelain. *Nuclear Instruments and Methods in Physics Research*, Vol. 244, pp. 409-411.

Choi, C.Y. 1975. *Chinese migration and settlement in Australia.* Sydney University Press, Sydney.

Cochran, M.D. and M.C. Beaudry. 2006. Material culture studies and historical archaeology. In D. Hicks and M.C. Beaudry (eds.) *The Cambridge companion to historical archaeology*, pp. 191-204. Cambridge University Press, Cambridge.

Comber, J. 1991. *Palmer Goldfield, Heritage Sites Study, Volume I.* Unpublished report to Queensland Department of Environment and Heritage.

Copeland, R. 2000 (1982). *Blue and white transfer-printed pottery.* Shire Publications, Princes Risborough.

Corcoran, A.M. 1993. *Chinese export porcelain in Australia.* Unpublished Master's Thesis. University of Sydney, Sydney.

Costello, J.G. and M.L. Maniery. 1988. *Rice bowls in the Delta, artifacts recovered from the 1915 Asian community of Walnut Grove, California.* Occasional Paper 16, Institute of Archaeology, University of California, Los Angeles.

Coysh, A.W. and R.K., Henrywood. 1982. *The dictionary of blue and white printed pottery 1780-1880.* The Antique Collectors' Club, Woodbridge.

Coysh, A.W. and R.K., Henrywood. 1989. *The dictionary of blue and white printed pottery 1780-1880 Volume II.* The Antique Collectors' Club, Woodbridge.

Crawford, J.D. 1877. *Notes by Mr. Crawford on Chinese immigration in the Australian colonies, September 1877.* Great Britain Foreign Office Confidential Prints, London.

Crook, P. 2000. Shopping and historical archaeology: exploring the contexts of urban consumption. *Australasian Historical Archaeology*, Vol. 18, pp. 17-28.

Crook, P. 2008. *Superior quality' exploring the nature of cost, quality and value in historical archaeology.* Unpublished PhD Thesis. La Trobe University, Bundoora.

Crook, P., L. Ellmoos and T. Murray. 2005. *Keeping up with the McNamaras, a historical archaeological study of the Cumberland and Gloucester Streets site, The Rocks, Sydney.* Volume 8 of the Archaeology of the Modern City Series, Historic Houses Trust of New South Wales, Sydney.

Crook, P., S. Lawrence and M. Gibbs. 2002. The role of artefact catalogues in Australian historical archaeology: a framework for discussion. *Australasian Historical Archaeology*, Vol. 20, pp. 26-38.

Curthoys, A. 2001. 'Chineseness' and Australian identity. In H. Chan, A. Curthoys and N. Chiang (eds.) *The overseas Chinese in Australasia: history, settlement and interactions*, pp. 16-29. Interdisciplinary Group for Australian Studies, National Taiwan University, Taipei and Centre for the Study of the Chinese Southern Diaspora, The Australian National University, Canberra.

Davies, L. 2012. *Ceramics from Lawrence Chinese Camp, preliminary study of ceramic remains excavated from the house lot of Sam Chew Lain.* Unpublished Honours Thesis. University of Otago, Dunedin.

Deetz, J. 1977. *In small things forgotten.* Anchor Books, New York.

Department of planning and community development, Victoria (DPCDV). 2012. *Chinese brickmaking kiln.* Viewed 29 October 2012. <http://www.dpcd.vic.gov.au/heritage/archaeology/Archaeology/chinese-brickmaking-kiln>

Diehl, M., J.A. Walters and J.H. Thiel. 1998. Acculturation and the composition of the diet of Tucson's overseas Chinese gardeners at the turn of the century. *Historical Archaeology*, Vol. 32, No. 4, pp. 19-33.

Dieringer, E. and B. Dieringer. 2001. *White ironstone china plate identification guide 1840-1890.* Schiffer Publishing, Atglen.

Douglas, R.K. 1895. *Society in China.* A.D. Innes and Co., London.

Drennan, R.D. 1996. *Statistics for archaeologists, a commonsense approach.* Plenum Press, New York.

Dunk, M. 2010. *Made in China. An analysis of the artefact assemblage from Atherton Chinatown, North Queensland.* Unpublished Honours Thesis. La Trobe University, Bundoora.

Esposito, V. 2008. *Ceramics Report, Hibernian Hotel, Portion 16, and Hennessy's Hut, Portion 265, Parish of Majura, County of Murray, New South Wales.* Unpublished report prepared for Archeo Analysis, Moruya, NSW.

Esposito, V. 2012. *Beyond the total: identifying inter- and intra-site variation through ceramic artefacts at Chinese goldmining settlements in southeast New South Wales, mid-nineteenth to early-twentieth centuries.* Unpublished PhD Thesis. The Australian National University, Canberra.

Etter, P. A. 1980. The West Coast Chinese and opium smoking. In R.L. Schuyler (ed.) *Archaeological perspectives on ethnicity in America*, pp. 97-101. Baywood Publishing, New York.

Evans, J. 1975. *Pattern, a study of ornament in Western Europe from 1180 to 1900, Volume II.* Hacker Art Books, New York.

Evans, W.S. Jr. 1980. Food and fantasy: material culture of the Chinese in California and the West, circa 1850-1900. In R.L. Schuyler (ed.) *Archaeological perspectives on ethnicity in America*, pp. 89-96 Baywood, New York.

Finch, I. 2010. *Konnyaku.* Viewed 30 November 2010. <http://www.chinese-porcelain-art.com/Antique_Japanese_Porcelain>

Fitts, R.K. 1999. 'The Archaeology of middle-class domesticity and gentility in Victorian Brooklyn. *Historical Archaeology*, Vol. 33, No. 1, pp. 39-62.

Ford, G. 1995. *Australian pottery: the first 100 years.* Salt Glaze Press, Wodonga.

Fournier, R. 1986. *Illustrated dictionary of pottery decoration.* Prentice Hall, New York.

Freestone, I.C., N.D. Meeks, and A.P. Middleton. 1985. Retention of phosphate in buried ceramics: an electron microbeam approach. *Archaeometry*, Vol. 27, No. 2, pp. 161-177.

Gammage, B. 1986. *Narrandera Shire.* Self-published for Narrandera Shire Council.

Gant-Thompson, C. 2008. *Unravelling Kiandra: tracing the threads of European and Chinese mining settlement in the landscape.* Unpublished Master's Thesis. The Australian National University, Canberra.

Gao, M. 2012. *Sojourners, Where is home?* Paper presented at the Australian Historical Association Conference, July 2012, Adelaide.

Gaughwin, D. 1995. Chinese settlement sites in north east Tasmania: an archaeological view. In P. Macgregor (ed.) *Histories of the Chinese in Australasia and the South Pacific, Proceedings of an international public conference held at the Museum of Chinese Australian History, Melbourne, 8-10 October 1993*, pp. 230-245. Museum of Chinese Australian History, Melbourne.

Gifford, J.C. 1951. The type-variety method of ceramic classification as an indicator of cultural phenomena. *American Antiquity*, Vol. 25, No. 3, pp. 341-347.

Gill, M.S. and T. Rehren. 2011. Material characterization of ceramic tile mosaic from two 17th century Islamic monuments in Northern India. *Archaeometry*, Vol. 53, No. 1, pp.22-36.

Godden, G.A. 1991 (1964). *Encyclopaedia of British pottery and porcelain marks.* Barrie and Jenkins, London.

Godden, G.A. 1999. *Godden's guide to ironstone stone and granite wares.* Antique Collectors' Club, Woodbridge.

Graham, Malcolm. 1908 (reprinted 2000). *Cup and saucer land.* Staffordshire and Stoke on Trent Archive Service, Stoke on Trent.

Graham, Marjorie. n.d. *Printed ceramics in Australia.* ASHA Occasional Paper No.2. The Australian Society for Historical Archaeology, University of Sydney, Sydney.

Grave, P., L. Lisle and M. Maccheroni. 2005. Multivariate comparison of ICP-OES and PIXE/PIGE analysis of east Asian storage jars. *Journal of Archaeological Science*, Vol. 32, pp. 885-896.

Greenwood, R.S. 1980. The Chinese on Main Street. In R.L. Schuyler (ed.) *Archaeological perspectives on ethnicity in America,* pp. 113-123. Baywood, New York.

Griffin, J.D. 200. *The Don Pottery.* Doncaster Museum Service, Doncaster.

Grimshaw, R.W. 1971 (4th edition revised). *The chemistry and physics of clays and allied ceramic materials.* Ernest Benn, London.

Grimwade, G. 1995. Of gods, timber and maize: cultural heritage management at Cedar Camp, Atherton Chinatown, Queensland. In P. Macgregor (ed.) *Histories of the Chinese in Australasia and the South Pacific, Proceedings of an international public conference held at the Museum of Chinese Australian History, Melbourne, 8-10 October 1993*, pp. 307-319. Museum of Chinese Australian History, Melbourne.

Grimwade, G. 2003. Gold, gardens, temples and feasts: Chinese temple, Croydon, Queensland. *Australasian Historical Archaeology*, Vol. 21, pp. 50-57.

Grimwade, G. 2008. Crispy roast pork: using Chinese Australasian pig ovens. *Australasian Historical Archaeology*, Vol. 26, pp. 21-28.

Grimwade, G. and M. Rowney. 2004. *Rusty's Markets re-development site archaeological report*. Unpublished report prepared by Gordon Grimwade and Associates for Gilligan's Development Pty. Ltd, Yungaburra.

Grimwade, G., M. Rowney and I. McCrae. 2007. *Salvage Archaeology Cooktown Sewerage Scheme*. Unpublished report prepared by Gordon Grimwade and Associates, for Cook Shire Council, Yungaburra.

Gulland, W.G. 1928 (5th edition). *Chinese porcelain*. Chapman and Hall, London.

Hall, E.T., F. Schweizer and P.A. Toller. 1973. X-ray fluorescence analysis of museum objects: a new instrument. *Archaeometry*, Vol. 15, pp. 53-78.

Ham, S.W., I. Shim, Y.E. Lee, J.Y., Kang and K. Koh. 2002. An archaeochemical microstructural study on Koryo inlaid celadon. *Bulletin of Korean Chemical Society*, Vol. 23, No. 1, pp. 1531-1540.

Hamer, F. and J. Hamer. 1997 (4th edition). *The potter's dictionary of materials and techniques*. A and C Black, London.

Hancock, R.G.V. 2000. Elemental analysis. In E. Ciliberto and G. Spoto (eds.) *Modern analytical methods in art and archaeology*, pp. 11-20. John Wiley and Sons, New York.

Hargrave, L.L. and W. Smith. 1936. A method for determining the texture of pottery. *American Antiquity*, Vol. 2, No. 1, pp. 32-36.

Harrison, B. 1995. *Later ceramics in South-East Asia*. Oxford University Press, London.

Harrison-Barbet, A. 1994. *Thomas Holloway Victorian philanthropist*. Royal Holloway, University of London, London.

Harrison-Hall, J. 1997. Chinese porcelain from Jingdezhen. In I. Freestone and D. Gaimster (eds.) *Pottery in the making*, pp. 194-199. British Museum Press, London.

Hayes, S. 2008. *Being middle class, an archaeology of gentility in nineteenth-century Australia*. Unpublished PhD Thesis. La Trobe University, Bundoora.

Heffernan, K.J. 1996. Introduction and overview. In K.J. Heffernan and L.M. Smith (eds.) *Archaeological investigations at Kiandra, Kosciusko National Park, II: excavations at Chinese emporia, mining races and Matthews' cottage, 1996*, pp. 1-12. Department of Archaeology and Anthropology, The Australian National University, Canberra.

Hellmann, V. and J. Yang. 1997 (reprinted 2007). Previously undocumented Chinese artifacts. In A. Praetzellis and M. Praetzellis (eds.) *Historical archaeology of an overseas Chinese community in Sacramento, California, Volume 1, Archaeological Excavations*, pp. 155-197. Anthropological Center, Sonoma State University Academic Foundation, Inc. Rohnert Park, California.

Hetherington, A.L. 1921. *The pottery and porcelain factories of China*. Kegan Paul, Trench, Trubner and Co., London.

Historical Census and Colonial Data Archive (HCCDA) 2010a. Census of New South Wales 1871—Nationality. Braidwood Registry District— Jembaicumbene Gold Field. Viewed 28 October 2010. <http://hccda.anu.edu.au/pages/NSW-1871-census-03_365>

Historical Census and Colonial Data Archive (HCCDA) 2010b. Census of New South Wales 1871—Nationality. Cooma Registry District— Kiandra Gold Field—Village of Kiandra and Environs. Viewed 28 October 2010. <http://hccda.anu.edu.au/pages/NSW-1871-census-03_376>

Ho, C. 1988. *Minnan blue-and-white wares, an archaeological survey of kiln sites of the 16th-19th centuries in southern Fujian, China*. BAR International Series 428, Oxford.

Hodgson, W. 1905. *How to identify old Chinese porcelain*. Methuen and Co, London.

Hoh, S. 1933. Pottery Industry in Shek-Waan, Kwangtung. *Lingnan Science Journal*, Vol. 12, pp. 57-64.

Holme, W.A. n.d. *Diary, 22 Oct. 1860 - 31 Jan. 1861*. Copy in Mitchell Library, Sydney. Original held by Mrs S. Mitchell, Casino.

Hommel, R.P. 1937. *China at work*. John Day Company, New York.

Honey, W.B. 1946. *The ceramic art of China and other countries of the Far East*. Faber and Faber, London.

Hsiao, K.C. 1960. *Rural China, imperial control in the nineteenth century*. University of Washington Press, Seattle.

Hueneke, K. 1987. *Kiandra to Kosciusko*. Tabletop Press, Canberra.

Hughes, G.B. n.d. *English and Scottish earthenware 1660-1860*. Abbey Fine Arts, London.

Hui, P. 1997. The development of Hong Kong Chinese business in the mid-19th to the early 20th century: a transnational perspective. *China Information*, Vol. 12, pp. 114-134. Viewed 15 November 2011. <http://cin.sagepub.com/content/12/1-2/114.refs.html>

Ioannou, N. 1988. 200 years of Australian clay culture. *Pottery in Australia*, Vol. 27, No. 2, pp. 65-80.

Jack, I. 2001. Some less familiar aspects of the Chinese in 19th-century Australia. In H. Chan, A. Curthoys and N. Chiang (eds.) *The overseas Chinese in Australasia: history, settlement and interactions*, pp. 44-53. Interdisciplinary Group for Australian Studies, National Taiwan University, Taipei and Centre for the Study of the Chinese Southern Diaspora, The Australian National University, Canberra.

Jack, I., K. Holmes and R. Kerr. 1984. Ah Toy's garden: a Chinese market-garden on the Palmer River Goldfield, North Queensland. *Australian Historical Archaeology*, Vol. 2, pp. 51-58.

Jackson, A. 1992. The Victorian perception and acquisition of Japanese culture. *Journal of Design History*, Vol. 5, No. 4, pp. 245-256.

Kaplan, S. M. 1952. Toward a classification of Chinese glazes a preliminary report. *Far Eastern Ceramic Group*, Vol. 4, No. 2, pp. 780-791.

Kelloway, S. 2008. *King of Irrawang, chemical analyses of colonial ceramics*. Unpublished Honours Thesis. University of Sydney, Sydney.

Kelly, H.E., A.A. Kowalsky and D.E. Kowalsky. 2001. *Spongeware 1835-1935 makers, marks, and patterns*. Schiffer Publishing, Atglen.

Koh Choo, C.K., Y.E. Lee, I.W. Shim, W.K. Choo, G.H. Kim, W.Y. Huh and S.C. Chun. 2004. Compositional and microstructural study of Koryo celadon and whiteware excavated from Sori kiln in Kyonggi Province. *Archaeometry*, Vol. 46, No. 2, pp. 247-265.

Kondo, S. 1923. The development of ceramic technology and science in Japan. *Journal of the American Ceramic Society*, Vol. 6, No. 1, pp. 212-218.

Kovel, R. and T. Kovel. 1986. *Kovels' new dictionary of marks*. Random House Reference, New York.

Kowalsky, A.A. and D.E. Kowalsky. 1999. *Encyclopedia of marks on American, English and European earthenware, ironstone, and stoneware 1780-1980*. Schiffer Publishing, Atglen.

Kuwayama, G. 1989. The significance of Chinese ceramics in the East and West. In S. Kotz (ed.) *Imperial taste: Chinese ceramics from the Percival David Foundation*, pp. 105-114. Chronicle Books, San Francisco.

Laird, C.N. 1918. The potteries at Shek Waan, near Canton, China. *The Journal of Industrial and Engineering Chemistry*, Vol. 10, No. 7, pp. 568-571.

Lau, M. 1979. Introduction. *Exhibition of Shiwan Wares Fung Ping Shan Museum*, pp. 10-12. University of Hong Kong, Hong Kong.

Lawrence Cheney, S. 1995. *No abiding city: the archaeology and history of an ephemeral mining settlement*. Unpublished PhD Thesis. La Trobe University, Bundoora.

Lawrence, S., 1999. Towards a feminist archaeology of households: gender and household structure on the Australian goldfields. In P.M. Allison (ed.) *The archaeology of household activities*, pp.121-141. Routledge, London.

Lawrence, S., A. Brooks and J. Lennon. 2009. Ceramics and status in regional Australia. *Australasian Historical Archaeology*, Vol. 27, pp. 67-78.

Lawrence, S. and P. Davies. 2011. *An archaeology of Australia since 1788*. Springer, New York.

Lentz, F.B. 1920. The world's ancient porcelain center. *National Geographic*, pp. 391-406. Viewed 24 March 2010. <http://gotheborg.com/natgeograph/index.shtml>

Leung, P.L., S.J. Stokes, T. Chen and D. Qin. 2000. A study of ancient Chinese porcelain wares of the Song-Yuan Dynasties from Cizhou and Ding kilns with energy dispersive x-ray fluorescence. *Archaeometry*, Vol. 42, No. 1, pp. 129-140.

Lindberg, G. 1947. Porcellanous ware – porcelain. *Ethnos*, Vol. 12, N. 3. pp. 95-105.

Little, W.L. 1969. *Staffordshire blue*. Crown Publishers, New York.

Lockett, T.A. 1996. *Pearlware: origins and types*. Viewed 31 March 2007. <http://www.thepotteries.org/features/pearlware.htm>

Low Choy, D. 2012. *Sojourners, settlers, selectors and subjects: interpreting a Queensland Chinese family history through a palimpsest approach*. Paper presented at the Australian Historical Association Conference, July 2012, Adelaide.

Lucas, M.T. 1994. À la russe, à la pell-mell, or à la practical: ideology and compromise at the late nineteenth-century dinner table. *Historical Archaeology*, Vol. 28, No. 4, pp. 80-93.

Lydon, J. 1996. *'Many inventions' historical archaeology and the Chinese in the Rocks, Sydney, 1890-1930.* Unpublished Master's Thesis. The Australian National University, Canberra.

Lydon, J. 1999. *'Many inventions' the Chinese in the Rocks, Sydney, 1890-1930.* Monash Publications, Clayton.

Macgregor, P. 2012. Lowe Kong Meng and Chinese engagement in the international trade of colonial Victoria. *Provenance*, Vol. 11. Viewed 1 November 2012. <http://prov.vic.gov.au/publications/provenance/lowe-kong-meng>

McCarthy, J. 1988. The new gold mountain: Chinese non-settlement in Northern Australia. In J. Birmingham, D. Bairstow and A. Wilson (eds.) *Archaeology and colonisation: Australia in a world context: selected papers from the Seventh Annual Conference of the Australian Society for Historical Archaeology*, pp. 139-147. Sydney, September 1987.

McCarthy, J. 1995. Tales from the Empire City: Chinese miners in the Pine Creek region, Northern Territory, 1872-1915. In P. Macgregor (ed.) *Histories of the Chinese in Australasia and the South Pacific, Proceedings of an international public conference held at the Museum of Chinese Australian History, Melbourne, 8-10 October 1993*, pp. 191-202. Museum of Chinese Australian History, Melbourne.

McCrae, I. 2001. *Canton to Cooktown: a study of Chinese ceramic shipping containers.* Unpublished Thesis. Armidale: The Teaching and Learning Centre, University of New England, Armidale.

McGowan, B. 1996. *Bungonia to Braidwood.* Self-published, Canberra.

McGowan, B. 2004a. Reconsidering race, the Chinese experience on the goldfields of southern New South Wales. *Australian Historical Studies*, Vol. 36, pp. 312-331.

McGowan, B. 2004b. The Chinese on the Braidwood goldfields: historical and archaeological opportunities. *Journal of Australian Colonial History*, Vol. 6, pp. 35-58.

McGowan, B. 2010. *Tracking the dragon, a history of the Chinese in the Riverina.* Museum of the Riverina, Wagga Wagga.

Majewski, T. and M.J., O'Brien. 1987. The use and misuse of nineteenth-century English and American ceramics in archaeological analysis. In M.B. Schiffer (ed.) *Advances in archaeological method and theory, Vol. 11*, pp. 97-209. Academic Press, New York.

Markus, A. 2001. Government control of Chinese immigration to Australia, 1855-1975. In H. Chan, A. Curthoys and N. Chiang (eds.) *The overseas Chinese in Australasia: history, settlement and interactions*, pp. 69-81. Interdisciplinary Group for Australian Studies, National Taiwan University, Taipei and Centre for the Study of the Chinese Southern Diaspora, The Australian National University, Canberra.

Martinez, J. 2011. *The Darwin KMT and the politics of speaking English.* Paper presented at Dragons Tails 2011, 2nd Australasian conference on overseas Chinese history and heritage, Chinese Museum, Melbourne.

Matlach, M. 2010. *S. Maw, Son & Sons Ltd.* Viewed 23 March 2011. <http://cosgb.blogspot.com/2010/10/s-maw-son-sons-ltd.html>

Meacham, S. 2003. Treasures of a life long past. *The Sydney Morning Herald*, 7 March, 2003.

Medley, M. 1976. *The Chinese potter*. Phaidon, Oxford.

Miller, G.L. 1980. Classification and economic scaling of 19th century ceramics. *Historical Archaeology*, Vol. 14, pp. 1-40.

Miller, G.L. 1984. George M. Coates, pottery merchant of Philadelphia. *Winterthur Portfolio*, Vol. 19, No. 1, pp. 37-49.

Miller, G.L. 1991. A revised set of CC index values for classification and economic scaling of English ceramics from 1787 to 1880. *Historical Archaeology*, Vol. 25, pp. 1-25.

Miller, G.L. 2004. *Smear glaze or dyed body wares.* Viewed 10 October 2006. <http://www.greatestjournal.com/community/potterynews/>

Moir, R. 1997. *Exploring creamware-pearlware-whiteware-ironstone categories through form-specific attributes: basic concepts in ceramic analyses and interpretation.* Unpublished round table discussion paper. Society of Historical Archaeology conference, January 1997, Corpus Christi, Texas.

Morrow, T.A. 2009. *Stoneware body and soul, a social interpretation of the Chinese stoneware record in British Columbia 1858-1958.* Unpublished Master's Thesis. University of British Columbia, Okanagan.

Motono, E. 2000. *Conflict and cooperation in Sino-British business 1860-1911.* MacMillan Press, Basingstoke.

Moye, D.G. 1959. *Historic Kiandra.* The Cooma-Monaro Historical Society, Cooma.

Mueller, F.W. 1987. Asian Tz'u: porcelain for the American market. In Great Basin Foundation (ed.) *Wong Ho Leun: An American Chinatown, Volume 1, Archaeology*, pp. 259-311. Great Basin Foundation, San Diego.

Muir, A.L. 2008. *Kitchen Ch'ing: Chinese archaeological ceramics in Victoria.* Unpublished Master's Thesis. La Trobe University, Bundoora.

Myles, V. 2004. *Viewing the blacksmith through his art: an archaeometallurgic portrait of iron working in Kiandra, NSW.* Unpublished Honours Thesis. The Australian National University, Canberra.

National Council of the Pottery Industry (NCPI). 1921-1922. *Pottery industry in Japan*, digitised 2007. Viewed 11 January 2010. <http://www.archive.org/details/informationrelat00natiuoft>

Neale, G. 2005. *Miller's encyclopedia of British transfer-printed pottery patterns 1790-1930.* Miller's, London.

New Zealand Historical Ceramics Database (NZHCD). 2012. Viewed from November 2007. <www.bickler.co.nz/china>

Niemeier, J. 1995. The changing role of the See Yup temple in Melbourne, 1866-1993. In P. Macgregor (ed.) *Histories of the Chinese in Australasia and the South Pacific, Proceedings of an international public conference held at the Museum of Chinese Australian History, Melbourne, 8-10 October 1993*, pp. 327-337. Museum of Chinese Australian History, Melbourne.

Nikko China. 2010. *The history of Nikko china.* Viewed 11 January 2010. <http://nikko-china.com/>

Noël Hume, I. 1976. *A guide to artifacts of colonial America.* Alfred A Knopf, New York.

Noël Hume, I. 2001. *If these pots could talk.* Chipstone Foundation, Milwaukee.

Noritake Co. 2010. *History of Noritake.* Viewed 11 January 2010. <http://www.noritake.co.jp/eng/company/about/history.html>

North American Pioneer Chinese Virtual Museum (NAPCVM). 2010. Viewed 4 November 2010. <http://www.chinesecol.com/treasure11.html>

O'Hoy, D.R. 1989. *Bendigo Pottery.* Bendigo Art Gallery, Bendigo.

Olsen, J.W. 1978. A study of Chinese ceramics excavated in Tucson. *The Kiva*, Vol. 44, No. 1, pp. 1-50.

Orton, C., P. Tyers and A. Vince. 1993. *Pottery in archaeology.* Cambridge University Press, Cambridge.

Osborne, D.A. 2008. *The Archaeology of a Riverina Chinatown at Narrandera.* Unpublished Honours Thesis. The Australian National University, Canberra.

Owen, J.V. 2001. Geochemical and mineralogical distinctions between Bonnin and Morris (Philadelphia, 1770-1772) porcelain and some contemporary British phosphatic wares. *Geoarchaeology: An International Journal*, Vol. 16, No. 7, pp.785-802.

Pastron, A.G., Gross, R. and D. Garaventa, 1981. Ceramics from Chinatown's tables: an historical archaeological approach to ethnicity and Appendix C. In A.G. Pastron, J. Prichett and M. Ziebarth (eds.) *Behind the seawall: historical archaeology along the San Francisco waterfront*, 3 Volumes, pp. 365-469 and 653-682. Report prepared by Archeo-Tec for the San Francisco Clean Water Program.

Peake, T.H. 2004. *An illustrated guide to Brownfield porcelain and majolica 1971-1900, Vol. 2.* Self-published.

Pettigrew, J. 2003. *Design for tea, tea wares from the Dragon Court to afternoon tea.* Sutton Publishing, Stroud.

Pickford, N., M. Hatcher and D. Freedman. 2000. *The legacy of the Tek Sing, China's Titanic – its tragedy and its treasure.* Granta Editions, Chesterton.

Piper, A. 1988. Chinese diet and cultural conservatism in nineteenth-century southern New Zealand. *Australian Historical Archaeology*, Vol. 6, pp. 34-42.

Pollard, A.M. and H. Hatcher. 1986. The chemical analysis of oriental ceramic body compositions: Part 2-greenwares. *Journal of Archaeological Science*, Vol. 13, pp. 261-287.

Pollard, A.M. and H. Hatcher. 1994. The chemical analysis of oriental ceramic body compositions: Part 1: wares from North China. *Archaeometry*, Vol. 36, No. 1, pp. 41-62.

Praetzellis, A. 1999. The archaeology of ethnicity: an example from Sacramento, California's early Chinese district. In G. Egan (ed.) *Old and new worlds*, pp.127-135. Oxbow, Oxford.

Praetzellis, A. and M. Praetzellis. 1992. Faces and facades: Victorian ideology in early Sacramento. In A.E. Yentsch and M.C. Beaudry (eds.) *The art and mystery of historical archaeology, essays in honor of James Deetz*, pp. 75-99. CRC Press, Boca Raton.

Praetzellis, A. and M. Praetzellis. 1997. *Historical archaeology of an overseas Chinese community in Sacramento, California, Volume 1: archaeological excavations*. Anthropological Studies Centre, Sonoma State University Academic Foundation, Inc. Rohnert Park.

Praetzellis, A. and M. Praetzellis. 2001. Mangling symbols of gentility in the Wild West: case studies in interpretive archaeology. *American Anthropologist*, Vol. 103, No. 3, pp. 645-654.

Praetzellis, A., M. Praetzellis and M. Brown. 1987. Artifacts as symbols of identity: an example from Sacramento's gold rush era Chinese community. In E. Staski (ed.) *Living in cities, current research in urban archaeology,* pp. 38-47. Society for Historical Archaeology, Pleasant Hill.

Preshaw, G.O. 1860. Wild Kiandra. In N. Keesing 1981 (ed.) *History of the Australian gold rushes by those who were there*, pp. 243-248. Angus and Robertson, Sydney.

Quellmalz, C.R. 1976. Late Chinese provincial export wares. *Oriental Art*, Vol. 22, pp. 289-298.

Rains, K. 2005. *Intersections: the overseas Chinese social landscape of Cooktown, 1873-1935*. Unpublished PhD Thesis. University of Queensland, Brisbane.

Reed, S.J.B. 1996. *Electron microprobe analysis and scanning electron microscopy in geology*. Cambridge University Press, Cambridge.

Ritchie, N.A. 1986. *Archaeology and history of the Chinese in southern New Zealand during the nineteenth century: a study of acculturation, adaptation and change*. Unpublished PhD Thesis. University of Otago, Dunedin.

Rye, O.S. 1981. *Pottery technology principles and reconstruction*. Taraxacum, Washington.

Samford, P.M. 1997. Response to a market: dating English underglaze transfer-printed wares. *Historical Archaeology*, Vol. 31, No. 2, pp. 1-30.

Sando, R.A. and D.L. Felton. 1993. Inventory records of ceramics and opium from a nineteenth century Chinese store in California. In P. Wegars (ed.) *Hidden heritage: historical archaeology of the overseas Chinese*, pp. 151-176. Baywood Publishing, New York.

Searle, J. and L.M. Smith. 2001. The Oven. In Smith L.M. (ed.) *Archaeological investigations at the Kiandra Chinese Camp, Kosciuszko National Park, NSW, a report on the ANU field school, February 2001*. The Australian National University, Canberra.

Shaoqing, C. 2002 (translated by D. Campbell). On the overseas Chinese secret societies of Australia. *New Zealand Journal of Asian Studies*, Vol. 4, No. 1, pp. 30-45.

Shennan, S. 1988. *Quantifying archaeology*. Edinburgh University Press. Edinburgh.

Shepard, A.O. 1956. *Ceramics for the archaeologist*. Carnegie Institute of Washington, Washington D.C.

Shipkowitz, D. and I. Shipkowitz. 2002. *The ABC's of ABC ware*. Schiffer Publishing, Atglen.

Singer, F. and S.S. Singer. 1963. *Industrial ceramics*. Chapman and Hall, London.

Skinner, G.W. 1964. Marketing and social structure in rural China: Part 1. *The Journal of Asian Studies*, Vol. 24, No. 1, pp. 3-43.

Smeaton, T.D. 1865. The great Chinese invasion. In N. Keesing 1981 (ed.) *History of the Australian Gold Rushes by Those Who were there*, pp. 118-121. Angus and Robertson, Sydney.

Smith, L.M. 1996. Brief history of Kiandra. In K.J. Heffernan and L.M. Smith (eds.) *Archaeological investigations at Kiandra, Kosciusko National Park, II: excavations at Chinese emporia, mining races and Matthews' cottage, 1996*, pp. 13-18. Department of Archaeology and Anthropology, The Australian National University, Canberra.

Smith, L.M. 1997. *The Chinese of Kiandra, New South Wales*. Report to the Heritage Office of the NSW, Department of Urban Development and Planning, Sydney.

Smith, L.M. 1998. *Cold hard cash: A study of Chinese ethnicity in archaeology at Kiandra, New South Wales*. Unpublished Master's Thesis. The Australian National University, Canberra.

Smith, L.M. 2006. *Hidden dragons, the archaeology of mid to late nineteenth-century Chinese communities in southeastern New South Wales*. Unpublished PhD Thesis. The Australian National University, Canberra.

Snodgrass, G. 2005. *Bohemian and Czechoslovakian ceramics and porcelain*. Viewed June 2010. <http://www.collectorscircle.com/bohemian/bohemian_china.html>

Snyder, J.B. 1997. *Romantic Staffordshire ceramics*. Schiffer Publishing, Atglen.

Snyder, J.B. 2004 (1992). *Flow blue*. Schiffer Publishing, Atglen.

Speltz, A. 1959. *The styles of ornament*. Dover Publications, New York.

Spence, J. 1977. Ch'ing. In K.C. Chang (ed.) *Food in Chinese culture*, pp. 259-294. Yale University Press, New Haven.

Spier, R.F.G. 1958. Food habits of nineteenth-century California Chinese. *California Historical Society Quarterly*, Vol. 37, pp. 79-86.

Stanin, Z. 2004. From Li Chun to Yong Kit: a market garden on the Lodden, 1851-1912. *Journal of Colonial History*, Vol. 6, pp. 15-34.

StatSoft, Inc. 2010. Electronic Statistics Textbook. Tulsa. Viewed April 2010. <http://www.statsoft.com/textbook/>

Steele, H. 1993. The manganese/cobalt ratio in nineteenth and twentieth century Asian porcelain. In P. Wegars (ed.) *Hidden heritage: historical archaeology of the overseas Chinese*, pp. 307-314. Baywood Publishing, New York.

Stenger, A. 1993. Sourcing and dating Asian porcelains by elemental analysis. In P. Wegars (ed.) *Hidden heritage: historical archaeology of the overseas Chinese*, pp. 315-331. Baywood Publishing, New York.

Stitt, I. 1974. *Japanese ceramics of the last 100 years*. Crown, New York.

Stoltzfus, D. and J. Snyder. 1997. *White ironstone, a survey of its many forms*. Schiffer Publishing, Atglen.

Summerhayes, G. 2000. *Lapita interaction*. Terra Australis 15, ANH Publications and The Centre for Archaeological Research, The Australian National University, Canberra.

Sussman, L. 1985. *The Wheat pattern an illustrated survey*. National Historic Parks and Sites Branch, Parks Canada, Quebec.

Tayler, J.B. 1930. The Hopei Pottery industry and the problem of modernisation. *The Chinese Social and Political Science Review*, Vol. 14, No. 2, pp. 184-209.

The Argus (Victoria: 1848-1957), 13 December 1866. The new Chinese Joss-house on Emerald-Hill, p.5. Viewed 3 November 2012. <http://nla.gov.au/nla.news-article5781017>

The Empire (Sydney, NSW: 1850-1875), 23 November 1861. Tumut, p.3. Viewed 17 January 2012. <http://nla.gov.au/nla.news-article60484220>

The Maitland Mercury and Hunter River General Advertiser (NSW: 1843-1893), 27 April 1861. Sydney news, p.6. Viewed 11 November 2012. <http://nla.gov.au/nla.news-article18681561>

The Mercury (Hobart, Tasmania: 1860- 954), 27 May 1864. Victoria, p.3. Viewed 3 November 2012. <http://nla.gov.au/nla.news-article8826436>

The Overseas Chinese Affairs Office of the State Council (TOCAOSC). 2006. *Common knowledge about Chinese culture*. Higher Education Press.

The Southeast Asian Ceramic Society (ed.) *Nonya ware and Kitchen Ch'ing*. Oxford University Press, West Malaysia Chapter, Kuala Lumpur.

The Sydney Morning Herald (NSW: 1842-1954), 25 October 1859. Chinese Joss-house, p.2. Viewed 16 March 2011. <http://nla.gov.au/nla.news-article13032205>

The Sydney Morning Herald (NSW: 1842-1954), 10 January 1860. Country news, p.2. Viewed 12 November 2012. <http://nla.gov.au/nla.news-article13035383>

The Sydney Morning Herald (NSW: 1842-1954), 21 August 1860. Our gold fields, p.7. Viewed 4 December 2009. < http://nla.gov.au/nla.news-article13044642>

The Sydney Morning Herald (NSW: 1842-1954), 5 November 1860. Braidwood gold-fields, p.8. Viewed 10 March 2011. < http://nla.gov.au/nla.news-article13048187>

The Sydney Morning Herald (NSW: 1842-1954), 11 May 1861. Kiandra, p.7. Viewed 25 May 2010. <http://nla.gov.au/nla.news-article13066968>

The Sydney Morning Herald (NSW: 1842-1954), 31 August 1861. Kiandra, p.5. Viewed 28 May 2010. <http://nla.gov.au/nla.news-article1305824>

The Sydney Morning Herald (NSW: 1842-1954), 7 February 1863. Death of a Chinaman under suspicious circumstances, p.7. Viewed 10 March 2011. < http://nla.gov.au/nla.news-article13073782>.

The Sydney Morning Herald (NSW: 1842-1954), 19 July, 1865. Random notes by a wandering reporter, p.5. Viewed 10 March 2011.
< http://nla.gov.au/nla.news-article31124787>

The Sydney Morning Herald (NSW: 1842-1954), 6 October 1865. The Chinese in Victoria: their manners, customs, &c., p. 2. Viewed 16 March 2011.
< http://nla.gov.au/nla.news-article13119929>

The Sydney Morning Herald (NSW: 1842-1954), 22 November 1866. Braidwood, p.5. Viewed 15 December 2011.
<http://nla.gov.au/nla.news-article13142796>

The Sydney Morning Herald (NSW: 1842-1954), 26 April 1867. A day in Canton, p.6. Viewed 16 March 2011.
<http://nla.gov.au/nla.news-article13142449>

The Sydney Morning Herald (NSW: 1842-1954), 8 January 1869. The progress of Mongolian colonisation, p.5. Viewed 10 March 2011.
< http://nla.gov.au/nla.news-article13187496>

The Sydney Morning Herald (NSW: 1842-1954), 11 April 1871. The Chinese murder at Kiandra, p.2. Viewed 7 April 2011.
< http://nla.gov.au/nla.news-article13222814>

The Sydney Morning Herald (NSW: 1842-1954), 10 July 1877. Braidwood, p.5. Viewed 10 March 2011. <http://nla.gov.au/nla.news-article13396737>

The Sydney Morning Herald (NSW: 1842-1954), 14 September 1886. Country news, Kiandra, p.7. Viewed 4 December 2009.
< http://nla.gov.au/nla.news-article13641378>

Thompson, C. and A. Wilson. 1987. A guide to the identification of nineteenth century ceramic body types. *The Australian Society for Historical Archaeology Research Bulletin*, Vol. 5, pp. 1-4.

Tite, M.S. 2008. Ceramic production, provenance and use – a review. *Archaeometry*, Vol. 50, No. 2, pp. 216-231.

Tite, M.S., I.C. Freestone and M. Bimson. 1984. A technological study of Chinese porcelain of the Yuan Dynasty. *Archaeometry*, Vol. 26, No. 2, pp. 139-154.

Trueangel, J. A. 1996. Surface archaeological features. In Heffernan, K. J. and L.M. Smith (eds.) *Archaeological investigations at Kiandra, Kosciusko National Park, II: excavations at Chinese emporia, mining races and Matthews' cottage, 1996*, pp. 23-29. Department of Archaeology and Anthropology, The Faculties, Australian National University, Canberra.

Underhill, A.P., G.M. Feinman, L.M. Nicholas, H. Fang, F. Luan, H. Yu and F. Cai. 2008. Changes in regional settlement patterns and the development of complex societies in southeastern Shandong, China. *Journal of Anthropological Archaeology*, Vol. 27, pp. 1-29.

Vader, J. and B. Murray. 1979. *Antique bottle collecting in Australia.* Summit Books, Sydney.

Van Buskirk, W.H. 2002. *Late Victorian flow blue and other ceramic wares: a selected history of potteries and shapes.* Schiffer Publishing, Atglen.

Voss, B.L. and R. Allen. 2008. Overseas Chinese archaeology: historical foundations, current reflections, and new directions. *Historical Archaeology*, Vol. 42, No. 3, pp. 5-28.

Votes and Proceedings of the New South Wales Legislative Assembly: Report on Chinese 1878-1879, Police Department, Inspector General's Office, Sydney.

Wakefield, H. 1962. *Victorian pottery.* Herbert Jenkins, London.

Wall, D.D. 1999. Examining gender, class and ethnicity in nineteenth century New York City. *Historical Archaeology*, Vol. 33, No. 1, pp. 102-117.

Ward, R. 2006. *Ceramics report 109-113 George Street, Parramatta*, Report for Casey and Lowe Pty. Ltd. Viewed 6 May 2007.
<http://www.caseyandlowe.com.au/pdf/109/v2section8 1.pdf>

Webber, K., I. Hoskins and J. McCann. 2003. *What's in store? A history of retailing in Australia.* Powerhouse Publishing, Sydney.

Wegars, P. 1988. The Asian comparative collection. *Australasian Historical Archaeology*, Vol. 6, pp. 43-48.

Wegars, P. 1993. Besides Polly Bemis: historical and artifactual evidence for Chinese women in the West, 1848-1930. In P. Wegars (ed.) *Hidden heritage: historical archaeology of the overseas Chinese*, pp. 229-254. Baywood Publishing, New York.

Wegars, P. 2003. From old gold mountain to new gold mountain: Chinese archaeological sites, artefact repositories, and archives in western North America and Australasia. *Australasian Historical Archaeology*, Vol. 21, pp. 70-83.

Wegars, P. 2012. *Chinese artifact illustrations, terminology, and selected bibliography.* Asian American Comparative collection, University of Idaho, Moscow.

Weigand, P.G., G. Harbottle and E.V. Sayre. 1977. Turquoise sources and source analysis: Mesoamerica and the Southwestern U.S.A. In T.K. Earle and J.E. Ericson (eds.) *Exchange systems in prehistory,* pp. 15-34. Academic Press, New York.

Willetts, W. 1981. Introduction. In The Southeast Asian Ceramic Society (ed.) *Nonya ware and Kitchen Ch'ing,* pp. 1-16. Oxford University Press, West Malaysia Chapter, Kuala Lumpur.

Williams, B. 2004. Opium pipe tops at the Market Street Chinese community in San Jose. *Proceedings of the Society for California Archaeology*, Vol. 17, pp. 219-227.

Williams, C.A.S. 1960. *Encyclopedia of Chinese symbolism and art motives*. Julian Press, New York.

Williams, P. 1978. *Staffordshire romantic transfer patterns*. Fountain House East, Jeffersontown.

Wilson, G. 1999. Ceramics and tobacco pipes artefact report. In Godden Mackay Logan (eds.) *Cumberland/Gloucester Streets Site, The Rocks, Archaeological Investigation Report*, Volume 4, Specialist Artefact Reports, Part 1, pp. 205-366. Prepared for the Sydney Cove Authority, Sydney.

Wilton, J. 2004. *Golden threads: The Chinese in regional New South Wales 1850-1950*. New England Regional Art Museum, Armidale.

Wood, N. 1978. *Oriental glazes*. Pitman, London.

Wood, N. 1999. *Chinese glazes*. A and C Black, London.

Worthy, L.H. 1982. Classification and interpretation of late-nineteenth- and early-twentieth-century ceramics. In R.S. Dickens Jr. (ed.) *Archaeology of urban America, the search for pattern and process*, pp. 329-360. Academic Press, New York.

Wylie, J. and R.E. Fike. 1993. Chinese opium smoking techniques and paraphernalia. In P. Wegars (ed.) *Hidden heritage: historical archaeology of the overseas* Chinese, pp. 255-303. Baywood Publishing, New York.

Yamin, R. 2001. Alternative narratives: respectability at New York's Five Points. In A. Mayne and T. Murray (eds.) *The archaeology of urban landscapes explorations in slumland*, pp. 154-171. Cambridge University Press, Cambridge.

Yang, M.M. 1994. *Gifts, favors, and banquets, the art of social relationships in China*. Cornell University Press, New York.

Yanyi, G. 1987. Raw materials for making porcelain and the characteristics of porcelain wares in north and south China in ancient times. *Archaeometry*, Vol. 29. No. 1, pp. 3-19.

Yap, C.T. 1986. A non-destructive scientific technique of detecting modern fake reproduction porcelains. *Oriental Art*, Vol. 32, pp. 48-50.

Yap, C.T. 1988. A quantitative spectrometric analysis of trace concentrations of manganese and cobalt in ceramics and the significance of As/Co and Mn/Co ratios. *Journal of Archaeological Science*, Vol. 15, pp. 173-177.

Yap, C.T. and Y. Hua. 1994. A study of Chinese porcelain raw materials for Ding, Xing, Gongxian and Dehua wares. *Archaeometry*, Vol. 36, No. 1, pp. 63-76.

Yap, C.T. and S.M. Tang. 1985. Energy dispersive x-ray fluorescence analysis of Chinese porcelains using Am-241. *Archaeometry*, Vol. 27, No. 1, pp. 61-63.

Yentsch, A. 1990. Minimum vessel lists as evidence of change in folk and courtly traditions of food use. *Historical Archaeology*, Vol. 24, No. 3, pp. 24-53.

Yin, M., Rehren, T. and J. Zheng. 2011. The earliest high-fired glazed ceramics in China: the composition of the proto-porcelain from Zhejiang during the Shang and Zhou periods (c.1700-221 BC). *Journal of Archaeological Science*, Vol. 38, pp. 2352-2365.

Yit, R.L. 2005. *The archaeology of Chinese suburban settlement*. Unpublished Honours Thesis. The Australian National University, Canberra.

Young, L. 2003. *Middle-class culture in the nineteenth-century: America, Australia and Britain.* Palgrave Macmillan, Basingstoke.

Young, S. 1956. An analysis of Chinese blue-and-white. *Oriental Art*, Vol. 2, pp. 43-47.

Yu, K.N. and J.M. Miao. 1996. Non-destructive analysis of Jingdezhen blue and white porcelains. *Archaeometry*, Vol. 38, No .2, pp. 257-262.

www.ingramcontent.com/pod-product-compliance
Lightning Source LLC
LaVergne TN
LVHW071207210926
848482LV00015B/745